K M GEORGE
OCTOBER 1972.

£6.45

GW00724549

# Group Representation Theory

## PART A

*Ordinary Representation Theory*

# PURE AND APPLIED MATHEMATICS

A Series of Monographs and Textbooks

COORDINATOR OF THE EDITORIAL BOARD

*S. Kobayashi*

UNIVERSITY OF CALIFORNIA AT BERKELEY

1. KENTARO YANO. Integral Formulas in Riemannian Geometry (1970)
2. S. KOBAYASHI. Hyperbolic Manifolds and Holomorphic Mappings (1970)
3. V. S. VLADIMIROV. Equations of Mathematical Physics (A Jeffrey, editor; A. Littlewood, translator) (1970)
4. B. N. PSHENICHNYI. Necessary Conditions for an Extremum (L. Neustadt, translation editor; K. Makowski, translator) (1971)
5. L. NARICI, E. BECKENSTEIN, and G. BACHMAN. Functional Analysis and Valuation Theory (1971)
6. D. S. PASSMAN. Infinite Group Rings (1971)
7. L. DORNHOFF. Group Representation Theory (in two parts). Part A: Ordinary Representation Theory. Part B: Modular Representation Theory (1971, 1972)

*In Preparation:*

W. BOOTHBY and G. L. WEISS (eds.). Symmetric Spaces: Short Courses Presented at Washington University

Y. MATSUSHIMA. Differentiable Manifolds (E. J. Taft, editor; E. T. Kobayashi, translator)

ARARAT BABAKHANIAN. Cohomological Methods in Group Theory

# Group Representation Theory *(in 2 parts)*

## PART A
*Ordinary Representation Theory*

## LARRY DORNHOFF
*Department of Mathematics*
*University of Illinois*
*Urbana, Illinois*

1971

MARCEL DEKKER, INC.,    New York

COPYRIGHT © 1971 BY MARCEL DEKKER, INC.

ALL RIGHTS RESERVED

No part of this work may be reproduced or utilized in any form
or by any means, electronic or mechanical, including xerography,
photocopying, microfilm, and recording, or by any information
storage and retrieval system, without the written permission of the
publisher.

MARCEL DEKKER, INC.
95 Madison Avenue, New York, New York 10016

LIBRARY OF CONGRESS CATALOG CARD NUMBER: 74–176305
ISBN NO.: 0–8247–1147–5

PRINTED IN THE UNITED STATES OF AMERICA

# *Preface*

This book grew out of a year course in representation theory of finite groups given at the University of Illinois during 1969–70. Its primary purpose is to provide a readable account of several major applications of representation theory to the structure of finite groups. The book should serve well as a text for a graduate course in representation theory and be useful for individual study by graduate students and mathematicians wishing to familiarize themselves with the subject. Part A separately should be useful as a text for a course in ordinary representation theory, with applications; exercises have been included in Part A to aid in this purpose.

This book presupposes knowledge of only several basic topics in algebra. These include the Sylow theorems and the structure of finite abelian groups from group theory; the structure of finitely generated modules over principal ideal domains; the notion of tensor product of modules over noncommutative rings; and occasionally basic Galois theory of fields. In Part A, we begin by developing the structure theory for semi-simple rings with unit. In Part B, we build the broad algebraic foundation needed for modular representation theory without assuming any knowledge of valuation theory. Indeed, we begin with a basic study of chain conditions for modules and the radical of a ring with unit. Readers of Part B with strong algebraic backgrounds will be able to skip most of Sections 39–50. Except for this, readers of Part B will find it necessary to read all of Sections 39–65 before studying the applications in Sections 66–72.

The material in Part A comes from a variety of books and research papers, no one source of predominant importance. References to sources and related results are given in the individual sections. Interdependency between the sections is greater than appears on the surface; but references

iii

to earlier results are usually precise enough for a reader interested in a specific section to read that section and be led back to the exact prerequisites needed.

One need only glance at the contents of Part B to see the predominant role of Richard Brauer in the development of modular representation theory. Indeed, the term "Brauer theory" is coming into wide use instead of "modular theory." Our presentation of modular representation theory owes much to Professor Walter Feit. His year-long course at Yale during 1967–68 and subsequent lecture notes, *Representations of Finite Groups*, are of great influence in our arrangement of the basic modular theory and requisite algebra. We only present the theory needed for the important applications in Sections 66–72. These theoretical results themselves are due primarily to Brauer, with important contributions from J. A. Green and D. G. Higman. Indeed, Feit's notes contain much more theory than we need here and include much of Feit's own work. If the present *Representations of Finite Groups* is followed by a sequel surveying extensively the applications of modular representation theory, Feit will have written a definitive work on the subject. We hope that our present modest Part B would then still be useful as a slower-paced introduction to Feit's more concise and extensive work.

I wish to express special thanks to Professor D. S. Passman and Professor J. L. Alperin, whose advice and encouragement led to the publication of this book. I am also indebted to the National Science Foundation for research support.

*Urbana, Illinois*                                                          LARRY DORNHOFF
*September 1971*

# Contents

# Contents of Part B

# §1

## Introduction

We adopt a firm convention that all rings discussed are associative, with identity. Most of the groups discussed are finite.

**Definitions**    Let $G$ be a group, $R$ a commutative ring (with 1), so the set $\text{Mat}_n(R)$ of $n \times n$ matrices over $R$ is also a ring. The set of all invertible matrices in $\text{Mat}_n(R)$ forms a group $GL(n, R)$ under matrix multiplication. A *matrix representation of $G$ over $R$* is a group homomorphism

$$\varphi: \quad G \to GL(n,R).$$

The *kernel* of $\varphi$ is, of course, $K = \{g \in G \mid \varphi(g) = \text{identity matrix}\}$, and is a normal subgroup of $G$. If $K = \{1\}$, then $\varphi$ is an isomorphism of $G$ into $GL(n,R)$, and we say we have a *faithful* matrix representation; the group $G$ is concretely realized as a set of matrices.

We point out some special cases:

(1)   If $R = k$ is a field of characteristic zero, then the theory of representations over $k$ is called *ordinary representation theory*, and is the topic of Part A.

(2)   If $R = k$ is a field of characteristic $p \neq 0$, then the theory of representations over $k$ is called *modular representation theory* (or *Brauer theory*) and is covered in Part B. This theory differs from the ordinary theory principally when $p$ divides the order of $G$.

(3)   If $R$ is not a field, representations over $R$ are still important. Representations over certain $R$'s are studied in modular representation theory, and representations over some other $R$'s are interesting in their own right. Representations over integral domains are called *integral representations*.

We recall some basic algebra.

1

**Definition**    An additive abelian group $M$ is called a *module* over the ring $A$ (*left A-module*) if we have a multiplication $a \cdot m \in M$, for $a \in A$, $m \in M$, such that $(a + a')m = am + a'm$, $a(a'm) = (aa')m$, $1m = m$. $M$ is called *free*, with *basis* $\{m_i\}_{i \in I}$, if each element of $M$ has a unique expression

$$a_1 m_{i_1} + \cdots + a_n m_{i_n}, \qquad i_1, \ldots, i_n \in I, \quad a_1, \ldots, a_n \in A.$$

**Definition**    Let $R$ be a commutative ring, and let $A$ be a ring which is also an $R$-module and satisfies $r(ab) = (ra)b = a(rb)$, all $a, b \in A$, $r \in R$. $A$ is called an *R-algebra*.

**Definition**    Let $G$ be a group, $R$ a commutative ring, $A$ the free $R$-module with basis the elements of $G$. We have a multiplication in $A$:

$$\left( \sum_{i=1}^{m} r_i g_i \right)\left( \sum_{j=1}^{n} s_j h_j \right) = \sum_{i=1}^{m} \sum_{j=1}^{n} (r_i s_j)(g_i h_j), \qquad \text{all } r_i, s_j \in R, \quad g_i, h_j \in G.$$

$A = RG$ is an $R$-algebra, called the *group ring* or *group algebra* of $G$ over $R$.

    *Notation*    Let $A$ be a ring, $V$ and $W$ $A$-modules. We denote $\text{Hom}_A(V, W) = \{f \colon V \to W \mid f(av) = af(v) \text{ and } f(v + v') = f(v) + f(v'),$ all $a \in A$, $v, v' \in V\}$, $\text{End}_A(V) = \text{Hom}_A(V, V)$. $\text{End}_A(V)$ is a ring; if $A$ is commutative, then $\text{Hom}_A(V, W)$ is an $A$-module and $\text{End}_A(V)$ is an $A$-algebra.

    If $R$ is a commutative ring, $V$ an $R$-module, we denote by $GL(V)$ the group of invertible elements of $\text{End}_R(V)$. If $V$ is $R$-free of dimension $n$, then we know

$$GL(V) \cong GL(n, R),$$

the isomorphism depending on the choice of basis for $V$. $n = \dim_R V$ is independent of the choice of basis of $V$.

**Definition**    Let $G$ be a group, $R$ a commutative ring, $V$ a finite-dimensional free $R$-module. A *representation of $G$ in $V$* is a group homomorphism

$$T \colon G \to GL(V).$$

Clearly a representation affords a matrix representation, by choosing a basis of $V$. $\dim_R V$ is the *degree* of $T$.

**Lemma 1.1** *Studying representations of the group G over R is equivalent to studying R-free RG-modules.*

*Proof* Suppose we are given a representation

$$\varphi: G \to GL(V),$$

$V$ an $R$-free $R$-module. By defining $g \cdot v = \varphi(g)v$, $V$ becomes an $RG$-module.

Conversely, given an $R$-free $RG$-module $V$, $g \in G$, define $\psi(g): V \to V$ by $\psi(g)v = gv$. Then each $\psi(g)$ is in $GL(V)$, and the fact that $V$ is an $RG$-module shows $\psi(g_1g_2)v = g_1g_2 \cdot v = g_1(g_2v) = g_1 \cdot \psi(g_2)v = \psi(g_1)\psi(g_2)v$, so $\psi(g_1g_2) = \psi(g_1)\psi(g_2)$, $\psi$ is a representation. $\psi$ is called the *representation afforded by V*.

*Remark* Of course if $R = k$ is a field, all $kG$-modules are $k$-free (i.e., are $k$-vector spaces). So our task is then to study all $kG$-modules.

**Definition** Two matrix representations $\varphi$, $\psi$ of $G$ in $\mathrm{Mat}_n(R)$ are *equivalent* if there is a matrix $M$ in $\mathrm{Mat}_n(R)$ such that

$$\psi(g) = M^{-1}\varphi(g)M, \qquad \text{all } g \in G.$$

**Lemma 1.2** *Let V and W be RG-modules which are n-dimensional and free as R-modules. Then V and W are isomorphic if and only if they afford equivalent matrix representations.*

*Proof* Assume first that $V$ and $W$ are isomorphic and let $\Gamma: V \to W$ be an $RG$-isomorphism. Let $V$ have basis $\{v_1, \ldots, v_n\}$, $W$ basis $\{w_1, \ldots, w_n\}$. Let $T: G \to GL(V)$, $T': G \to GL(W)$ be the representations afforded by $V, W$ and let $\varphi: G \to GL(n,R)$, $\varphi': G \to GL(n,R)$ be the matrix representations afforded by $V$ and $W$ in the given bases. Since $\Gamma$ is an $RG$-isomorphism, for any $g \in G$ we have

$$g\Gamma(v_i) = \Gamma(gv_i) \qquad \text{or} \qquad T'(g)\Gamma(v_i) = \Gamma(T(g)v_i).$$

$\Gamma$ is invertible so $\Gamma^{-1}T'(g)\Gamma = T(g)$. If $M$ is the matrix of $\Gamma$ in the given bases, this means

$$M^{-1}\varphi'(g)M = \varphi(g), \qquad \text{as desired.}$$

Conversely, if the matrix representations are equivalent, arguing backward shows that $V$ and $W$ are isomorphic.

*Notation*  $|G|$ denotes the order of the group (or set) $G$, $|G:H|$ the index of the subgroup $H$ in the group $G$. $K \lhd G$ means that $K$ is a normal subgroup of $G$.

We give two examples of representations.

*Example 1.1*  $RG$ is itself an $R$-free $RG$-module. If $G$ is finite, then $\dim_R RG = |G|$. Let $n = |G|$, $G = \{1 = x_1, x_2, \ldots, x_n\}$. Then for any $x \in G$, $xx_i$ is some $x_j$. Hence the matrix of $x$ in the basis $G$ has form

$$
\begin{array}{cc}
 & \begin{array}{ccccc} 1 & \cdots & & x_i & \cdots \end{array} \\
\begin{array}{c} 1 \\ \vdots \\ x \\ \vdots \\ \vdots \\ x_j \\ \vdots \\ \vdots \end{array} &
\left(\begin{array}{ccccc}
0 & \cdots & & 0 & \cdots \\
\vdots & & & \vdots & \\
1 & & & \vdots & \\
0 & & & \vdots & \\
\vdots & & & 0 & \\
0 & \cdots & 0 & 1 & \\
\vdots & & & \vdots & \\
0 & & & 0 &
\end{array}\right)
\end{array} .
$$

This is a *permutation matrix*, with one 1 in each row and each column. This representation is called the *regular representation* of $G$, and can be constructed for any finite group.

*Example 1.2*  The smallest nonabelian group has order 6 (the symmetric group on three symbols). It is generated by elements $x, y$ with relations $x^2 = 1$, $y^3 = 1$, $yx = xy^2$. Let $\omega$ be a cube root of 1. Then

$$
\begin{pmatrix} 0 & 1 \\ 1 & 0 \end{pmatrix}^2 = \begin{pmatrix} 1 & 0 \\ 0 & 1 \end{pmatrix}, \qquad
\begin{pmatrix} \omega & 0 \\ 0 & \omega^2 \end{pmatrix}^3 = \begin{pmatrix} 1 & 0 \\ 0 & 1 \end{pmatrix}
$$

and

$$
\begin{pmatrix} \omega & 0 \\ 0 & \omega^2 \end{pmatrix}\begin{pmatrix} 0 & 1 \\ 1 & 0 \end{pmatrix} = \begin{pmatrix} 0 & 1 \\ 1 & 0 \end{pmatrix}\begin{pmatrix} \omega & 0 \\ 0 & \omega^2 \end{pmatrix}^2 .
$$

Hence we may construct a matrix representation of our group by mapping

$$x \to \begin{pmatrix} 0 & 1 \\ 1 & 0 \end{pmatrix}, \qquad y \to \begin{pmatrix} \omega & 0 \\ 0 & \omega^2 \end{pmatrix}.$$

Note that this can only be done over $\mathbf{C}$ = complex numbers, or some other ring or field containing nontrivial cube roots of 1.

It is easy to check that this representation is faithful.

**Definitions**    Let $A$ be a ring, $M$ an $A$-module $\neq 0$. $M$ is *reducible* if there is a proper submodule $0 \neq N \neq M$; otherwise $M$ is *irreducible*. $M$ is *completely reducible* (or *semisimple*) if $M$ is a direct sum of irreducible $A$-modules.

A representation $G \to GL(V)$, $V$ a free $R$-module, is *reducible, irreducible,* or *completely reducible,* according to whether $V$ is reducible, irreducible, or completely reducible as $RG$-module.

We now interpret these terms matrix-theoretically, assuming that $R = k$ is a field.

If $V$ is *reducible* with submodule $W$, we can write $V = W \oplus W'$, $W'$ a $k$-subspace of $V$, and choose bases $\{v_1, \ldots, v_m\}$ of $W$, $\{v_{m+1}, \ldots, v_n\}$ of $W'$. In the basis $\{v_1, \ldots, v_m, v_{m+1}, \ldots, v_n\}$ of $V$, we see that matrices of elements of $G$ have form

$$\begin{pmatrix} A & B \\ 0 & C \end{pmatrix},$$

where $A$ is an $m \times m$ matrix, $B$ an $m \times (n - m)$ matrix, $C$ an $(n - m) \times (n - m)$ matrix, and 0 the $(n - m) \times m$ zero matrix.

If $V$ is completely reducible, let $V = W_1 \oplus \cdots \oplus W_t$, the $W_i$ irreducible submodules of $V$. By choosing bases of the $W_i$ and combining them, we see that the matrices of elements of $G$ have the form

$$\begin{pmatrix} A_1 & 0 & \cdots & 0 \\ & & & \vdots \\ 0 & A_2 & & 0 \\ \vdots & & & \\ 0 & & 0 & A_t \end{pmatrix}.$$

We see that the presence of submodules enables us to greatly simplify matrix representations.

*Example 1.3*   If $G \neq 1$, then the regular representation of $G$ is always reducible, because $RG$ always has the 1–dimensional submodule with basis element $\sum_{g \in G} g$.

*Example 1.4*   If $G$ is the nonabelian group of order 6, we shall show that the representation constructed in Example 1.2 is irreducible, taking $k = \mathbf{C}$.

Let $V$ be a 2–dimensional $\mathbf{C}$-vector space with basis $\{v_1, v_2\}$. The matrices

$$\varphi(x) = \begin{pmatrix} 0 & 1 \\ 1 & 0 \end{pmatrix}, \qquad \varphi(y) = \begin{pmatrix} \omega & 0 \\ 0 & \omega^2 \end{pmatrix}$$

show that $xv_1 = v_2$, $xv_2 = v_1$, $yv_1 = \omega v_1$, $yv_2 = \omega^2 v_2$. Clearly $x$ does not fix the subspace $\mathbf{C}v_2$. Other 1–dimensional subspaces have form $\mathbf{C}(v_1 + \lambda v_2)$, $\lambda \in \mathbf{C}$. If such a subspace is fixed by $G$, then $x(v_1 + \lambda v_2) = v_2 + \lambda v_1 \in \mathbf{C}(v_1 + \lambda v_2)$, say $v_2 + \lambda v_1 = \alpha(v_1 + \lambda v_2)$. This forces $\alpha = \lambda$, $1 = \lambda \alpha$, so $\lambda = \pm 1$. But $y(v_1 + v_2) = \omega v_1 + \omega^2 v_2 \notin \mathbf{C}(v_1 + v_2)$ and $y(v_1 - v_2) = \omega v_1 - \omega^2 v_2 \notin \mathbf{C}(v_1 - v_2)$, proving no 1-dimensional subspace is a $\mathbf{C}G$-submodule.

# §2

## Theory of Semisimple Rings

In this section we follow Lang [1, Chapter XVII].

**Lemma 2.1** (Schur's lemma)    *Let $A$ be a ring, $V$ and $W$ irreducible $A$-modules. Then*

$$Hom_A(V,W) = (0) \text{ if } V \not\cong W;$$

$$= a \text{ division ring, if } V = W.$$

*Proof* Suppose $0 \neq f \in \mathrm{Hom}_A(V,W)$. Then $\ker f \neq V$, so $\ker f = (0)$, and $\mathrm{im}\, f \neq (0)$, so $\mathrm{im}\, f = W$, using the irreducibility of $V$ and $W$. Hence $f$ is an isomorphism and must have an inverse, proving that $\mathrm{Hom}_A(V,V) = \mathrm{End}_A(V)$ is a division ring.

**Lemma 2.2**    *Let $A$ be any ring, $V = V_1 \oplus \cdots \oplus V_n$ and $W = W_1 \oplus \cdots \oplus W_m$ $A$-modules. Let $\varepsilon_j : V_j \to V$ be the natural injection $\varepsilon_j : x \to x$, and let $\pi_i : W \to W_i$ be the natural projection. Then:*

(i) *If for each $i$ and $j$, $\varphi_{ij} \in \mathrm{Hom}_A(V_j, W_i)$, then we can define a $\varphi \in \mathrm{Hom}_A(V,W)$ by*

$$\varphi(v_1 + \cdots + v_n) = \begin{pmatrix} \varphi_{11} & \cdots & \varphi_{1n} \\ \vdots & & \vdots \\ \varphi_{m1} & \cdots & \varphi_{mn} \end{pmatrix} \begin{pmatrix} v_1 \\ \vdots \\ v_n \end{pmatrix}$$

$$= \underbrace{\varphi_{11}(v_1) + \cdots + \varphi_{1n}(v_n)}_{\in W_1} + \cdots + \underbrace{\varphi_{m1}(v_1) + \cdots + \varphi_{mn}(v_n)}_{\in W_m},$$

*all $v_i \in V_i$.*

(ii) *Conversely, if $\varphi \in \mathrm{Hom}_A(V,W)$, then $\varphi_{ij} = \pi_i \circ \varphi \circ \varepsilon_j \in \mathrm{Hom}_A(V_j, W_i)$, and*

7

$$\varphi(v_1 + \cdots + v_n) = \begin{pmatrix} \varphi_{11} & \cdots & \varphi_{1n} \\ \vdots & & \vdots \\ \varphi_{m1} & \cdots & \varphi_{mn} \end{pmatrix} \begin{pmatrix} v_1 \\ \vdots \\ v_n \end{pmatrix}.$$

(iii) $\quad \mathrm{Hom}_A(V,W) \cong \begin{pmatrix} \mathrm{Hom}_A(V_1,W_1) & \cdots & \mathrm{Hom}_A(V_n,W_1) \\ \vdots & & \\ \mathrm{Hom}_A(V_1,W_m) & \cdots & \mathrm{Hom}_A(V_n,W_m) \end{pmatrix}$

*as additive groups.*

(iv)    *In particular, if* $V^{(n)} = V \oplus \cdots \oplus V$ *(n copies), then* $\mathrm{End}_A(V^{(n)}) \cong \mathrm{Mat}_n(\mathrm{End}_A(V))$ *as rings.*

*Proof*    (i) is clear.

$$\begin{pmatrix} \varphi_{11} & \cdots & \varphi_{1n} \\ \vdots & & \vdots \\ \varphi_{m1} & \cdots & \varphi_{mn} \end{pmatrix} \begin{pmatrix} v_1 \\ \vdots \\ v_n \end{pmatrix} = \sum_{i,j} \varphi_{ij}(v_j)$$

$$= \sum_{i,j} \pi_i \circ \varphi \circ \varepsilon_j(v_j)$$

$$= \sum_i \pi_i \varphi(v_1 + \cdots + v_n) = \varphi(v_1 + \cdots + v_n)$$

since $\sum_i \pi_i = 1$. Parts (iii) and (iv) follow from (i) and (ii).

**Lemma 2.3**    *Let $A$ be a ring, $V$ an $A$-module. Then the following are equivalent.*

(1)    *$V$ is a sum of irreducible submodules.*
(2)    *$V$ is completely reducible.*
(3)    *If $W$ is any submodule of $V$, then $V = W \oplus W'$ for some submodule $W'$.*

*Remark*    Zorn's lemma is used to obtain several maximal objects in the following proof.

*(1) $\Rightarrow$ (2)*    If $V = \sum_{i \in I} V_i$, the $V_i$ irreducible, let $J$ be a maximal subset of $I$ such that $V' = \sum_{j \in J} V_j$ is direct. If $i \notin J$, then $V_i \cap V'$ is a submodule of $V_i$ and so is 0 or $V_i$. If it is 0, then $\sum_{j \in J \cup \{i\}} V_j$ is direct, contradicting the maximality of $J$. Hence it is $V_i$, so all $V_i \subseteq V'$, $V' = V$, done.

$(2) \Rightarrow (3)$   Let $V = \oplus \sum_{i \in I} V_i$, the $V_i$ irreducible. Choose a maximal subset $J$ of $I$ such that $W + (\oplus \sum_{j \in J} V_j)$ is direct. For any other $V_i$, as in the previous proof, we see that $V_i \cap (W + (\sum_{j \in J} V_j))$ is not 0. Hence all $V_i \subseteq W \oplus (\oplus \sum_{j \in J} V_j)$, and we may take $W' = \oplus \sum_{j \in J} V_j$.

$(3) \Rightarrow (1)$   *Claim 1:* If $V$ satisfies (3), so does any submodule $V_0$ of $V$. *Proof:* Let $W$ be any submodule of $V_0$, so $V = W \oplus W'$ by (3). We see that then $V_0 = W \oplus (W' \cap V_0)$ for if $v_0 \in V_0$, $v_0 = x + y$, $x \in W$, $y \in W'$, then $y = v_0 - x \in V_0$, $y \in W' \cap V_0$.

*Claim 2:* Any submodule $0 \neq V_0$ of $V$ contains an irreducible submodule. *Proof:* Choose $0 \neq v_0 \in V_0$, $W_0$ a maximal submodule of $V_0$ such that $v_0 \notin W_0$. By Claim 1, $V_0 = W_0 \oplus W_1$. If $W_1$ is not irreducible, then by Claim 1, $W_1 = W_2 \oplus W_3$, $V_0 = W_0 \oplus W_2 \oplus W_3$, $v_0 \notin W_0 = (W_0 + W_2) \cap (W_0 + W_3)$, so $v_0 \notin W_0 + W_2$ or $v_0 \notin W_0 + W_3$, a contradiction to the maximality of $W_0$. Hence $W_1$ is irreducible, done.

*Conclusion:* If (3) holds for $V$, let $V'$ be the sum of all irreducible submodules of $V$. By (3), $V = V' \oplus W$, some $W$. If $W \neq 0$, then by Claim 2 there is an irreducible $W_0 \subseteq W$. But $W_0 \subseteq V'$, a contradiction. Hence, $W = 0$, $V' = V$, (1) holds.

*Remark*   It is easy to see that submodules and factor modules of completely reducible modules are completely reducible.

**Lemma 2.4**   *Let $A$ be a ring, $V$ a completely reducible $A$-module, $B = \text{End}_A(V)$. $V$ is a $B$-module under $\varphi \cdot v = \varphi(v)$, any $\varphi \in B, v \in V$. Then for any $v \in V$ and any $f \in \text{End}_B(V)$, there is an $a \in A$ with $av = f(v)$.*

*Proof*   By complete reducibility of $V$, $V = Av \oplus W$, some submodule $W$. Let $\pi: V \to Av$ be the projection, so $\pi \in \text{End}_A(V) = B$. Hence, $f(v) = f(\pi v) = \pi f(v) \in \pi(V) = Av$, done.

**Theorem 2.5** (Jacobson density theorem)   *Let $V$ be an irreducible $A$-module, $B = \text{End}_A(V)$, $f \in \text{End}_B(V)$, $v_1, \ldots, v_n \in V$. Then there is an $a \in A$ such that $av_i = f(v_i)$, all $i$.*

*Proof*   Denote $V^{(n)} = V \oplus V \oplus \cdots \oplus V$ ($n$ copies), and define $f^{(n)}: V^{(n)} \to V^{(n)}$ by $f^{(n)}(x_1 + \cdots + x_n) = f(x_1) + \cdots + f(x_n)$, all $x_i$, $f(x_i) \in i$th summand $V$. Denote $B' = \text{End}_A(V^{(n)})$. Given any $\varphi \in B'$, we use Lemma 2.2 to write

$$\varphi(w_1 + \cdots + w_n) = \varphi_{11}(w_1) + \cdots + \varphi_{1n}(w_n) + \cdots + \varphi_{nn}(w_n)$$

$$= \begin{pmatrix} \varphi_{11} & \cdots & \varphi_{1n} \\ \vdots & & \vdots \\ \varphi_{n1} & \cdots & \varphi_{nn} \end{pmatrix} \begin{pmatrix} w_1 \\ \vdots \\ w_n \end{pmatrix},$$

where all $\varphi_{ij} \in \text{End}_A(V) = B$. Hence we see $f^{(n)}(\varphi(w_1 + \cdots + w_n)) = \varphi(f^{(n)}(w_1 + \cdots + w_n)), f^{(n)} \in \text{End}_{B'}(V^{(n)})$.

By Lemma 2.4, then, there is $a \in A$, $a(v_1 + \cdots + v_n) = f^{(n)}(v_1 + \cdots + v_n) = f(v_1) + \cdots + f(v_n)$, all $v_i$, $f(v_i) \in i$th summand. Thus, $av_i = f(v_i)$ for all $i$.

**Corollary 2.6**    *Let $A$ be a ring with a faithful irreducible module $V$, and let $D = \text{End}_A(V)$, a division ring. If $V$ is a finite-dimensional $D$-vector space, then $A \cong \text{End}_D(V)$.*

*Proof*    Let $\{v_1, \ldots, v_n\}$ be a $D$-basis of $V$. By Theorem 2.5, if $f \in \text{End}_D(V)$ then there is an $a \in A$, $av_i = f(v_i)$, all $i$; this means $av = f(v)$ for all $v \in V$, so we have a map $A \to \text{End}_D(V)$ which is onto. Since $V$ is faithful, this map has kernel 0.

**Definition**    A ring $A$ is *semisimple* if $A$ is itself a completely reducible $A$-module.

**Lemma 2.7**    *If $A$ is semisimple, then every $A$-module is completely reducible.*

*Proof*    Let $V$ be any $A$-module. We can find a free $A$-module $F$ with submodule $W$, $F/W \cong V$. $A$ is completely reducible, so $F$ and $F/W \cong V$ are also completely reducible.

**Definition**    A left ideal of $A$ is *simple* if it is irreducible as $A$-module.

**Definition**    A ring $A$ is *simple*, if it is semisimple and has only one isomorphism class of simple left ideals.

**Lemma 2.8**    *Let $A$ be a semisimple ring, $L$ a simple left ideal, $V$ an irreducible $A$-module. Then $L \cong V$ or $LV = 0$.*

*Proof*    $ALV = LV$, a submodule of $V$, is 0 or $V$. If it is $V$, choose

$v \in V$, $Lv \neq 0$. $Lv$ is an $A$-submodule, so $Lv = V$. The map $L \to V$ given by $a \to av$ is an $A$-isomorphism since $L$ is irreducible as $A$-module.

**Theorem 2.9**    *Let $A$ be a semisimple ring. Then:*

   *(1)   $A$ has only finitely many nonisomorphic simple left ideals $L_1, \ldots, L_s$.*

   *(2)   If $A_i$ is the sum of all left ideals of $A$ isomorphic to $L_i$, then $A_i$ is a 2-sided ideal in $A$ and a simple ring.*
   *(3)   $A = \oplus \sum_{i=1}^{s} A_i$.*
   *(4)   If $A_i$ has unit element $e_i$, then $1 = e_1 + \cdots + e_s$ and $A_i = Ae_i$. $i \neq j$ implies $A_i A_j = 0$.*

   *Proof*   Let $\{L_i\}_{i \in I}$ be a set of representatives of all isomorphism classes of simple left ideals, and define $A_i$, all $i \in I$, as in (2). By Lemma 2.8, $i \neq j$ implies $A_i A_j = 0$. Of course $A = \sum_{i \in I} A_i$, so $A_i \subseteq A_i A = A_i A_i \subseteq AA_i = A_i$, proving that each $A_i$ is a 2-sided ideal.
   Write $1 = \sum_{i \in I} e_i$, $e_i \in A_i$, so all but finitely many $e_i$ are 0, say $1 = \sum_{i=1}^{s} e_i$, where $e_1, \ldots, e_s \neq 0$. If $x \in A_k$, $k \notin \{1, \ldots, s\}$, then $x = 1x = (e_1 + \cdots + e_s)x = 0$, proving (1).
   If $0 = x_1 + \cdots + x_s$, $x_i \in A_i$, then $0 = e_j x_1 + \cdots + e_j x_s = e_j x_j = (e_1 + \cdots + e_s)x_j = 1x_j = x_j$, all $j$, so (3) holds. We thus see $e_j$ is a unit in $A_j$, so (4) has been checked. Any simple left ideal of $A_i$ is a simple left ideal of $A$, and so by construction is isomorphic to some $L_i$. Hence (2) holds.

**Theorem 2.10**    *With the notation of Theorem 2.9, let $V \neq 0$ be an $A$-module. Then $V = A_1 V \oplus \cdots \oplus A_s V = e_1 V \oplus \cdots \oplus e_s V$, and $A_i V$ is the sum of all irreducible submodules of $V$ which are isomorphic to $L_i$.*

   *Proof*   Let $V_i$ be the sum of all irreducible submodules of $V$ which are isomorphic to $L_i$. If $W$ is any irreducible submodule of $V$, then $AW = W$, and $A = \sum_i \sum_{L \cong L_i} L$ so $LW = W$ for some simple left ideal $L$. By Lemma 2.8, $W \cong L \cong L_i$.
   This implies $V = V_1 + \cdots + V_s$. $e_i V_j = 0$ if $i \neq j$ and $e_i V_j = V_j$ if $i = j$, so $0 = v_1 + \cdots + v_s$ implies $0 = e_i v_1 + \cdots + e_i v_s = e_i v_i = (e_1 + \cdots + e_s) v_i = v_i$, and the sum of the $V_i$ is direct. Thus, $V_i = e_i V = A_i V$, done.

**Corollary 2.11**    *If $A$ is a semisimple ring, then every irreducible $A$-module is isomorphic to a simple left ideal of $A$.*

**Corollary 2.12**    *A simple ring A has only one isomorphism class of irreducible modules.*

**Lemma 2.13**    *Let A be a ring, $\psi \in \mathrm{End}_A(A)$. Then for some $a \in A$ we have $\psi(x) = xa$, all $x \in A$.*

*Proof*   $\psi(x) = \psi(x1) = x\psi(1)$. Let $a = \psi(1)$.

**Theorem 2.14**    *Let A be a simple ring. Then A is a finite direct sum of simple left ideals. The only 2–sided ideals of A are (0) and A. If L and M are simple left ideals, then for some $a \in A$, $La = M$; also, $LA = A$.*

*Proof*   $A$ is semisimple, so $A = \oplus \sum_{j \in J} L_j$, the $L_j$ simple left ideals. Let $1 = \sum_{j=1}^{m} \beta_j$, $\beta_j \in L_j$. Then $A = A1 = \sum_{j=1}^{m} A\beta_j = \sum_{j=1}^{m} L_j$.

If $L$ and $M$ are simple left ideals, let $A = L \oplus L'$, $L'$ a left ideal, $\pi \colon A \to L$ the projection. By Corollary 2.12, there is an $A$-isomorphism $\sigma \colon L \to M$. By Lemma 2.13, there is an $a \in A$, $\sigma \cdot \pi(x) = xa$, for all $x \in A$; in particular, $\sigma(x) = xa$ for $x \in L$, so $M = La$.

This shows every $M$ is in $LA$, so $LA = A$.

If $0 \neq I$ is a 2-sided ideal, choose a simple left ideal $L \subseteq I$. Then $I = IA \supseteq LA = A$, so $I = A$.

**Corollary 2.15**    *Let A be a simple ring, L a simple left ideal of A. If V is any irreducible A-module, $LV = V$ and V is faithful.*

*Proof*   $LV = L(AV) = (LA)V = AV = V$, using Theorem 2.14. If $a \in A$ satisfies $aV = 0$, then $AaAV = AaV = A \cdot 0 = 0$. But $AaA$ is a 2–sided ideal in $A$, so this means $AaA \neq A$, $AaA = 0$, $a = 0$.

*Remark*   If $D$ is a division ring, $D^{\mathrm{op}}$ denotes the *opposite* division ring to $D$; that is, $D^{\mathrm{op}}$ has the same underlying set and addition as $D$, but a multiplication $\circ$ where $x \circ y = yx$. We explain here why, if $\dim_D V = n$, then

$$\mathrm{End}_D(V) \cong \mathrm{Mat}_n(D^{\mathrm{op}}).$$

$V$ is isomorphic to the direct sum of $n$ copies of $D$ as $D$-module, so by Lemma 2.2 we have

$$\mathrm{End}_D(V) \cong \mathrm{Mat}_n(\mathrm{End}_D(D));$$

we shall show $\text{End}_D(D) \cong D^{\text{op}}$. If $\psi \in \text{End}_D(D)$, then by Lemma 2.13, $\psi(x) = xd$ for some $d \in D$; we denote $\psi = \psi_d$, so $\text{End}_D(D) = \{\psi_d | d \in D\}$. For any $x \in D$, $(\psi_c \psi_d)(x) = \psi_c(\psi_d(x)) = \psi_c(xd) = (xd)c = x(dc) = \psi_{dc}(x)$, so $\text{End}_D(D) \cong D^{\text{op}}$.

**Theorem 2.16**     *Let $A$ be a simple ring, $V$ an irreducible $A$-module, $D = \text{End}_A(V)$. Then*

$$A \cong \text{End}_D(V) \cong \text{Mat}_n(D^{\text{op}}), \quad \text{some } n.$$

*Proof*   By Corollary 2.15, $V$ is faithful. Hence by Corollary 2.6, it is enough to show that $V$ is finite-dimensional as $D$-vector space. If not, let $\{v_1, \dots, v_n, \dots\}$ be infinitely many $D$-linearly independent elements of $V$, and let $I_t = \{a \in A \mid av_1 = av_2 = \cdots = av_t = 0\}$. Certainly $I_1 \supset I_2 \supset \cdots$. By the Jacobson density theorem, $I_1 \neq I_2 \neq \cdots$. All the $I_j$ are left ideals of $A$. Let $I_t = J_t \oplus I_{t+1}$, so $A \supseteq J_1 \oplus J_2 \oplus \cdots$, all $J_i \neq 0$. Choose $J_0$ so $A = J_0 \oplus J_1 \oplus J_2 \oplus \cdots$, and choose $m$ so that $1 = \sum_{i=0}^m a_i$, $a_i \in J_i$. Then $A = A1 \subseteq \sum_{i=0}^m A a_i \subseteq J_0 \oplus \cdots \oplus J_m$, proving that $J_{m+1} = 0$, a contradiction.

**Theorem 2.17** (Wedderburn's theorem)     *Let $A$ be a semisimple ring (with $1$). Then $A = A_1 \oplus \cdots \oplus A_r$, the $A_i$ rings with $A_i A_j = 0$ when $i \neq j$. Moreover, $A_i \cong \text{Mat}_{n_i}(D_i^{\text{op}})$, $D_i$ a division ring, $D_i = \text{End}_A(V_i)$, $V_i$ some irreducible $A$-module. Each irreducible $A$-module is isomorphic to exactly one of the $V_i$, and $A_i V_j = (0)$ if $i \neq j$, $A_i V_i = V_i$, $\dim_{D_i}(V_i) = n_i$.*

*Proof*   All but the last fact is contained in Lemma 2.1, Theorems 2.9 and 2.16, and Corollaries 2.11 and 2.12. The fact $\dim_{D_i}(V_i) = n_i$ is seen in the proof of Corollary 2.6.

**Theorem 2.18**     *(a) Let $D$ be a division ring, $V$ an $n$-dimensional $D$-vector space, $A = \text{End}_D(V)$. Then $A$ is a simple ring with irreducible module $V$, and $D = \text{End}_A(V)$.*

*(b) Let $D$ and $E$ be division rings, such that $\text{Mat}_m(D) \cong \text{Mat}_n(E)$. Then $m = n$ and $D \cong E$.*

*Proof*   (a) For any $0 \neq v \in V$ we certainly have $Av = V$, so $V$ is an irreducible $A$-module. Let $\{v_1, \dots, v_n\}$ be a $D$-basis of $V$, and define

$$\varphi : A \to V^{(n)} = V \oplus \cdots \oplus V \quad (n \text{ copies})$$

by $\varphi: a \to (av_1, \ldots, av_n)$. We see that ker $\varphi = 0$ and $\varphi$ is onto, so $A \cong V^{(n)}$ as $A$-modules. This proves $A$ is a sum of simple left ideals, so $A$ is semisimple. Let $\pi_i: V^{(n)} \to V$ be the projection onto the $i$th summand. If $L$ is any simple left ideal of $A$, then $\varphi(L) \neq (0)$, so for some $j$, $\pi_j \circ \varphi(L) \neq 0$. $\pi_j \circ \varphi: L \to V$ must be an isomorphism since $L$ and $V$ are irreducible. We have seen that $L \cong V$ for any simple left ideal of $A$, so $A$ is a simple ring.

Since it is a vector space, $V$ is certainly a completely reducible $D$-module, $A = \operatorname{End}_D(V)$. For any $f \in \operatorname{End}_A(V)$ and any $0 \neq v \in V$, there is by Lemma 2.4 a $d \in D$ such that $f(v) = dv$. If $w \in V$, then $w = av$ for an $a \in A$; hence

$$f(w) = f(av) = af(v) = a(dv) = d(av) = dw,$$

proving that $f = d \in D$, $D = \operatorname{End}_A(V)$.

(b)   If $A \cong \operatorname{Mat}_m(D) \cong \operatorname{Mat}_n(E)$, let $V$ be an irreducible $A$-module. By part (a), we have both $D^{op} \cong \operatorname{End}_A(V)$ and $E^{op} \cong \operatorname{End}_A(V)$, implying that $D \cong E$. Also, $m = \dim_D(V) = \dim_E(V) = n$.

EXERCISES

**1**   Let $A$ be a ring, $V$ a completely reducible $A$-module. Show that if $\operatorname{End}_A(V)$ is a division ring, then $V$ is irreducible. (This is sort of a converse of Schur's lemma.)

**2**   Show that a semisimple commutative ring is a direct sum of fields.

**3**   Let $A$ be a ring. $e \in A$ is called an *idempotent* if $e^2 = e \neq 0$. Two idempotents $e_1, e_2$ are *orthogonal* if $e_1 e_2 = e_2 e_1 = 0$. A *central idempotent* is an idempotent in the center of $A$.

(a)   Prove that if $A = V_1 \oplus \cdots \oplus V_n$, the $V_i$ submodules of $A$ as $A$-module, then $1 = e_1 + \cdots + e_n$, the $e_i$ pairwise orthogonal idempotents with $V_i = Ae_i$.

(b)   Conversely, show that if $1 = e_1 + \cdots + e_n$, the $e_i$ pairwise orthogonal idempotents, then $A = Ae_1 \oplus \cdots \oplus Ae_n$.

(c)   Prove that if $A = I_1 \oplus \cdots \oplus I_n$, the $I_j$ two-sided ideals of $A$, then $1 = e_1 + \cdots + e_n$, the $e_j$ pairwise orthogonal central idempotents with $I_j = Ae_j$.

(d)   Conversely, show that if $1 = e_1 + \cdots + e_n$, the $e_i$ pairwise orthogonal central idempotents, then $A = Ae_1 \oplus \cdots \oplus Ae_n$ where each $Ae_i$ is a two-sided ideal of $A$.

(e)   Show that if $e \neq 1$ is any idempotent in $A$, then $\{e, 1 - e\}$ is a pair of orthogonal idempotents and $A = Ae \oplus A(1 - e)$.

# §3

## Semisimple Group Algebras

**Theorem 3.1** (Maschke's theorem)     *Let $G$ be a finite group of order $n, k$ a field such that* char $k \nmid n$. *Then the group algebra $kG$ is a semisimple ring.*

*Proof*   Let $V$ be a $kG$-submodule of $kG$; we will find a $kG$-submodule $W$, $kG = V \oplus W$. At least we can find a $k$-subspace $N$, $kG = V \oplus N$. Let $\pi: kG \to V$ be the projection onto $V$ with kernel $N$; $\pi$ is only $k$-linear, not necessarily $kG$-linear.

Define $\pi^*: kG \to kG$ by

$$\pi^*(x) = \frac{1}{n} \sum_{g \in G} g^{-1} \pi(gx).$$

(By our hypothesis, $1/n$ exists in $k$.) We will prove that $V = \pi^*(kG)$, and that

$$kG = V \oplus (1 - \pi^*)(kG) \text{ as } kG\text{-modules.}$$

If $x \in V$, then

$$\pi^*(x) = \frac{1}{n} \sum_{g \in G} g^{-1} \pi(gx) = \frac{1}{n} \sum_{g \in G} g^{-1} gx = \frac{1}{n} \cdot nx = x.$$

For any $y \in kG$, $g \in G$, $\pi(gy) \in V$, so

$$\pi^*(y) = \frac{1}{n} \sum_{g \in G} g^{-1} \pi(gy) \in V.$$

We have proved $\pi^*$ is a projection of $kG$ onto $V$, for it is now easy to see $(\pi^*)^2 = \pi^*$ and

$$kG = \pi^*(kG) \oplus (1 - \pi^*)(kG) = V \oplus (1 - \pi^*)(kG).$$

16

It remains to prove that $W = (1 - \pi^*)(kG)$ is a $kG$-submodule. Suppose $h \in G$, $x \in kG$. Then

$$h^{-1}\pi^*(hx) = \frac{1}{n} \sum_{g \in G} h^{-1}g^{-1}\pi(ghx) = \frac{1}{n} \sum_{y \in G} y^{-1}\pi(yx) = \pi^*(x),$$

so

$$\pi^*(hx) = h\pi^*(x).$$

Any element $(1 - \pi^*)(x)$ of $(1 - \pi^*)(kG)$ therefore satisfies $h(1 - \pi^*)x = hx - h\pi^*(x) = hx - \pi^*(hx) = (1 - \pi^*)(hx) \in (1 - \pi^*)(kG)$, and we are done.

**Theorem 3.2** *Let $G$ be a finite group of order $n$, and let $k$ be an algebraically closed field of characteristic not dividing $n$. Then*

$$kG \cong \mathrm{Mat}_{n_1}(k) \oplus \cdots \oplus \mathrm{Mat}_{n_s}(k),$$

*where $n = n_1^2 + \cdots + n_s^2$. $kG$ has exactly $s$ nonisomorphic irreducible modules, of dimensions $n_1, n_2, \ldots, n_s$. $s$ is the number of conjugacy classes of $G$.*

*Proof* By Maschke's theorem, $kG$ is semisimple. Hence, by Wedderburn's theorem,

$$kG \cong \mathrm{Mat}_{n_1}(D_1^{\mathrm{op}}) \oplus \cdots \oplus \mathrm{Mat}_{n_s}(D_s^{\mathrm{op}}),$$

the $D_i$ division rings, $D_i = \mathrm{End}_{kG}(V_i)$, the $V_i$ irreducible $kG$-modules, $k \subset D_i^{\mathrm{op}} \subset kG$. Thus $D_i$ is finite-dimensional over $k$, $k$ in the center of $D_i$. We claim each $D_i = k$. For if $d \in D_i$, then $k(d)$ is a field, finite-dimensional over $k$. $k$ is algebraically closed, so $k(d) = k$, $d \in k$, $D_i = k$, $D_i^{\mathrm{op}} = k$. We now know that

$$kG \cong \mathrm{Mat}_{n_1}(k) \oplus \cdots \oplus \mathrm{Mat}_{n_s}(k).$$

Taking the dimension of both sides over $k$,

$$n = n_1^2 + \cdots + n_s^2.$$

It remains to show that $s$ is the number of conjugacy classes of $G$. Let $Z$ be the center of $kG$. The center of $\mathrm{Mat}_{n_i}(k)$ is the set of scalar matrices, and so has dimension 1 over $k$; therefore $\dim_k Z = s$.

For each conjugacy class $\mathscr{C}_i$ of $G$, let $C_i = \sum_{x \in \mathscr{C}_i} x \in kG$. If $g \in G$, then $g^{-1}C_i g = C_i$ so all $C_i \in Z$. The $C_i$ are obviously $k$-linearly independent elements of $kG$.

If $\sum_{g \in G} a_g g \in Z$ for some $a_g \in k$, then for any $h \in G$, $\sum a_g g = h^{-1}(\sum a_g g)h = \sum a_g h^{-1}gh$. This means $a_g = a_{h^{-1}gh}$, and implies that $\sum a_g g$ is a $k$-linear combination of the $C_i$. $\{C_i\}$ is a $k$-basis of $Z$, so $s = \dim_k Z = $ number of $C_i$'s = number of conjugacy classes of $G$.

*Example*   Let $G$ be the nonabelian group of order 6. In Examples 1.2 and 1.4, we constructed an irreducible $\mathbf{C}G$-module of degree 2. $G$ has 3 conjugacy classes, so the degrees of the irreducible $\mathbf{C}G$-modules are $n_1, n_2, n_3 = 2$, where $6 = n_1^2 + n_2^2 + 2^2$; hence $n_1 = n_2 = 1$.

The two one-dimensional representations of $G$ over $\mathbf{C}$ are as follows; one is the trivial representation of $G$ in which $g \to 1$, all $g \in G$. $G$ has a normal subgroup $N$ of index 2 (in the notation of Example 1.2, $N$ is the cyclic group generated by $y$, $|N| = 3$). The other 1–dimensional representation $\varphi$ of $G$ is defined by $\varphi(g) = 1$ if $g \in N$, $\varphi(g) = -1$ if $g \in G - N$; it thus has kernel $N$.

We have therefore constructed *all* the irreducible representations of $G$ over $\mathbf{C}$.

EXERCISE

Let $\langle x \rangle$ be the cyclic group of order $p$ generated by $x$, where $p$ is a prime. Let $F$ be a field of characteristic $p$. Show that

$$x \to \begin{pmatrix} 1 & 1 \\ 0 & 1 \end{pmatrix}$$

defines a two-dimensional representation of $\langle x \rangle$ over $F$ which is reducible but not completely reducible. (This shows the hypothesis $p \nmid |G|$ is necessary in Maschke's theorem.)

# §4

## Splitting Fields and Absolutely Irreducible Modules

*Remark*  We saw in Theorem 3.2 that when $k$ is algebraically closed of characteristic not dividing the order of $G$ and $V$ is an irreducible $kG$-module, then the division ring $\mathrm{End}_{kG}(V)$ is actually $k$ itself. This equation $\mathrm{End}_{kG}(V) = k$ occurs for some other fields $k$; when it occurs is the subject of our next discussion.

Recall that if $A$ is an algebra over a field $k$, and $K$ is an extension field of $k$, then we may form a new algebra $A^K = K \otimes_k A$. The reason is as follows: $A$ and $K$ are $k$-modules. The map

$$K \times A \times K \times A \to K \otimes_k A$$

defined by

$$(\alpha, a, \beta, b) \to \alpha\beta \otimes ab$$

is 4–multilinear, and so factors through the tensor product to give a $k$-linear map

$$\varphi : K \otimes A \otimes K \otimes A \to K \otimes A$$

satisfying

$$\varphi : \alpha \otimes a \otimes \beta \otimes b \to \alpha\beta \otimes ab.$$

Composing this with the natural map

$$\otimes : (K \otimes A) \times (K \otimes A) \to K \otimes A \otimes K \otimes A$$

gives us a bilinear map

$$(K \otimes A) \times (K \otimes A) \to K \otimes A$$

satisfying

19

$$(\alpha \otimes a, \beta \otimes b) \to \alpha\beta \otimes ab;$$

this last map is the multiplication making $K \otimes A$ into a ring with unit $1 \otimes 1$, and hence an algebra over $K$.

If $V$ is an $A$-module and hence a $k$-vector space, then the $K$-vector space $V^K = K \otimes_k V$ becomes an $A^K$-module under a module action satisfying $(\alpha \otimes a)(\beta \otimes v) = \alpha\beta \otimes av$, all $\alpha \otimes a \in A^K$, all $\beta \otimes v \in V^K$.

We emphasize that $A^K$ and $V^K$ are not very mysterious; $A$, $V$ are $k$-linear combinations of their basis elements, and we now allow the coefficients to be chosen from $K$ instead of just $k$. For example, if $G$ is a group then $(kG)^K \cong KG$ by an isomorphism $\alpha \otimes g \to \alpha g$, all $\alpha \in K$, $g \in G$.

**Definition**     Let $A$ be a $k$-algebra, $V$ an irreducible $A$-module. $V$ is *absolutely irreducible* if, for every extension field $K$ of $k$, $V^K$ is an irreducible $A^K$-module. $k$ is a *splitting field* for $A$ if every irreducible $A$-module is absolutely irreducible.

**Lemma 4.1**     *Let $k$ be a field, $A$ a finite-dimensional $k$-algebra with a faithful irreducible $A$-module $V$. Then $D = \mathrm{End}_A(V)$ is a division ring containing $k$ in its center, and $D$ is finite-dimensional over $k$. Also, $A \cong \mathrm{Mat}_m(D^{\mathrm{op}})$, where $m = \dim_D(V)$.*

*Proof*   $D$ is a division ring by Schur's lemma, clearly with $k$ in its center. $\dim_k V$ is finite and $D \subset \mathrm{End}_k(V)$, so $\dim_k D$ is finite. $D \supset k$ so $\dim_D V$ is finite; hence $A \cong \mathrm{End}_D(V) \cong \mathrm{Mat}_m(D^{\mathrm{op}})$ by Corollary 2.6.

**Theorem 4.2**     *Let $A$ be a finite-dimensional $k$-algebra, $V$ an irreducible $A$-module. Then $V$ is absolutely irreducible if and only if $\mathrm{End}_A(V) = k$.*

*Proof, Step 1*   We may assume that $V$ is a faithful $A$-module.

*Proof of Step 1*   For any $a \in A$, define $a_V \in \mathrm{End}_k(V)$ by $a_V v = av$, all $v \in V$, and let $B = \{a_V | a \in A\}$. $B$ is a subalgebra of $\mathrm{End}_k(V)$ and the map $a \to a_V$ is a ring homomorphism from $A$ onto $B$, with kernel the ideal $I = \{a \in A | aV = 0\}$. $V$ is a faithful $B$-module, and $\mathrm{End}_A(V) = \mathrm{End}_B(V)$; to prove Step 1 it is enough to show $V$ is absolutely irreducible as $A$-module if and only if $V$ is absolutely irreducible as $B$-module, for we then simply replace $A$ by $B$.

Let $F$ be an extension field of $k$; we will show $V^F$ irreducible as $A^F$-module if and only if irreducible as $B^F$-module. Let $V$ have $k$-basis

$\{v_1, \ldots, v_n\}$, $I$ $k$-basis $\{a_1, \ldots, a_r\}$, and $A$ $k$-basis $\{a_1, \ldots, a_r, \ldots, a_m\}$; then $\dim_k B = \dim_k A - \dim_k I$ and so $B$ has $k$-basis $\{(a_{r+1})_V, \ldots, (a_m)_V\}$. $A^F$ has $F$-basis $\{1 \otimes a_1, \ldots, 1 \otimes a_m\}$ and $B^F$ has $F$-basis $\{1 \otimes (a_{r+1})_V, \ldots, 1 \otimes (a_m)_V\}$.

If $i \le r$, then $(1 \otimes a_i)(1 \otimes v_j) = 1 \otimes a_i v_j = 1 \otimes 0 = 0$; if $i > r$, then $(1 \otimes a_i)(1 \otimes v_j) = 1 \otimes a_i v_j = 1 \otimes (a_i)_V v_j = (1 \otimes (a_i)_V)(1 \otimes v_j)$, proving that $A^F$ and $B^F$ induce the same elements of $\mathrm{End}_F(V^F)$, so $V^F$ is irreducible as $A^F$-module if and only if it is irreducible as $B^F$-module.

*Step 2*   For any extension field $F$ of $k$, $V^F$ is a faithful $A^F$-module.

*Proof of Step 2*   Let $A$ have $k$-basis $\{a_1, \ldots, a_m\}$, $V$ $k$-basis $\{v_1, \ldots, v_n\}$, where $a_i v_j = \sum_{l=1}^n \alpha_{ijl} v_l$, all $\alpha_{ijl} \in k$. Hence $(\sum_{i=1}^m \beta_i a_i) v_j = \sum_{i=1}^m \sum_{l=1}^n \beta_i \alpha_{ijl} v_l$. $V$ faithful means that, if $\sum_{i=1}^m \beta_i \alpha_{ijl} = 0$ for all $j$ and all $l$, then all $\beta_i = 0$.

Using the bases $\{1 \otimes a_i\}$ of $A^F$, $\{1 \otimes v_j\}$ of $V^F$ we similarly see $V^F$ will be faithful if we show that $\gamma_i \in F$, $\sum_{i=1}^m \gamma_i \alpha_{ijl} = 0$ for all $j$ and $l$ implies all $\gamma_i = 0$.

The system of $n^2$ equations $\sum_{i=1}^m \alpha_{ijl} X_i = 0$ has no nontrivial solution over $k$, so contains $m$ linearly independent equations. In this set of $m$ equations, the determinant of coefficients is invertible. This fact remains true in $F$, so the system has no nontrivial solution in $F$ either, proving that all $\gamma_i = 0$.

*Step 3*   If $V$ is absolutely irreducible, then $\mathrm{End}_A(V) = k$.

*Proof of Step 3*   By Steps 1 and 2, if $K$ is the algebraic closure of $k$ then $V^K$ is a faithful irreducible $A^K$-module. In Lemma 4.1 applied to $A^K$ and $V^K$, $D$ must be $K$ itself; hence we have

$$\mathrm{End}_{A^K}(V^K) = K \qquad \text{and} \qquad A^K \cong \mathrm{Mat}_n(K),$$

where

$$n = \dim_K(V^K) = \dim_k V.$$

Applying Lemma 4.1 to $A$ and $V$, we have

$$D = \mathrm{End}_A(V), \qquad A \cong \mathrm{Mat}_m(D^{\mathrm{op}}), \qquad m = \dim_D V.$$

Now

$$n^2 = \dim_K A^K = \dim_k A = m^2(\dim_k D).$$

But from above,

$$n = \dim_k V = (\dim_k D)(\dim_D V) = m(\dim_k D).$$

Hence $\dim_k D = n/m = n^2/m^2$, forcing

$$\dim_k D = 1, \qquad k = D = \text{End}_A(V).$$

*Step 4, Conclusion*   If $\text{End}_A(V) = k$, then $V$ is absolutely irreducible.

*Proof*   Let $F$ be any extension field of $k$. By Lemma 4.1 $A = \text{Mat}_n(k)$, so $A^F = F \otimes_k \text{Mat}_n(k) \cong \text{Mat}_n(F)$. $V^F$ is an $A^F$-module with $\dim_F V^F = \dim_k V = n$, and so must be irreducible.

**Definition**       If $G$ is a finite group and $k$ a field, $k$ is a *splitting field for* $G$ if $k$ is a splitting field for $kG$. We see that $k$ is a splitting field for $kG$ if and only if $\text{End}_{kG}(V) = k$ for all irreducible $kG$-modules $V$.

**Corollary 4.3**       *Let $k$ be a field, $G$ a finite group, char $k \nmid |G|$. The following are equivalent:*

(1)   *$k$ is a splitting field for $G$.*
(2)   *$kG$ is a direct sum of full matrix rings over $k$.*

*Proof* $(1) \Rightarrow (2)$   Theorem 3.2 and the last sentence before this Corollary.

*Proof* $(2) \Rightarrow (1)$   If $k$ were not a splitting field, we would have $kG \cong \oplus \sum \text{Mat}_{n_i}(D_i^{\text{op}})$, some $D_i = \text{End}_{kG}(V_i) \neq k$; this would contradict Theorem 2.18 (b).

**Corollary 4.4**       *If $k$ is a splitting field for $G$ and char $k \nmid |G|$, then*

$$kG \cong \text{Mat}_{n_1}(k) \oplus \cdots \oplus \text{Mat}_{n_s}(k),$$

*where $n = n_1^2 + \cdots + n_s^2$. $kG$ has exactly $s$ nonisomorphic irreducible modules, of dimensions $n_1, n_2, \ldots, n_s$. $s$ is the number of conjugacy classes of $G$.*

*Proof*   Same as the proof of Theorem 3.2, except that we already know all $D_i = k$.

*Remark*   The *exponent* of a group $G$ is the smallest integer $m$ such that $x^m = 1$, all $x \in G$. We will show in §17 that if $F$ is any subfield of **C** containing all $m$th roots of 1, where $m$ is the exponent of $G$, then $F$ is a splitting field for $G$.

*Example* If $\langle x \rangle$ is a cyclic group of order 3 and **Q** = rational numbers, one can easily show that

$$x \rightarrow \begin{pmatrix} -1 & -1 \\ 1 & 0 \end{pmatrix}$$

is an irreducible representation of $\langle x \rangle$ over **Q**. This cannot be absolutely irreducible, as over **C** all irreducible representations of $\langle x \rangle$ have degree 1 (only solutions of $3 = \sum (\text{degree})^2$).

## EXERCISE

Let $G$ be a finite group, $k$ a field with char $k \nmid |G|$, and let $V_1, \ldots, V_r$ be a full set of nonisomorphic irreducible $kG$-modules. Show that $k$ is a splitting field for $G$ if and only if

$$|G| = \sum_{i=1}^{r} (\dim_k V_i)^2.$$

# §5

## Characters

**Definition**    Let $k$ be a field, $G$ a group, $f: G \to k$ a function. $f$ is a *class function* if it is constant on conjugacy classes, i.e., if $f(x^{-1}yx) = f(y)$, all $x, y \in G$. If $G$ has $h$ conjugacy classes, then the class functions on $G$ form a $k$-vector space $Cf(G,k)$ of $k$-dimension $h$.

Recall that similar matrices have the same trace, and the matrices of the same linear transformation in two different bases are similar. Hence for any representation $T: G \to GL(V)$ and any $x \in G$, the trace $\chi(x) = \text{tr } T(x)$ is well defined.

**Definition**    Let $k$ be a field, $G$ a group, $V$ a finite-dimensional $k$-vector space, $T: G \to GL(V)$ a representation. The function $\chi: G \to k$ defined by

$$\chi(x) = \text{tr } T(x), \qquad \text{all } x \in G$$

is called a *character;* it is an *irreducible character* if the representation $T$ is irreducible.

**Lemma 5.1**    *Characters are class functions. Isomorphic $kG$-modules or equivalent representations afford the same character.*

*Proof*  $\chi(x^{-1}yx) = \text{tr } T(x^{-1}yx) = \text{tr } T(x)^{-1}T(y)T(x) = \text{tr } T(y) = \chi(y)$. Isomorphic modules afford equivalent representations, and equivalent representations have similar matrices.

**Lemma 5.2**    *Let $T$ and $U$ be irreducible representations afforded, respectively, by the irreducible $kG$-modules $V$ and $W$. For suitable bases, let $T$ afford the matrix representation $\varphi(x) = (a_{ij}(x))$, $U$ the matrix representation $\psi(x) = (b_{ij}(x))$. Then:*

24

(a)  If $V \ncong W$, then for all $i,j,k,l$,

$$\sum_{x \in G} b_{ij}(x^{-1})a_{kl}(x) = 0.$$

(b)  If $k$ is a splitting field for $G$ and char $k \nmid |G|$, then

$$\sum_{x \in G} a_{ij}(x^{-1})a_{kl}(x) = \delta_{jk}\delta_{il}\frac{|G|}{\deg T}.$$

*Proof*  For any $f \in \mathrm{Hom}_k(V,W)$, let $f$ have matrix $(c_{ij})$ in the chosen bases of $V$ and $W$. Define $f^*: V \to W$ by

$$f^* = \sum_{x \in G} U(x^{-1})fT(x).$$

For any $y \in G$, then $f^*T(y) = \sum_{x \in G} U(y)U(y^{-1}x^{-1})fT(xy) = U(y)f^*$, so for any $v \in V$, $f^*(yv) = f^*T(y)v = U(y)f^*(v) = yf^*(v)$. This proves that $f^* \in \mathrm{Hom}_{kG}(V,W)$.

(a)  If $V \ncong W$, then by Schur's lemma $f^* = 0$ for any $f$. Choose $(c_{ij})$ to be $(\delta_{ir}\delta_{js})$, $r$ and $s$ fixed. Taking the $i,j$th entry in the definition of $f^*$,

$$0 = \sum_{x \in G}\sum_{m,n} b_{im}(x^{-1})c_{mn}a_{nj}(x) = \sum_{x \in G} b_{ir}(x^{-1})a_{sj}(x).$$

(b)  Here $\mathrm{End}_{kG}(V) = \{aI \mid a \in k\}$, $I$ the identity map. Hence for any $f \in \mathrm{End}_k(V)$, $f^* = a_f I$, some $a_f \in k$. Also, $\mathrm{tr}\, f^* = \sum_{x \in G}\mathrm{tr}\,(T(x)^{-1}fT(x)) = |G|\mathrm{tr}\, f$. If $f$ has matrix $(\delta_{ir}\delta_{js})$ for fixed $r,s$, then $\mathrm{tr}\, f = \delta_{rs}$. Hence $|G|\delta_{rs} = |G|\mathrm{tr}\, f = \mathrm{tr}\, f^* = \mathrm{tr}(a_f I) = a_f(\deg T)$. Setting $r = s$ proves $\deg T \neq 0$ in $k$, since $|G| \neq 0$; therefore $|G|/\deg T \in k$. We have shown

$$\sum_{x \in G} T(x^{-1})fT(x) = f^* = a_f I = \frac{|G|\delta_{rs}}{\deg T} I.$$

Taking the $i,j$th matrix entry in this equation, we obtain (b).

**Theorem 5.3** (Orthogonality Relations)    *Let $\chi$ and $\zeta$ be the characters afforded by the irreducible kG-modules $V$ and $W$. Then:*

(1)  *For any field $k$, $V \ncong W$ implies*

$$\sum_{x \in G} \chi(x)\zeta(x^{-1}) = 0.$$

(2) *If $k$ is a splitting field for $G$ and char $k \nmid |G|$, then $\sum_{x \in G} \chi(x)\chi(x^{-1}) = |G|$.*

(3) *Let $k$ be a splitting field for $G$ with char $k \nmid |G|$. Assume that $G$ has exactly $h$ conjugacy classes $\mathscr{C}_1, \ldots, \mathscr{C}_h$. Denote $h_i = |\mathscr{C}_i|$, choose $x_i \in \mathscr{C}_i$, and denote the irreducible characters of $G$ by $\chi_1, \ldots, \chi_h$. Then*

$$\sum_{m=1}^{h} \chi_m(x_i)\chi_m(x_j^{-1}) = \frac{\delta_{ij}|G|}{h_j} = \delta_{ij}|C_G(x_j)|.$$

(Here, of course, $C_G(x_j) = $ *centralizer* of $x_j$ in $G = \{x \in G | xx_j = x_j x\}$.)

*Proof (1)* Let $V$ afford the matrix representation $(a_{ij}(x))$, $W$ the representation $(b_{ij}(x))$. Then

$$\sum_{x \in G} \chi(x)\zeta(x^{-1}) = \sum_{x \in G} \sum_{l} a_{ll}(x) \sum_{m} b_{mm}(x^{-1})$$

$$= \sum_{l,m} \sum_{x \in G} a_{ll}(x)b_{mm}(x^{-1}) = 0,$$

by Lemma 5.2(a).

*Proof (2)* Let $V$ afford $(a_{ij}(x))$. Using Lemma 5.2(b), we have

$$\sum_{x \in G} \chi(x)\chi(x^{-1}) = \sum_{l,m} \sum_{x \in G} a_{ll}(x)a_{mm}(x^{-1})$$

$$= \sum_{l,m=1}^{\dim V} \delta_{lm} \frac{|G|}{\dim V} = \dim V \cdot \frac{|G|}{\dim V} = |G|.$$

*Proof (3)* We introduce two $h \times h$ matrices $B$ and $C$ by

$$b_{ij} = \frac{h_j}{|G|} \chi_i(x_j^{-1}), \qquad c_{ij} = \chi_j(x_i).$$

Using (1) and (2), the $r,s$th entry of $BC$ is

$$\sum_{t=1}^{h} b_{rt}c_{ts} = \sum_{t=1}^{h} \frac{h_t}{|G|} \chi_r(x_t^{-1})\chi_s(x_t) = \delta_{rs}.$$

With $I$ the identity matrix, this says $BC = I$, so also $CB = I$. The $i,j$th entry of $CB$ is therefore

$$\delta_{ij} = \sum_{m=1}^{h} \chi_m(x_i) \frac{h_j}{|G|} \chi_m(x_j^{-1}), \qquad \text{proving (3).}$$

*Remark*  All the orthogonality relations can be deduced from a very few; see Nagao [2].

**Definitions**   Let $V$ be a $k$-vector space,

$$( , ): V \times V \to k$$

a bilinear form. $( , )$ is *symmetric* if $(x,y) = (y,x)$, all $x,y \in V$. If $( , )$ is symmetric, denote $V^{\perp} = \{v \in V | (v,w) = 0, \text{ all } w \in V\}$; $( , )$ is called *nonsingular* if $V^{\perp} = \{0\}$. A basis $\{v_1, \ldots, v_n\}$ of $V$ is called *orthogonal* if $i \neq j$ implies $(v_i, v_j) = 0$; it is *orthonormal*, if it is orthogonal and $(v_i, v_i) = 1$, all $i$.

**Definition**   An *algebraic number field* is a finite extension field $K$ of the rational numbers $\mathbf{Q}$. Any element of $K$ is an *algebraic number*. $\alpha \in K$ is an *algebraic integer* if $\alpha$ is a root of a monic polynomial with integer coefficients.

**Lemma 5.4**   *Let $K$ be a finite extension field of the rational numbers $\mathbf{Q}$, and let $\alpha \in K$. Then the following are equivalent:*

- (a)   $\alpha$ *is an algebraic integer.*
- (b)   *If $f(X)$ is the monic polynomial irreducible over $\mathbf{Q}$ and satisfied by $\alpha$, then $f(X)$ is a polynomial over the integers $\mathbf{Z}$.*
- (c)   $\mathbf{Z}[\alpha]$ *is a finitely generated $\mathbf{Z}$-submodule of $K$. ($\mathbf{Z}[\alpha]$ is the subring of $K$ generated by $\mathbf{Z}$ and $\alpha$.)*

*Proof (a) $\Rightarrow$ (b)*   Let $\alpha$ be a zero of the monic polynomial $h(X)$ over $\mathbf{Z}$, so $h(X) = f(X)g(X)$, some polynomial $g(X)$ over $\mathbf{Q}$. We write

$$f(X) = \frac{aF(X)}{b}, \qquad g(X) = \frac{cG(X)}{d},$$

where $a,b,c,d \in \mathbf{Z}$, and $F(X)$ and $G(X)$ are *primitive* polynomials over $\mathbf{Z}$; i.e., the G.C.D. of the coefficients of $F(X)$ is 1, and similarly for $G(X)$.

We claim that $F(X)G(X)$ is also primitive. If not, let $p$ be a prime dividing all coefficients of $F(X)G(X)$, and let — denote taking the residue mod $p$. In the field $\mathbf{Z}/p\mathbf{Z}$, we have $0 = \overline{F(X)G(X)} = \overline{F(X)}\,\overline{G(X)}$, a contradiction since $0 \neq \overline{F(X)}$, $0 \neq \overline{G(X)}$.

$h(X) = f(X)g(X)$ implies that

$$bdh(X) = acF(X)G(X).$$

Therefore primitivity of $h(X)$ and $F(X)G(X)$ implies $bd = ac$, $h(X) = F(X)G(X)$. We now see that $F(X)$ and $G(X)$ are monic, so $f(X) = F(X)$ is a polynomial over $\mathbf{Z}$.

*Proof* $(b) \Rightarrow (c)$   Let $f(X) = X^m + a_1 X^{m-1} + \cdots + a_m$, all $a_i \in \mathbf{Z}$. Then $\alpha^m = -a_1\alpha^{m-1} - \cdots - a_m$, and we see $\mathbf{Z}[\alpha] = \mathbf{Z} \oplus \mathbf{Z}\alpha \oplus \cdots \oplus \mathbf{Z}\alpha^{m-1}$.

*Proof* $(c) \Rightarrow (a)$   Let $\mathbf{Z}[\alpha] = \mathbf{Z}f_1(\alpha) + \cdots + \mathbf{Z}f_n(\alpha)$, where the $f_i(\alpha) \in \mathbf{Z}[\alpha]$ so $f_i(X)$ is a polynomial over $\mathbf{Z}$. Choose $N$ greater than the degrees of all the $f_i(X)$, and $\alpha^N \in \mathbf{Z}[\alpha]$ so $\alpha^N = a_1 f_1(\alpha) + \cdots + a_n f_n(\alpha)$, $a_i \in \mathbf{Z}$. This gives a monic polynomial over $\mathbf{Z}$ satisfied by $\alpha$.

**Corollary 5.5**   *The algebraic integers in an algebraic number field $K$ form a subring of $K$.*

*Proof*   If $\alpha, \beta \in K$ are algebraic integers, we saw that $\mathbf{Z}[\alpha]$ is generated by $1, \alpha, \ldots, \alpha^{m-1}$ and $\mathbf{Z}[\beta]$ is generated by $1, \beta, \ldots, \beta^{n-1}$. Then $\mathbf{Z}[\alpha, \beta]$ is generated by $\{\alpha^i \beta^j\}$, $i < m$, $j < n$. The $\mathbf{Z}$-submodules $\mathbf{Z}[\alpha + \beta]$ and $\mathbf{Z}[\alpha\beta]$ are therefore finitely generated, so $\alpha + \beta$ and $\alpha\beta$ are algebraic integers.

**Corollary 5.6**   *The only algebraic integers which are rational numbers are the ordinary integers.*

*Proof*   If $r \in \mathbf{Q}$, then the irreducible polynomial for $r$ over $\mathbf{Q}$ is $X - r = 0$. By Lemma 5.4(b), $r \in \mathbf{Z}$.

**Theorem 5.7**   *Let $k$ be a splitting field for $G$ with char $k \nmid |G|$, and let $Cf(G,k)$ be the $k$-vector space of class functions $f: G \to k$. By defining*

$$(f_1, f_2) = (f_1, f_2)_G = \frac{1}{|G|} \sum_{x \in G} f_1(x) f_2(x^{-1}),$$

$( \, , \, )_G$ *becomes a symmetric nonsingular bilinear form $Cf(G,k) \times Cf(G,k) \to k$, called the* inner product. *If $\chi_1, \ldots, \chi_h$ are the irreducible characters of $G$, then $\chi_1, \ldots, \chi_h$ form an orthonormal basis for $Cf(G,k)$.*

*Proof*   The orthogonality relations (1), (2) of Theorem 5.3 show $(\chi_i, \chi_j) = \delta_{ij}$. Using this, we see that $\chi_1, \ldots, \chi_h$ are $k$-linearly independent. $h = $ number of conjugacy classes by Corollary 4.4, so $\{\chi_1, \ldots, \chi_h\}$ is a basis of $Cf(G,k)$. $(\chi_i,\chi_j) = \delta_{ij}$ now implies that $(\,,)$ is symmetric nonsingular, with orthonormal basis $\{\chi_1, \ldots, \chi_h\}$.

**Theorem 5.8**     *For any* $\alpha \in \mathbf{C} =$ *complex numbers, denote* $\bar{\alpha} =$ *complex conjugate of* $\alpha$. *Let* $\chi_1, \ldots, \chi_h$ *be the irreducible characters of* $G$ *over* $\mathbf{C}$. *Then:*

(i)   $\chi_i(x^{-1}) = \overline{\chi_i(x)}$, *all* $x \in G$.

(ii)  *All* $\chi_i(x)$ *are sums of* $|G|$*th roots of* 1, *and hence are algebraic integers.*

(iii) *If* $0 \neq f = \sum_{i=1}^{h} r_i \chi_i$, *the* $r_i \in \mathbf{R} =$ *real numbers, then* $(f,f) > 0$; *that is, on the real vector space with basis* $\{\chi_1, \ldots, \chi_h\}$, $(\,,)$ *is a* positive definite *bilinear form.*

*Proof*  Let $\varphi_i: G \to \mathrm{Mat}_{n_i}(\mathbf{C})$ be the matrix representation with character $\chi_i$. $\varphi_i(x)$ is similar to its Jordan form, so $\chi_i(x) = \sum_{j=1}^{n_i} \varepsilon_{ij}$, $\{\varepsilon_{ij}\}$ the eigenvalues of $\varphi_i(x)$. $\varphi_i(x)^{|G|} = \varphi_i(x^{|G|}) = \varphi_i(1) = I$ has as its eigenvalues the set of $|G|$th powers of the $\varepsilon_{ij}$, so all $\varepsilon_{ij}$ are $|G|$th roots of 1; this proves (ii). Roots $\gamma$ of 1 satisfy $\gamma^{-1} = \bar{\gamma}$, so

$$\chi_i(x^{-1}) = \sum \varepsilon_{ij}^{-1} = \sum \overline{\varepsilon_{ij}} = \overline{\chi_i(x)}.$$

That proves (i). For (iii),

$$(f,f) = \left(\sum_i r_i \chi_i, \sum_j r_j \chi_j\right) = \sum_{i,j} r_i r_j (\chi_i,\chi_j) = \sum_{i,j} r_i r_j \delta_{ij} = \sum_i r_i^2 > 0.$$

**Lemma 5.9**     *Let* $k$ *be a splitting field of characteristic zero for* $G$. *Let* $\mathscr{C}_1, \ldots, \mathscr{C}_h$ *be the conjugacy classes of* $G$, *denote* $h_i = |\mathscr{C}_i|$, *and choose representatives* $x_i \in \mathscr{C}_i$. *Let* $\chi_1, \ldots, \chi_h$ *be the irreducible characters of* $G$, *with* $n_i = \deg \chi_i$. *Then the numbers*

$$\omega_{ij} = \frac{h_i \chi_j(x_i)}{n_j}, \qquad i,j = 1, \ldots, h,$$

*are algebraic integers.*

*Proof*  Let $C_i = \sum_{x \in \mathscr{C}_i} x \in kG$. We saw in the proof of Theorem 3.2 that $\{C_1, \ldots, C_h\}$ is a basis of the center of $kG$. Hence there exist integers $c_{ijm}$ such that

$$C_i C_j = \sum_{m=1}^{h} c_{ijm} C_m.$$

If $\chi_l$ is the character of the representation $T_l$, let $V_l$ be the corresponding irreducible $kG$-module, and extend $T_l: G \rightarrow \mathrm{End}_{kG}(V_l)$ to all of $kG$ by $k$-linearity. Then

$$T_l(C_i)T_l(C_j) = \sum_m c_{ijm}T_l(C_m). \qquad (*)$$

All $C_i \in$ center of $kG$, so for all $x \in kG$, $T_l(C_i)$ and $T_l(x)$ commute. For any $v \in V_l$, this means

$$T_l(C_i) \cdot xv = T_l(C_i)T_l(x)v = T_l(x)T_l(C_i)v = x \cdot T_l(C_i)v,$$

so $T_l(C_i) \in \mathrm{End}_{kG}(V_l) = k$. Hence for some $\omega_{ij} \in k$,

$$T_l(C_i) = \omega_{il}I, \qquad I = \text{identity map}.$$

Taking traces in this equation, $h_i\chi_l(x_i) = \omega_{il}n_l$, so the $\omega_{ij}$ are as stated in the Lemma. The equation (*) becomes

$$\omega_{il}\omega_{jl} = \sum_m c_{ijm}\omega_{ml}.$$

This may be rewritten

$$0 = \sum_{m=1}^{h} (c_{ijm} - \delta_{jm}\omega_{il})\omega_{ml}.$$

If we denote $\mathscr{C}_1 = \{1\}$, then $h_1 = 1$ and $\omega_{11} = 1 \cdot \chi_l(1)/n_l = 1$.

For fixed $i$ and $l$, our system of equations

$$0 = \sum_m (c_{ijm} - \delta_{jm}\omega_{il})X_m$$

has a nontrivial solution, since $\omega_{11} = 1 \neq 0$. Therefore

$$0 = \det(c_{ijm} - \delta_{jm}\omega_{il}) = \det \begin{pmatrix} c_{i11} - \omega_{il} & \cdots\cdots & c_{i1h} \\ c_{i21} & c_{i22} - \omega_{il} & \\ \vdots & & \\ c_{ih1} & \cdots\cdots & c_{ihh} - \omega_{il} \end{pmatrix}.$$

This means $\omega_{il}$ is a characteristic root of the matrix $(c_{ijm})_{j,m=1\ldots h}$. The entries of this matrix are integers, so $\omega_{il}$ satisfies a monic polynomial over the integers and is an algebraic integer.

**Lemma 5.10**    *Let*  $T: G \to \text{Mat}_t(\mathbf{C})$  *be a representation affording the character*  $\chi$ . *Then for any*  $x \in G$ ,

$$|\chi(x)| \le t;$$

*equality holds only if*  $T(x) = \alpha I$ , *some*  $\alpha \in \mathbf{C}$ .

*Proof*  Let  $T(x)$  have characteristic roots  $\varepsilon_1, \ldots, \varepsilon_t$ , so  $\chi(x) = \varepsilon_1 + \cdots + \varepsilon_t$ . Therefore
$$|\chi(x)| = |\varepsilon_1 + \cdots + \varepsilon_t| \le |\varepsilon_1| + \cdots + |\varepsilon_t| = t,$$
and equality holds only if  $\varepsilon_1 = \varepsilon_2 = \cdots = \varepsilon_t = \alpha$ , say. If equality holds, then the matrix  $\varphi(x)$  afforded by  $T(x)$  satisfies its characteristic equation  $(X - \alpha)^t = 0$ , and also satisfies  $X^{|G|} = 1$ . Therefore  $\varphi(x)$  is a zero of

$$\text{G.C.D.}((X - \alpha)^t, X^{|G|} - 1) = X - \alpha.$$

EXERCISE

Show that if  $\chi$  is an irreducible complex character of  $G$ , and  $x \in G$  has order 2, then  $\chi(x)$  is an ordinary integer and

$$\chi(x) \equiv \chi(1) \qquad (\text{mod } 2).$$

# §6

## Burnside's $p^a q^b$ Theorem

We pause to recall some basic abstract finite group theory.

*Sylow's theorem* asserts that if $|G| = p^a m$, where $p$ is a prime and $p \nmid m$, then: (1) $G$ has a subgroup (*Sylow p-subgroup*) of order $p^a$; (2) any two Sylow $p$-subgroups of $G$ are conjugate; (3) the number of Sylow $p$-subgroups of $G$ is $1 + kp$ for some integer $k$.

A group $G$ is *solvable* if there is a sequence

$$G_0 = 1 \subset G_1 \subset G_2 \subset \cdots \subset G_m = G,$$

where all $G_i \lhd G_{i+1}$ and all $G_{i+1}/G_i$ are abelian. It is easy to see that subgroups and factor groups of solvable groups are solvable, and if $G$ has a normal subgroup $N$ such that $N$ and $G/N$ are solvable then $G$ is solvable.

A group $G$ is *simple* if its only normal subgroups are $G$ and 1. Clearly the only simple, solvable finite groups are cyclic of prime order. Determination of the noncyclic simple groups is the most important unsolved problem of finite group theory.

A finite group is a *p-group* if its order is a power of the prime $p$.

**Lemma 6.1**    *Every finite p-group has a nontrivial center.*

*Proof*   Let $|G| = p^m$, and let $G$ have conjugacy classes $\mathscr{C}_1, \ldots, \mathscr{C}_h$, $x_i \in \mathscr{C}_i$. We know that $|\mathscr{C}_i| = |G|/|C_G(x_i)|$. For each $i$, $C_G(x_i)$ is a subgroup of $G$ and so has order a power of $p$; hence $|\mathscr{C}_i|$ is always a power of $p$, say $|\mathscr{C}_i| = p^{a_i}$, $a_i \geq 0$. $\mathscr{C}_1 = \{1\}$, so $|\mathscr{C}_1| = p^0 = 1$. $G = \cup_{i=1}^h \mathscr{C}_i$ so $p^m = |G| = \sum_i |\mathscr{C}_i| = 1 + \sum_{i=2}^h |\mathscr{C}_i|$. If all $|\mathscr{C}_i| > 1$ for $i > 1$ this means $0 \equiv 1 \pmod{p}$, a contradiction. So for some $i > 1$, $\mathscr{C}_i = \{x_i\}$ only; this means $x_i \in$ center of $G = Z(G)$.

**Corollary 6.2**    *Finite p-groups are solvable.*

*Proof* By induction on $|G|$. By Lemma 6.1, $Z(G) \neq 1$, and $Z(G)$ is solvable since abelian. $G/Z(G)$ is solvable by induction, so $G$ is solvable.

The object of this section is to prove

**Theorem 6.3** (Burnside's $p^a q^b$ Theorem) (Burnside [2]). *Let* $|G| = p^a q^b$, *$p$ and $q$ primes. Then $G$ is solvable.*

*Remark* This theorem was only recently proved without group representation theory by Thompson; see Goldschmidt [1] for comments. The proof follows a couple of Lemmas.

**Lemma 6.4** *Let $\chi$ be an irreducible character of $G$ over $\mathbf{C}$, and suppose that $(|\mathscr{C}|, \deg \chi) = 1$ for some conjugacy class $\mathscr{C}$ of $G$. Then for any $x \in \mathscr{C}$, either $\chi(x) = 0$ or $|\chi(x)| = \deg \chi$.*

*Proof* Choose integers $s$ and $t$ such that $s|\mathscr{C}| + t\chi(1) = 1$. (Of course $\chi(1) = \deg \chi$.) The left side of the equation

$$\frac{s|\mathscr{C}|\chi(x)}{\chi(1)} + t\chi(x) = \frac{\chi(x)}{\chi(1)}$$

is an algebraic integer by Lemma 5.9, so $a = \chi(x)/\chi(1)$ is an algebraic integer. $\chi(x)$ is a sum of $\chi(1)$ roots of unity, and so is every algebraic conjugate of $\chi(x)$. Let $a = a_1, \ldots, a_m$ be all the algebraic conjugates of $a$. Then all $a_i$ are algebraic integers, all $|a_i| \leq 1$. $N(a) = a_1 a_2 \cdots a_m$, being invariant under all field automorphisms over $\mathbf{Q}$, is both rational and an algebraic integer, so $N(a)$ is 0 or $\pm 1$. If $N(a) = 0$, then $a = 0$ and $\chi(x) = 0$. If $N(a) = \pm 1$, then $|a| = 1$ so $|\chi(x)| = \deg \chi$.

**Lemma 6.5** *Assume $G$ has a conjugacy class $\mathscr{C} \neq \{1\}$ such that $|\mathscr{C}|$ is a power of a prime. Then $G$ is not a noncyclic simple group.*

*Proof* Let $|\mathscr{C}| = p^c$, and choose $x \in \mathscr{C}$; if $c = 0$, then $x \in Z(G) \lhd G$ so a noncyclic $G$ is not simple, done.

Let $\{\chi_1, \ldots, \chi_h\}$ be all the irreducible complex characters of $G$, where $\chi_1 = 1_G = $ the trivial character. If $G$ is simple, then $\chi_2, \ldots, \chi_h$ are all characters of faithful representations. By the orthogonality relations,

$$0 = \sum_{i=1}^{h} \chi_i(1)\chi_i(x) = 1 + \sum_{i=2}^{h} \chi_i(1)\chi_i(x).$$

We may arrange the $\chi_i$'s so that

$$\text{if} \quad 2 \leq i \leq h_0, \qquad \text{then} \quad p \nmid \chi_i(1);$$

$$\text{if} \quad h_0 + 1 \leq i \leq h, \qquad \text{then} \quad p \mid \chi_i(1).$$

If $2 \leq i \leq h_0$, then $1 = (p^c, \chi_i(1))$, so by Lemma 6.4, $\chi_i(x) = 0$ or $|\chi_i(x)| = \chi_i(1)$. But $|\chi_i(x)| = \chi_i(1)$ would imply by Lemma 5.10 that $x \in Z(G)$, a contradiction to the simplicity of $G$. So $\chi_i(x) = 0$ for $2 \leq i \leq h_0$.

Let $\chi_i(1) = pm_i$ for $i > h_0$. Our orthogonality relation becomes

$$0 = 1 + \sum_{i=h_0+1}^{h} \chi_i(1)\chi_i(x) = 1 + \sum_{i=h_0+1}^{h} pm_i\chi_i(x).$$

If $\beta = \sum_{i=h_0+1}^{h} m_i\chi_i(x)$, then $\beta$ is an algebraic integer. But $0 = 1 + p\beta$ so $\beta = -1/p$, the final contradiction. Therefore $G$ is not simple.

*Proof of Theorem 6.3*   We use induction on $|G| = p^a q^b$. Let $P$ be a Sylow $p$-subgroup of $G$, and choose $1 \neq x \in Z(P)$. Let $\mathscr{C}$ be the conjugacy class of $G$ containing $x$. $C_G(x) \supseteq P$, so $p \nmid |G : C_G(x)|$, and $|\mathscr{C}| = |G|/|C_G(x)| = q^{b'}$, some $b' \leq b$. By Lemma 6.5, $G$ is not simple, and so has a normal subgroup $N$, $1 \neq N \neq G$. By induction, starting with Lemma 6.2, $N$ and $G/N$ are solvable, so $G$ is solvable.

# §7

## Multiplicities, Generalized Characters, Character Tables

**Definitions**    Let $A$ be a ring, $V$ and $W$ $A$-modules. We say $W$ is a *constituent* of $V$ if $V$ has submodules $V_1 \subset V_2$ such that $W \cong V_2/V_1$. We say $W$ is a *component* of $V$ if $W$ is isomorphic to a direct summand of $V$. A chain

$$(0) = V_0 \subset V_1 \subset \cdots \subset V_m = V$$

is called a *composition series* of $V$ if each $V_{i+1}/V_i$ is irreducible. The $V_{i+1}/V_i$ are then called the *composition factors* of $V$. The *Jordan-Hölder theorem* asserts that if $V$ has two composition series, then both series have the same composition factors (up to isomorphism). Of course if $V$ is completely reducible with composition series, then every composition factor is a component, and $V$ is completely determined by listing its irreducible components. A proof of the Jordan-Hölder theorem appears in §38 of Part B.

**Lemma 7.1**    *Let $k$ be any field, $G$ a finite group, $V$ a finite-dimensional $kG$-module. Then the character of $V$ is the sum of the characters of its composition factors. Also, if $V_1, \ldots, V_l$ are $kG$-modules with characters $\theta_1, \ldots, \theta_l$, then the character of $V_1 \oplus \cdots \oplus V_l$ is $\theta_1 + \cdots + \theta_l$.*

*Proof*    Let

$$(0) = V_0 \subset V_1 \subset \cdots \subset V_m = V$$

be a composition series for $V$, and choose a $k$-basis of $V$ with its first $\dim_k V_1$ elements in $V_1$, the next $\dim_k V_2 - \dim_k V_1$ elements in $V_2 - V_1$, etc. The matrices of elements $x \in G$ in this basis have form

35

$$\varphi(x) = \begin{pmatrix} \varphi_1(x) & * & \cdots & & * \\ 0 & \varphi_2(x) & & & \vdots \\ \vdots & & & & * \\ 0 & 0 & \cdots & 0 & \varphi_m(x) \end{pmatrix}.$$

Here $\varphi_i(x)$ is the matrix representation afforded by $V_i/V_{i-1}$ and $\operatorname{tr}\varphi(x) = \sum_i \operatorname{tr}\varphi_i(x)$, proving the first part of the Lemma. The second part has a similar proof, or follows from the first and the Jordan-Hölder theorem.

**Definition**     Let $A$ be a ring, $V$ an $A$-module with a composition series, and $W$ an irreducible $A$-module. The number of composition factors of $V$ isomorphic to $W$ is the *multiplicity of $W$ in $V$*.

**Lemma 7.2**     *Let $V$ and $W$ be $kG$-modules with respective characters $\chi$ and $\theta$, where we assume* char $k = 0$. *Then $V \cong W$ if and only if $\chi = \theta$.*

*Proof*   By Lemma 5.1, $V \cong W$ implies $\chi = \theta$. Conversely, assume $\chi = \theta$. Let $K$ be the algebraic closure of $k$, $\chi_1, \ldots, \chi_h$ the irreducible $KG$-characters, $\varphi_1, \ldots, \varphi_t$ the irreducible $kG$-characters. By Theorem 5.3(1), $i \neq j$ implies $(\varphi_i, \varphi_j)_G = 0$.

Let $V_1, \ldots, V_t$ be the irreducible $kG$-modules affording $\varphi_1, \ldots, \varphi_t$, and let $U_1, \ldots, U_h$ be the irreducible $KG$-modules affording $\chi_1, \ldots, \chi_h$. The $KG$-module $V_i^K$ affords the same matrix representation as $V_i$, and so affords the character $\varphi_i$. $V_i^K$ is completely reducible, say

$$V_i^K \cong \underbrace{U_1 \oplus \cdots \oplus U_1}_{a_{i1}} \oplus \cdots \oplus \underbrace{U_h \oplus \cdots \oplus U_h}_{a_{ih}},$$

where $a_{ij} \geq 0$ is the number of copies of $U_j$. Then $\varphi_i = \sum_{j=1}^h a_{ij}\chi_j$, so

$$(\varphi_i, \varphi_i)_G = \left( \sum_j a_{ij}\chi_j, \sum_l a_{il}\chi_l \right)$$

$$= \sum_{j,l} a_{ij}a_{il}\delta_{jl} = \sum_j a_{ij}^2 > 0,$$

using the orthogonality relation $(\chi_i, \chi_j) = \delta_{ij}$.

Let $m_i$ be the multiplicity of $V_i$ in $V$, $n_i$ the multiplicity of $V_i$ in $W$. Then

$$(\varphi_j, \chi) = (\varphi_j, \sum_i m_i \varphi_i) = \sum_i m_i(\varphi_j, \varphi_i) = m_j(\varphi_j, \varphi_j),$$

so $m_j = (\varphi_j, \chi)/(\varphi_j, \varphi_j)$. Similarly, $n_j = (\varphi_j, \theta)/(\varphi_j, \varphi_j)$. But $\chi = \theta$, so $m_j = n_j$, all $j$. By complete reducibility of $V$ and $W$, this proves $V \cong W$.

**Definition**  Let $k$ be a field of characteristic zero, $\chi$ a $kG$-character, $\varphi$ an irreducible $kG$-character. By Lemma 7.2, $\chi$ and $\varphi$ determine the $kG$-modules affording them, say $V$ affording $\chi$, $W$ affording $\varphi$. We define the *multiplicity of $\varphi$ in $\chi$* to be the multiplicity of $W$ in $V$.

**Corollary 7.3**  *Let $k$ be a field with* char $k = 0$, $\chi$ *a $kG$-character, $\varphi$ an irreducible $kG$-character. Then the multiplicity of $\varphi$ in $\chi$ is $(\varphi, \chi)/(\varphi, \varphi)$.*

*Proof*  Last paragraph of previous proof.

**Definition**  Let $k$ be a splitting field of characteristic zero for $G$, and let $\chi_1, \ldots, \chi_h$ be the characters of the irreducible $kG$-modules $V_1, \ldots, V_h$. By Theorem 5.7, $Cf(G, k)$ has $k$-basis $\{\chi_1, \ldots, \chi_h\}$. With $\mathbf{Z}$ denoting the integers, the *ring of generalized characters of $G$* is

$$\mathbf{Z}\chi_1 \oplus \cdots \oplus \mathbf{Z}\chi_h = \{a_1\chi_1 + \cdots + a_h\chi_h| \text{ all } a_i \in \mathbf{Z}\}.$$

Note that a function $f: G \to k$ is a character if and only if $f = a_1\chi_1 + \cdots + a_h\chi_h$, all $a_i \in \mathbf{Z}$, all $a_i \geq 0$, some $a_j > 0$.

**Lemma 7.4**  *Let $k$ be a splitting field for $G$,* char $k = 0$, *and let $\chi_1, \ldots, \chi_h$ be the irreducible $kG$-characters. Then for any character $\chi = a_1\chi_1 + \cdots + a_h\chi_h$, the multiplicity of $\chi_i$ in $\chi$ is $a_i = (\chi, \chi_i)$.*

*Proof*  By Corollary 7.3, the multiplicity is $(\chi, \chi_i)/(\chi_i, \chi_i) = (\chi, \chi_i) = (\sum a_j\chi_j, \chi_i) = \sum a_j\delta_{ij} = a_i$.

**Lemma 7.5**  *Let $\mathbf{C} =$ complex numbers, $\bar{\alpha}$ the complex conjugate of $\alpha$. Let $\chi_1, \ldots, \chi_h$ be the irreducible $\mathbf{C}G$-characters, and define $\bar{\chi}_i$ by $\bar{\chi}_i(g) = \overline{\chi_i(g)}$, all $g \in G$. Then $\bar{\chi}_i$ is one of the $\chi_j$.*

*Proof*  Let $\varphi_i: G \to \mathrm{Mat}_{n_i}(\mathbf{C})$ be the matrix representation with $\chi_i(g) = \mathrm{tr}\, \varphi_i(g)$, all $g \in G$, and define $\psi_i: G \to \mathrm{Mat}_{n_i}(\mathbf{C})$ by $\psi_i(g) = {}^t\varphi_i(g^{-1})$. (${}^tA$ denotes the transpose of the matrix $A$.) Then

$$\psi_i(g)\psi_i(h) = {}^t\varphi_i(g^{-1}){}^t\varphi_i(h^{-1}) = {}^t(\varphi_i(h^{-1})\varphi_i(g^{-1}))$$

$$= {}^t\varphi_i(h^{-1}g^{-1}) = {}^t\varphi_i((gh)^{-1}) = \psi_i(gh),$$

so $\psi_i$ is a representation. Its character is

$$\operatorname{tr} {}^t\varphi_i(g^{-1}) = \operatorname{tr} \varphi_i(g^{-1}) = \chi_i(g^{-1}) = \overline{\chi_i(g)} = \bar{\chi}_i(g),$$

by Theorem 5.8(i).

$$(\bar{\chi}_i, \bar{\chi}_i) = \frac{1}{|G|} \sum_{x \in G} \bar{\chi}_i(g)\bar{\chi}_i(g^{-1}) = \overline{(\chi_i, \chi_i)} = 1,$$

so $\bar{\chi}_i$ is irreducible.

**Lemma 7.6**    *Let $k$ be a splitting field for $G$ with char $k \nmid |G|$. Let $\chi_1, \ldots,$
$\chi_h$ be the irreducible characters afforded by the irreducible $kG$-modules
$V_1, \ldots, V_h$, where $\dim_k V_i = n_i$. Let $\rho_G$ denote the character* (regular
character) *of $G$ afforded by the regular $kG$-module $kG$. Then:*

(1)    *Each $V_i$ has multiplicity $n_i$ in $kG$, so*

$$\rho_G = \sum_{i=1}^{h} n_i \chi_i.$$

(2)    *The values of $\rho_G$ are*

$$\rho_G(x) = |G| \qquad \text{if} \quad x = 1$$

$$= 0 \qquad \text{if} \quad x \neq 1.$$

(3)    *Assume also char $k = 0$. Then any character $\zeta$ of $G$ satisfying
$\zeta(x) = 0$, all $x \neq 1$, is a multiple of $\rho_G$.*

*Proof*   We know

$$kG \cong \operatorname{Mat}_{n_1}(k) \oplus \cdots \oplus \operatorname{Mat}_{n_h}(k),$$

where $\operatorname{Mat}_{n_i}(k)$ has as its only irreducible module $V_i$. $\dim_k V_i = n_i$ and
$\dim_k \operatorname{Mat}_{n_i}(k) = n_i^2$, so $V_i$ has multiplicity $n_i$ in $\operatorname{Mat}_{n_i}(k)$. Therefore (1)
holds.

In the basis $G$ of $kG$, the matrix of $1 \in G$ is the identity matrix, with
trace $|G|$, and the matrices of other $x \in G$ are permutation matrices of trace
zero, so (2) holds.

If char $k = 0$ and the character $\zeta$ satisfies $\zeta(x) = 0$ for all $x \neq 1$, then $\zeta = [\zeta(1)/\rho_G(1)]\rho_G = r\rho_G$, $r$ a rational number. $\rho_G = \sum_{i=1}^{h} n_i \chi_i = 1_G + \sum_{i=2}^{h} n_i \chi_i$, so $\zeta = r1_G + \sum_{i=2}^{h}(rn_i)\chi_i$. This implies that $r$ is an integer and $\zeta = r\rho_G$.

*Notation*    Here and elsewhere, $\chi_1 = 1_G$ denotes the *1-character* or *principal character* of the trivial $kG$-module $k$.

**Definition**      Let $k$ be a splitting field for $G$ of characteristic zero, $\chi$ and $\theta$ $kG$-characters. We say $\chi$ *is a constituent of* $\theta$ and write $\chi \subset \theta$, if $\theta = \chi + \chi'$ where $\chi'$ is either 0 or a character.

**Theorem 7.7**      *Let $k$ be a splitting field for $G$ with* char $k = 0$, *and let $\chi$ be an irreducible $kG$-character. Then $\chi(1) = \deg \chi$ divides $|G|$.*

*Proof*    Let $\chi = \chi_t$ where $\chi_1, \ldots, \chi_h$ are the irreducible $kG$-characters. Let $\mathscr{C}_1, \ldots, \mathscr{C}_h$ be the conjugacy classes of $G$ where $x_i \in \mathscr{C}_i$, and denote $h_i = |\mathscr{C}_i|$, $n_i = \chi_i(1)$. By Lemma 5.9, the following expression is an algebraic integer:

$$\sum_{i=1}^{h} \omega_{it}\chi_t(x_i^{-1}) = \sum_{i=1}^{h} \frac{h_i \chi_t(x_i)}{n_t}\chi_t(x_i^{-1}) = \frac{1}{n_t}\sum_{x \in G}\chi_t(x)\chi_t(x^{-1}) = \frac{|G|}{n_t}.$$

Being a rational number and an algebraic integer, we see $|G|/n_t$ is an integer; that is, $n_t \mid |G|$.

*Remark*    An alternate proof of Theorem 7.7 appears in Murnaghan [1]. The result will be sharpened in §22.

**Definition**      Let $G$ be a finite group, $k$ a splitting field of characteristic zero for $G$. Let $\chi_1, \ldots, \chi_h$ be the irreducible characters and $\mathscr{C}_1, \ldots, \mathscr{C}_h$ the conjugacy classes of $G$, where $x_i \in \mathscr{C}_i$. The *character table* of $G$ is the $h \times h$ matrix $(\chi_i(x_j))$. Since the $\chi_i$ are linearly independent, the rows of the matrix are linearly independent, and character tables are always nonsingular matrices.

*Example*    Let $G$ be the nonabelian group of order 6, generated by elements $x, y$ with $x^2 = 1$, $y^3 = 1$, $x^{-1}yx = y^2$. $G$ has the three conjugacy classes $\mathscr{C}_1 = \{1\}$, $\mathscr{C}_2 = \{y, y^2\}$, $\mathscr{C}_3 = \{x, xy, xy^2\}$. We computed the three irreducible representations of $G$ over **C**. One, of degree 2, has matrices

$$\varphi(x) = \begin{pmatrix} 0 & 1 \\ 1 & 0 \end{pmatrix}, \qquad \varphi(y) = \begin{pmatrix} \omega & 0 \\ 0 & \omega^2 \end{pmatrix}.$$

Let its character be $\chi_3$, so $\chi_3(1) = 2$, $\chi_3(x) = 0$, $\chi_3(y) = \omega + \omega^2 = -1$. Let $\chi_1 = 1_G$. The third character $\chi_2$ has degree 1 and kernel the subgroup generated by $y$, so $\chi_2(y) = 1$, $\chi_2(x) = -1$. The character table is

|          | $\mathscr{C}_1$ | $\mathscr{C}_2$ | $\mathscr{C}_3$ |
|----------|-----------------|-----------------|-----------------|
| $\chi_1$ | 1               | 1               | 1               |
| $\chi_2$ | 1               | 1               | $-1$            |
| $\chi_3$ | 2               | $-1$            | 0 .             |

The extent to which a finite group is determined by its character table is the subject of several interesting conjectures in Brauer [7]. Two of these conjectures are answered in Dade [1] and Saksonov [2]. Other interesting recent papers on character tables include Solomon [1], Weidman [1], and Brauer [11].

**Lemma 7.8**   Let $\theta$ be a **CG**-*character afforded by the representation T. Then for any $x \in G$, $\theta(x) = \theta(1)$ if and only if $x$ is in the kernel of T.*

*Proof*   If $x \in \ker T$, $T(x) = T(1)$ so $\theta(x) = \operatorname{tr} T(x) = \operatorname{tr} T(1) = \theta(1)$. Conversely, suppose $\theta(x) = \theta(1) = \deg T$. By Lemma 5.10, $T(x) = \alpha I$, some $\alpha$, so $\theta(x) = (\deg T)\alpha$. Therefore $\alpha = 1$, $T(x) = I = T(1)$, $x \in \ker T$.

**Definition**   The *kernel* of a character $\theta$ (over any field) is the kernel of the representation affording $\theta$. A character if *faithful* if its kernel is 1.

Existence of faithful irreducible or near-irreducible representations of finite groups is discussed in Weisner [1], Kochendörffer [1], Gaschütz [1], Žmud' [1], and Tazawa [1].

EXERCISES

1   Let $G$ be a group of odd order, and write $G - \{1\}$ as the union of $(|G| - 1)/2$ pairs $\{x_i, x_i^{-1}\}$. Let $1_G \neq \chi$ be an irreducible complex character of $G$, and show that $\chi \neq \bar{\chi}$. (*Hint:* Consider the equation

$$0 = \chi(1) + \sum_i (\chi(x_i) + \chi(x_i^{-1})).)$$

**2** Show that $g \in G$ is conjugate to $g^{-1}$ if and only if $\chi(g)$ is real for all complex characters $\chi$ of $G$.

(An element of a group is therefore called *real* if it is conjugate to its inverse.)

**3** Let $G$ be a finite group, $x \in G$, and let $H$ be a normal subgroup of $G$ such that $H \cap C_G(x) = 1$.

(a) Show that $|C_G(x)| \leq |C_{G/H}(xH)|$.

(b) Show that if $\chi$ is an irreducible complex character of $G$ whose kernel does not contain $H$, then $\chi(x) = 0$. (*Hint:* Which orthogonality relation gives an expression for $|C_G(x)|$? This little result is important in the Feit-Thompson proof that groups of odd order are solvable.)

**4** Let $G$ be a finite noncyclic simple group, $\chi$ an irreducible complex character of $G$ of prime degree $p$, $P$ a Sylow $p$-subgroup of $G$.

(a) Use Theorem 4.2 to show that $P$ is abelian.

(b) Use Lemma 6.4 to show that if $x \in P - \{1\}$, then $\chi(x) = 0$.

(c) Conclude that $|P| = p$.

**5** Let $\chi$ be an irreducible complex character of $G$ of degree $\chi(1) > 1$.

(a) Let $F = \mathbf{Q}(\varepsilon)$, $\varepsilon$ a primitive $|G|$th root of 1. If $g \in G$ and $S = \{h \in G| \langle h \rangle = \langle g \rangle\}$, show that the Galois group Gal $(F/\mathbf{Q})$ is transitive on the set $\{\chi(h)|h \in S\}$.

(b) In (a), use the fact the arithmetic mean of positive real numbers is at least the geometric mean to show that if $\chi(g) \neq 0$, then

$$\sum_{h \in S} |\chi(h)|^2 \geq |S|.$$

(c) Show that $\chi(x) = 0$ for some $x \in G$.

**6** There are exactly two nonisomorphic nonabelian groups of order 8. One is the *dihedral group* $D_8$, generated by elements $a$, $b$ with relations

$$a^2 = 1, \quad b^4 = 1, \quad a^{-1}ba = b^3.$$

The other is the *quaternion group* $Q_8$, generated by elements $c$, $d$ with relations

$$c^4 = 1, \quad d^2 = c^2, \quad c^{-1}dc = d^3.$$

(a)   List the conjugacy classes of $D_8$ and of $Q_8$. Verify that $D_8$ and $Q_8$ are not isomorphic.

(b)   Show that the center $Z(D_8)$ of $D_8$ is $\langle b^2 \rangle$ and the center $Z(Q_8)$ of $Q_8$ is $\langle d^2 \rangle$. Show that $D_8/Z(D_8)$ and $Q_8/Z(Q_8)$ are each elementary abelian of order 4.

(c)   Use (b) to find four one-dimensional representations of $D_8$ and four one-dimensional representations of $Q_8$ over **C**.

(d)   Using the orthogonality relations, complete the character tables of $D_8$ and $Q_8$. Show that $D_8$ and $Q_8$ have the same character table.

**7**   Show that the hypothesis char $k = 0$ is necessary in Lemma 7.2.

# §8

## Representations of Abelian Groups

**Definition** Clearly a one-dimensional representation over any field coincides with its character. Such representations and characters are called *linear*.

**Theorem 8.1** *Let $H$ be a finite abelian group of exponent $m$, and let $k$ be a field of characteristic $p$ (perhaps $p = 0$). Then:*

(1) *if $p|m$, then the number of distinct linear representations of $H$ in $k$ is less than $|H|$.*

(2) *if $p\nmid m$ and $k$ contains a primitive $m$th root of 1, then there are exactly $|H|$ distinct linear representations of $H$ in $k$, and $k$ is a splitting field for $H$.*

(3) *If $p\nmid m$ and $k$ does not contain a primitive $m$th root of 1, then there are less than $|H|$ distinct linear representations of $H$ in $k$, and $k$ is not a splitting field for $H$.*

*Proof* By the structure theorem for finite abelian groups, $H = H_1 \times \cdots \times H_t$, each $H_i$ a cyclic group of prime power order generated by an element $x_i$. Any representation $T: H \to k$ is completely determined by $\{T(x_1), \ldots, T(x_t)\}$. $T(x_i)^{|H_i|} = T(x_i^{|H_i|}) = T(1) = 1$, so there are at most $|H_i|$ possibilities for $T(x_i)$, and at most $|H| = \prod_i |H_i|$ possibilities for $\{T(x_1), \ldots, T(x_t)\}$.

(1) If $p|m$, then $p|\,|H_i|$ for some $i$, so $k$ contains only one $|H_i|$th root of 1. (*Reason:* if $|H_i| = p^a$, then $X^{p^a} - 1 = (X - 1)^{p^a}$.) Hence there are fewer than $|H|$ possibilities for $T$.

(2) In this case, all possibilities for $T$ occur. $kH$ is semisimple, so the $|H|$ possibilities for $T$ are all the irreducible representations of $H$ over $k$. They, being linear, are absolutely irreducible, so $k$ is a splitting field.

43

(3)   In this case, the $|H_i|$th roots of 1 are not present in $k$ for some $i$, so there are fewer than $|H|$ possibilities for $T$.

Suppose $k$ were a splitting field for $H$. $H$ has $|H|$ conjugacy classes, so $kH$ would have $|H|$ distinct irreducible modules $V_i$. $kH$ is semisimple, so each $V_i$ is a submodule of $kH$. By the previous paragraph some $n_i = \dim_k V_i > 1$ so $\sum_i n_i > |H| = \dim_k kH$, a contradiction.

**Definitions**      If $x, y \in G$, $y^x = x^{-1}yx$ is the *conjugate* of $y$ by $x$. The *commutator* of $x$ and $y$ is $[x,y] = x^{-1}y^{-1}xy$. The *commutator subgroup* $G'$ of $G$ is the subgroup generated by $\{[x,y] \,|\, x,y \in G\}$.

**Lemma 8.2**      $G' \lhd G$.

*Proof*  If      $x,y,w \in G$,      then      $w^{-1}[x,y]w = w^{-1}x^{-1}y^{-1}xyw = w^{-1}x^{-1}ww^{-1}y^{-1}ww^{-1}xww^{-1}yw = (x^{-1})^w(y^{-1})^w x^w y^w$.  Now  $(x^{-1})^w = (x^w)^{-1}$  since  $(x^{-1})^w x^w = w^{-1}x^{-1}ww^{-1}xw = 1$.  Hence  $w^{-1}[x,y]w = (x^w)^{-1}(y^w)^{-1}x^w y^w = [x^w,y^w] \in G'$. So $(G')^w \subseteq G'$, $(G')^w = G'$.

**Lemma 8.3**      (a) *For any group* $G$, $G/G'$ *is abelian*. (b) *If* $H \lhd G$ *and* $G/H$ *is abelian, then* $G' \subseteq H$.

*Proof*  (a)   If  $xG'$,  $yG' \in G/G'$,  then  $(xG')^{-1}(yG')^{-1}xG'yG' = x^{-1}y^{-1}xyG' = G' = 1$, so $xG'yG' = yG'xG'$.

(b)   If $x,y \in G$ then $xHyH = yHxH$, so $[x,y]H = H$. Therefore $[x,y] \in H$, proving $G' \subseteq H$.

**Corollary 8.4**      *If* $G$ *is a finite group and* $k$ *a splitting field for* $G$ *with* char $k \nmid |G|$, *then* $G$ *has exactly* $|G:G'|$ *linear representations. All of them have kernel containing* $G'$, *so they constitute a full set of representations of* $G/G'$.

*Proof*  If $T: G \to k$ is linear, then Im $T$ is a multiplicative subgroup of $k$, and hence abelian. $G/\ker T \cong \text{im } T$, so $\ker T \supseteq G'$ by Lemma 8.3. Of course any group homomorphism $G \to k$ with kernel containing $G'$ induces a homomorphism $G/G' \to k$, and conversely.

**Definition**      A group $G$ is *perfect* if $G = G'$. Noncyclic simple groups are perfect. By Corollary 8.4, any group $G$ is perfect if and only if $1_G$ is its only linear character over **C**.

# §9

## Induced Characters

**Definition**   Let $H$ be a subgroup of the group $G$. A *cross section* (*transversal, set of coset representatives*) of $H$ in $G$ is a set of elements $\{x_i\}$,

$$G = x_1 H \cup x_2 H \cup \cdots \qquad \text{(disjoint union)}.$$

If $G$ is finite, of course $|\{x_i\}| = |G{:}H|$. (We have defined *left* coset representatives. Right coset representatives can also be defined.)

**Definition**   Let $G$ be a group with subgroup $H$, $R$ a commutative ring, $V$ an $RG$-module. $RH \subset RG$ so $V$ is also an $RH$-module; when so considered, we denote $V$ by $V_H$, the *restriction of $V$ to $RH$*.

**Definition**   Let $G$ be a group with subgroup $H$, $R$ a commutative ring, $W$ an $RH$-module. Let $\{x_i\}$ be a cross section of $H$ in $G$. $RG$ has $R$-basis $G$, so

$$RG = Rx_1 H \oplus Rx_2 H \oplus \cdots .$$

Considering $RG$ as a right $RH$-module, this expresses $RG$ as a direct sum of right $RH$-submodules. We form the tensor product $W^G = RG \otimes_{RH} W$, a (left) $RG$-module called the *induced module*. Now $RG = \oplus \sum_i Rx_i H = \oplus \sum_i x_i RH$, so $W^G = \oplus \sum_i x_i RH \otimes W = \oplus \sum_i x_i \otimes W$, at least a direct sum as $R$-modules. If $x \in G$, then $x$ acts on $W^G$ as follows: let $xx_i = x_j h$, $h \in H$. Then if $w \in W$,

$$x(x_i \otimes w) = xx_i \otimes w = x_j h \otimes w = x_j \otimes hw \in x_j \otimes W.$$

**Lemma 9.1**   *Let $G$ be a finite group with subgroup $H$, $R$ a commutative ring, $W$ an $R$-free $RH$-module with $R$-basis $\{w_1, \ldots, w_m\}$. Let $\varphi$ be the matrix representation of $H$ afforded by $W$ in this basis. Define $\dot\varphi$ on all of $G$ by*

$$\dot{\phi}(g) = \varphi(g) \qquad if \quad g \in H$$

$$= the \ m \times m \ zero \ matrix, \qquad if \quad g \notin H.$$

Let $x_1, \ldots, x_t$ be a cross section of $H$ in $G$, so $t = |G:H|$. Then $W^G$ has R-basis

$$\{x_i \otimes w_j | 1 \leq i \leq t, 1 \leq j \leq m\}.$$

By arranging these basis elements in the order $x_1 \otimes w_1, \ldots, x_1 \otimes w_m,$ $x_2 \otimes w_1, \ldots, x_2 \otimes w_m, \ldots, x_t \otimes w_1, \ldots, x_t \otimes w_m,$ the matrix representation $\Phi(g)$ of $G$ afforded by $W^G$ is

$$\Phi(g) = \begin{pmatrix} \dot{\phi}(x_1^{-1}gx_1) & \dot{\phi}(x_1^{-1}gx_2) & \cdots & \dot{\phi}(x_1^{-1}gx_t) \\ \dot{\phi}(x_2^{-1}gx_1) & \dot{\phi}(x_2^{-1}gx_2) & \cdots & \dot{\phi}(x_2^{-1}gx_t) \\ \vdots & \vdots & & \vdots \\ \dot{\phi}(x_t^{-1}gx_1) & \dot{\phi}(x_t^{-1}gx_2) & \cdots & \dot{\phi}(x_t^{-1}gx_t) \end{pmatrix}.$$

In this block array, only one block in each row and each column is a nonzero block.

*Proof* Let $\varphi(h) = (\alpha_{ij}(h))$, where $hm_j = \sum_i \alpha_{ij}(h)m_i$. Let $g \in G$, and let $i,j$ be fixed. If $gx_i = x_lh$, $h \in H$, then

$$g(x_i \otimes m_j) = gx_i \otimes m_j = x_l \otimes hm_j$$

$$= \sum_{s=1}^{m} \alpha_{sj}(h)x_l \otimes m_s = \sum_{s=1}^{m} \alpha_{sj}(x_l^{-1}gx_i)x_l \otimes m_s,$$

which implies the result. It also implies only one nonzero block, the $l$th, in the $i$th column.

Finally, only one nonzero block in each row, for if $i \neq i'$ and $x_l^{-1}gx_i \in H$, then $x_i \in g^{-1}x_lH$ so $x_{i'} \notin g^{-1}x_lH$, $x_l^{-1}gx_{i'} \notin H$.

**Lemma 9.2** *Let $G$ be a finite group with subgroup $H$, $k$ a field with char $k \nmid |G|$, and let $W$ be a $kH$-module affording the character $\theta$. Then the character $\theta^G$ of the induced $kG$-module $W^G$ satisfies the formula*

$$\theta^G(x) = \frac{1}{|H|} \sum_{y \in G} \dot{\theta}(y^{-1}xy), \qquad where \quad \dot{\theta}(g) = \theta(g) \quad if \quad g \in H$$

$$= 0 \quad if \quad g \notin H.$$

*Proof* Taking $x_1, \ldots, x_t$ as a cross section of $H$ in $G$, and taking traces in Lemma 9.1, we see $\theta^G(x) = \sum_{i=1}^{t} \dot\theta(x_i^{-1}xx_i)$. For any $h \in H$ and any $x \in G$, we have $\dot\theta(h^{-1}xh) = \dot\theta(x)$, whether $x \in H$ or $x \notin H$. Hence

$$\dot\theta(x_i^{-1}xx_i) = \frac{1}{|H|} \sum_{h \in H} \dot\theta(h^{-1}x_i^{-1}xx_ih),$$

and

$$\theta^G(x) = \sum_{i=1}^{t} \dot\theta(x_i^{-1}xx_i) = \frac{1}{|H|} \sum_{i=1}^{t} \sum_{h \in H} \dot\theta(h^{-1}x_i^{-1}xx_ih)$$

$$= \frac{1}{|H|} \sum_{y \in G} \dot\theta(y^{-1}xy).$$

**Definitions** If $\theta$ is a character of a subgroup $H$ of $G$, $\theta^G$ is called the *induced character* of $G$. This powerful tool of representation theory was developed primarily by Frobenius [1].

More generally, if $\theta: H \to k$ is any class function, we define $\theta^G: G \to k$ by $\theta^G(x) = (1/|H|) \sum_{y \in G} \dot\theta(y^{-1}xy)$; clearly $\theta^G$ is a class function of $G$. If $\theta$ is a character, we have seen $\theta^G$ is a character; if $\theta$ is a generalized character of $H$, then $\theta^G$ is a generalized character of $G$.

If $V$ is a $kG$-module with character $\theta$, and $H$ is a subgroup of $G$, then we denote the character of $V_H$ by $\theta_H$, the *restriction of $\theta$ to $H$*. If $h \in H$, clearly $\theta_H(h) = \theta(h)$, so $\theta_H$ is an actual restriction function.

*Remark* Let $k$ be a field, $G$ a group with characters $\chi, \theta$ afforded by $kG$-modules $V$ and $W$. We saw that $V \oplus W$ has character $\chi + \theta$ in Lemma 7.1, so the sum of two characters is a character. More surprisingly, we now show

**Theorem 9.3** *If $\chi, \theta$ are characters of $kG$-modules $V$ and $W$, respectively, then the product $\chi\theta$ is also a character of $G$. (By $\chi\theta$, we mean the function $\tau: G \to k$ defined by $\tau(x) = \chi(x)\theta(x)$, all $x \in G$.)*

*Proof* Let $V$ and $W$ have $k$-bases $\{v_1, \ldots, v_m\}$ and $\{w_1, \ldots, w_n\}$, respectively. We form the $k$-vector space $V \otimes_k W$, with $k$-basis $\{v_i \otimes w_j | 1 \le i \le m, 1 \le j \le n\}$. We make $V \otimes W$ into a $kG$-module by defining $g(v \otimes w) = gv \otimes gw$, all $g \in G$, and extending to all of $kG$ by $k$-linearity. This is possible, because the map $V \times W \to V \otimes W$ defined by $(v, w) \to gv \otimes gw$ is bilinear.

We compute the character of $V \otimes_k W$. If

$$gv_j = \sum_{i=1}^{m} \alpha_{ij} v_i \quad \text{and} \quad gw_j = \sum_{i=1}^{n} \beta_{ij} w_i,$$

then

$$\chi(g) = \text{tr}(\alpha_{ij}) = \sum_{i=1}^{m} \alpha_{ii}, \qquad \theta(g) = \text{tr}(\beta_{ij}) = \sum_{i=1}^{n} \beta_{ii}.$$

We see that

$$g(v_r \otimes w_s) = gv_r \otimes gw_s = \sum_{i=1}^{m} \sum_{j=1}^{n} \alpha_{ir} \beta_{js}(v_i \otimes w_j)$$

$$= \alpha_{rr} \beta_{ss} v_r \otimes w_s + \text{other terms}.$$

Thus $\alpha_{rr}\beta_{ss}$ is the diagonal entry in the matrix of $g$ on $V \otimes W$. Therefore the character $\tau$ afforded by $V \otimes W$ is

$$\tau(g) = \sum_{r=1}^{m} \sum_{s=1}^{n} \alpha_{rr} \beta_{ss} = \left( \sum_{r=1}^{m} \alpha_{rr} \right)\left( \sum_{s=1}^{n} \beta_{ss} \right) = \chi(g)\theta(g).$$

   *Remark*   Theorem 9.3 implies that the set of generalized characters of $G$ over a splitting field $k$ of characteristic zero forms a commutative ring, under the above multiplication of functions.

**Theorem 9.4** (Frobenius)    *Let $G$ be a finite group with subgroups $H$ and $A$, $H \subseteq A \subseteq G$. Let $k$ be a field with char $k \nmid |G|$. Let $\chi$ be a class function on $H$, $\theta$ a class function on $G$. Then we have:*
   (a)  $(\chi^A)^G = \chi^G$    (*Transitivity of Induction*).
   (b)  $\chi^G \theta = (\chi \theta_H)^G$.
   (c)  $(\chi, \theta_H)_H = (\chi^G, \theta)_G$    (*Frobenius Reciprocity*).

   *Proof*   Denote $\tau = \chi^A$, and define $\dot{\chi}$, $\dot{\tau}$, $(\chi\theta_H)^{\boldsymbol{\cdot}}$ on $G$ as follows:

$$\dot{\chi}(g) = \chi(g) \quad \text{if} \quad g \in H \qquad \dot{\tau}(g) = \tau(g) \quad \text{if} \quad g \in A$$

$$= \ 0 \quad \text{if} \quad g \notin H; \qquad\qquad = \ 0 \quad \text{if} \quad g \notin A;$$

$$(\chi\theta_H)^{\boldsymbol{\cdot}}(g) = \chi(g)\theta(g) \quad \text{if} \quad g \in H$$

$$= \ 0 \qquad \text{if} \quad g \notin H.$$

*Proof of (a)* The following equation holds for all $g \in G$, whether $g \in A$ or $g \notin A$:

$$\dot{\tau}(g) = \frac{1}{|H|} \sum_{a \in A} \dot{\chi}(a^{-1}ga).$$

Hence, for any $g \in G$, we have

$$(\chi^A)^G(g) = \tau^G(g) = \frac{1}{|A|} \sum_{x \in G} \dot{\tau}(x^{-1}gx)$$

$$= \frac{1}{|A|}\frac{1}{|H|} \sum_{x \in G} \sum_{a \in A} \dot{\chi}(a^{-1}x^{-1}gxa).$$

As $x$ varies over $G$ and $a$ varies over $A$, the element $xa$ runs through all elements of $G$ $|A|$ times. Hence the above expression equals

$$\frac{1}{|H|} \sum_{y \in G} \dot{\chi}(y^{-1}gy) = \chi^G(g).$$

*Proof of (b)* The following equation holds for all $g \in G$, whether $g \in H$ or $g \notin H$:

$$(\chi\theta_H)^{\cdot}(g) = (\dot{\chi}\theta)(g).$$

Hence

$$(\chi\theta_H)^G(g) = \frac{1}{|H|} \sum_{x \in G} \dot{\chi}(x^{-1}gx)\theta(x^{-1}gx)$$

$$= \theta(g) \cdot \frac{1}{|H|} \sum_{x \in G} \dot{\chi}(x^{-1}gx)$$

$$= \theta(g)\chi^G(g) = (\chi^G\theta)(g).$$

*Proof of (c)* We have

$$(\chi^G, \theta)_G = \frac{1}{|G|} \sum_{g \in G} \chi^G(g)\theta(g^{-1})$$

$$= \frac{1}{|G|}\frac{1}{|H|} \sum_{g \in G} \sum_{x \in G} \dot{\chi}(x^{-1}gx)\theta(x^{-1}g^{-1}x)$$

$$= \frac{1}{|H|} \sum_{y \in G} \dot{\chi}(y)\theta(y^{-1}) = \frac{1}{|H|} \sum_{y \in H} \chi(y)\theta(y^{-1}) = (\chi, \theta_H)_H.$$

*Remark*   Frobenius reciprocity is very important. We pause to interpret it in terms of modules and characters.

Let $k$ be a splitting field of characteristic zero for $G$ and its subgroups. Let $W$ be an irreducible $kH$-module, $H$ a subgroup of $G$, and let $V$ be an irreducible $kG$-module. Assume that $W$ affords the character $\chi$, $V$ the character $\theta$. Then by Lemma 7.4, $(\chi,\theta_H)_H$ is the number of composition factors of $V_H$ isomorphic to $W$, and $(\chi^G,\theta)_G$ is the number of composition factors of $W^G$ isomorphic to $V$.

Frobenius reciprocity tells us that if

$$\theta_H = a\chi + \text{other irreducible } H\text{-characters},$$

then $a = (\chi,\theta_H)_H = (\chi^G,\theta)_G$ so

$$\chi^G = a\theta + \text{other irreducible } G\text{-characters}.$$

**Lemma 9.5**   *Let $k$ be a splitting field of characteristic zero for $G$, and let $\chi: G \to k$ be a character. If $(\chi,\chi)_G = 1$, then $\chi$ is irreducible.*

*Proof*   We know $\chi = a_1\chi_1 + \cdots + a_t\chi_t$, the $a_i$ integers and the $\chi_i$ irreducible characters with $(\chi_i,\chi_j) = \delta_{ij}$. So

$$1 = (\chi,\chi) = \left(\sum_i a_i\chi_i, \sum_j a_j\chi_j\right) = \sum_{i,j} a_i a_j \delta_{ij} = \sum_i a_i^2,$$

proving that one $a_i$ is 1, the others zero.

*Example*   Consider the alternating group $A_4$, of order 12. $A_4$ has a normal elementary abelian subgroup $N$ of order 4, say $N = \{1,x_1,x_2,x_1x_2\}$, and an element $y$ of order 3 such that $x_1^y = x_2$, $x_2^y = x_1x_2$, $(x_1x_2)^y = x_1$. $A_4 = G$ has 4 conjugacy classes, namely

$$\mathscr{C}_1 = \{1\}, \qquad \mathscr{C}_2 = \{x_1,x_2,x_1x_2\},$$

$$\mathscr{C}_3 = \{y,yx_1,yx_2,yx_1x_2\}, \qquad \mathscr{C}_4 = \{y^2,y^2x_1,y^2x_2,y^2x_1x_2\}.$$

Let $\lambda$ be the linear character of $N$ defined by

$$\lambda(1) = \lambda(x_1) = 1, \qquad \lambda(x_2) = \lambda(x_1x_2) = -1.$$

We will compute $\lambda^G$, over the field $k = \mathbf{C}$. Denote

$$\begin{aligned}
\dot{\lambda}(g) &= \lambda(g) && \text{if } g \in N \\
&= 0 && \text{if } g \notin N.
\end{aligned}$$

Then

$$\lambda^G(1) = \frac{1}{|N|} \sum_{g \in G} \lambda(g^{-1}1g) = \frac{|G|}{|N|} = 3.$$

$$\lambda^G(x_1) = \frac{1}{|N|} \sum_{g \in G} \lambda(g^{-1}x_1g) = \frac{1}{4}(4\lambda(x_1) + 4\lambda(x_2) + 4\lambda(x_1x_2))$$

$$= \frac{1}{4}(4 - 4 - 4) = -1.$$

Finally, $N \lhd G$, so for all $g \in G$ we have $g^{-1}yg \notin N$, $g^{-1}y^2g \notin N$. This implies $\lambda^G(y) = \lambda^G(y^2) = 0$. We now see that

$$(\lambda^G, \lambda^G)_G = \frac{1}{|G|} \sum_{y \in G} \lambda^G(g)\lambda^G(g^{-1})$$

$$= \frac{1}{12}(3^2 + (-1)^2 + (-1)^2 + (-1)^2) = 1,$$

so by Lemma 9.5, $\chi_4 = \lambda^G$ is an irreducible character of $G$.

G, however, has only four irreducible characters. Three of them are linear characters of $G/N$ since $|G/N| = 3$. If $\omega$ denotes a cube root of 1, we see that the character table of $G = A_4$ is

|  | $\mathscr{C}_1$ | $\mathscr{C}_2$ | $\mathscr{C}_3$ | $\mathscr{C}_4$ |
|---|---|---|---|---|
| $\chi_1$ | 1 | 1 | 1 | 1 |
| $\chi_2$ | 1 | 1 | $\omega$ | $\omega^2$ |
| $\chi_3$ | 1 | 1 | $\omega^2$ | $\omega$ |
| $\chi_4$ | 3 | $-1$ | 0 | 0 |

One can show that if $1_N$ denotes the 1-character of $N$, then $1_N^G = \chi_1 + \chi_2 + \chi_3$, so $1_N^G$ is not irreducible.

EXERCISES

1 The group $G = A_4$ has order 12, and we computed its complex

character table. $G$ has a normal subgroup $N$ of order 4. For each irreducible character $\theta$ of $N$, express $\theta^G$ in terms of the irreducible characters of $G$.

**2**   The *dihedral group* $D_{2m}$ of order $2m$ has generators $a$, $b$ and relations

$$a^m = 1, \qquad b^2 = 1, \qquad b^{-1}ab = a^{-1}.$$

Describe all of the irreducible characters of $D_{2m}$.

**3**   If $\chi$ and $\theta$ are irreducible complex characters of $G$, show that

$$(\chi\theta, 1_G) = 1 \qquad \text{if} \quad \chi = \bar{\theta}$$

$$\qquad\qquad = 0 \qquad \text{otherwise.}$$

**4** (Brauer [8])   Let $\theta$ be a faithful complex character of $G$ with exactly $r$ distinct values $a_1 = \theta(1)$, $a_2, \ldots, a_r$. Let $\chi$ be any irreducible complex character of $G$.

(a)   If $(\theta^n, \chi)_G = 0$ for some integer $n \geq 0$ and $A_i = \{g \in G \mid \theta(g) = a_i\}$, show that

$$0 = \sum_{j=1}^{r} a_j^n \sum_{g \in A_j} \overline{\chi(g)}.$$

(b)   Show that for some $0 \leq n < r$, $\chi$ is a constituent of $\theta^n$.

**5**   If $\chi$ is a faithful complex character of the subgroup $H$ of $G$, show that $\chi^G$ is a faithful character of $G$.

# §10

## Representations of Direct Products

**Theorem 10.1** (Burnside)    *Let $G$ be a finite group, $k$ a splitting field for $G$, $V$ an irreducible $kG$-module. For each $g \in G$, let $T(g) \in \operatorname{End}_k(V)$ be the representation of $g$ on $V$; that is,*

$$T(g)v = gv, \qquad \text{all} \quad v \in V.$$

*Then*

$$\operatorname{End}_k(V) = \sum_{g \in G} kT(g);$$

*that is, the set $\{T(g) | g \in G\}$ spans the $k$-vector space $\operatorname{End}_k(V)$, and hence contains $(\dim_k V)^2$ linearly independent elements.*

*Proof*    Let $A = \sum_{g \in G} kT(g)$; $A$ is a subalgebra of $\operatorname{End}_k(V)$ since $T(g)T(h) = T(gh)$. An $A$-submodule of $V$ would be a $kG$-submodule, so $V$ is an irreducible $A$-module. Since $k$ is a splitting field, $\operatorname{End}_{kG}(V) = k$; it is clear that any element of $\operatorname{End}_A(V)$ is in $\operatorname{End}_{kG}(V)$, so $\operatorname{End}_A(V) = k$.

Let $V$ have $k$-basis $\{v_1, \ldots, v_n\}$. If $E \in \operatorname{End}_k(V)$, then by the Jacobson density theorem 2.5 there is an $a \in A$ with $Ev_i = av_i$, all $i$. Hence $E = a$, $\operatorname{End}_k(V) = A$.

**Lemma 10.2**    *Let $G$ be a finite group, $k$ a field with* char $k \nmid |G|$. *Let $kG$ have nonisomorphic irreducible modules $V_1, \ldots, V_s$, and assume that $G$ has $h$ conjugacy classes. Then $s \leq h$.*

*Proof*    By Maschke's theorem, $kG = A_1 \oplus \cdots \oplus A_r$, the $A_i$ simple rings. Each $A_i$ has only one irreducible module, so $r = s$. Let $K$ be the algebraic closure of $k$, so $KG \cong (kG)^K = A_1^K \oplus \cdots \oplus A_s^K$. The $A_i^K$ are semisimple rings, so are direct sums of simple rings, and we see $KG$ is a sum of $t \geq s$ simple rings. By Theorem 3.2, $t = h$, done.

53

**Theorem 10.3**    *Let $G = H_1 \times H_2$ be a finite group, the direct product of subgroups $H_1$ and $H_2$. Let $k$ be a field.*

(a)    *If $T_1: H_1 \rightarrow \mathrm{End}_k(V_1)$ and $T_2: H_2 \rightarrow \mathrm{End}_k(V_2)$ are representations where $V_1, V_2$ are $k$-vector spaces, then there is a representation*

$$T: G \rightarrow \mathrm{End}_k(V_1 \otimes_k V_2)$$

*satisfying*

$$T(h_1 h_2)(v_1 \otimes v_2) = T_1(h_1)v_1 \otimes T_2(h_2)v_2, \qquad all \quad h_i \in H_i, \quad v_i \in V_i.$$

(b)    *If $k$ is a splitting field for $H_1$ and $H_2$, $V_1$ an irreducible $kH_1$-module, $V_2$ an irreducible $kH_2$-module, then $V_1 \otimes_k V_2$ is an irreducible $kG$-module under a module action satisfying*

$$(h_1 h_2)(v_1 \otimes v_2) = h_1 v_1 \otimes h_2 v_2, \qquad all \quad h_i \in H_i, \quad v_i \in V_i.$$

(c)    *If $k$ is a splitting field for $H_1$ and $H_2$ and char $k \nmid |G|$, then all irreducible $kG$-modules are of the form given in (b).*

(d)    *If $k$ is a splitting field for $H_1$ and $H_2$ and char $k \nmid |G|$, then all irreducible characters of $G$ are obtained as follows. Let $\{\chi_1, \ldots, \chi_m\}$ be the irreducible characters of $H_1$, $\{\theta_1, \ldots, \theta_n\}$ the irreducible characters of $H_2$. Then $G$ has exactly $mn$ irreducible characters $\psi_{ij}$ $(1 \leq i \leq m, 1 \leq j \leq n)$, satisfying $\psi_{ij}(h_1 h_2) = \chi_i(h_1)\theta_j(h_2)$.*

*Proof* (a)    The map $V_1 \times V_2 \rightarrow V_1 \otimes_k V_2$ defined by $(v_1, v_2) \rightarrow T_1(h_1)v_1 \otimes T_2(h_2)v_2$ is bilinear, so there is

$$T(h_1 h_2) \in \mathrm{End}_k(V_1 \otimes_k V_2), \qquad T(h_1 h_2)(v_1 \otimes v_2) = T_1(h_1)v_1 \otimes T_2(h_2)v_2.$$

Clearly $T(h_1 h_2) \cdot T(h'_1 h'_2) = T(h_1 h_2 \cdot h'_1 h'_2)$, so $T$ is a representation.

(b)    By (a), $V_1 \otimes_k V_2$ is a $kG$-module; we shall show it is irreducible. For any $f_1 \in \mathrm{End}_k(V_1)$, $f_2 \in \mathrm{End}_k(V_2)$ the map $(v_1, v_2) \rightarrow f_1(v_1) \otimes f_2(v_2)$ is bilinear, so there is an $f_1 \otimes f_2 \in \mathrm{End}_k(V_1 \otimes_k V_2)$, $(f_1 \otimes f_2)(v_1 \otimes v_2) = f_1(v_1) \otimes f_2(v_2)$. It is well known (or an easy exercise) that $\mathrm{End}_k(V_1 \otimes V_2) = \mathrm{End}_k(V_1) \otimes \mathrm{End}_k(V_2)$. For any $h_1 \in H_1$, $h_2 \in H_2$, $T(h_1 h_2)$ is the map $T_1(h_1) \otimes T_2(h_2)$.

By Theorem 10.1,

$$\mathrm{End}_k(V_1) = \sum_{h_1 \in H_1} k T_1(h_1) \qquad \text{and} \qquad \mathrm{End}_k(V_2) = \sum_{h_2 \in H_2} k T_2(h_2).$$

We see, then,

$$\mathrm{End}_k(V_1 \otimes V_2) = \mathrm{End}_k(V_1) \otimes \mathrm{End}_k(V_2)$$

$$= \sum_{h_1 \in H_1} \sum_{h_2 \in H_2} kT_1(h_1) \otimes T_2(h_2)$$

$$= \sum_{h_1 h_2 \in G} kT(h_1 h_2).$$

Certainly, then, $V_1 \otimes V_2$ has no proper $kG$-invariant subspace, and is irreducible.

(c)   Let the irreducible $kH_1$-modules be $V_{11}, V_{12}, \ldots, V_{1m}$, the irreducible $kH_2$-modules $V_{21}, \ldots, V_{2n}$, where $V_{1i}$ has character $\chi_i$, $V_{2j}$ has character $\theta_j$. Then by computing the trace of elements of $G$ on a basis, we see as in the proof of Theorem 9.3 that $W_{ij} = V_{1i} \otimes V_{2j}$ has character $\psi_{ij}$. And

$$\sum_{h_1 h_2 \in G} \psi_{ij}(h_1 h_2)\psi_{rs}(h_1^{-1}h_2^{-1}) = \sum_{h_1 h_2 \in G} \chi_i(h_1)\theta_j(h_2)\chi_r(h_1^{-1})\theta_s(h_2^{-1})$$

$$= \left( \sum_{h_1 \in H_1} \chi_i(h_1)\chi_r(h_1^{-1}) \right) \left( \sum_{h_2 \in H_2} \theta_j(h_2)\theta_s(h_2^{-1}) \right)$$

$$= \delta_{ir}\delta_{js}|H_1|\,|H_2|.$$

Hence the $mn$ characters $\psi_{ij}$ are all distinct, implying that the $W_{ij}$ are pairwise nonisomorphic.

If $\mathscr{C}_1, \ldots, \mathscr{C}_m$ are the conjugacy classes of $H_1$ and $\mathscr{D}_1, \ldots, \mathscr{D}_n$ are the conjugacy classes of $H_2$, then $\{\mathscr{C}_i \times \mathscr{D}_j\}$ are all the conjugacy classes of $G$, so $G$ has $mn$ conjugacy classes. By Lemma 10.2, the $W_{ij}$ are all of the irreducible $kG$-modules.

Part (d) of the Theorem follows from (c).

# §11

## Permutation Groups

**Definitions**    Let $\Omega$ be a set. The set of all one-to-one mappings of $\Omega$ onto itself forms a group, called the *symmetric group* on $\Omega$. In the case $|\Omega| = n$, any subgroup of this group is called a *permutation group of degree n*, and its elements are called *permutations*.

We can use a cycle notation for permutations; if $\Omega = \{1, 2, 3, 4, 5, 6, 7\}$, then $g = (14)(357)$ is the function which sends $1 \to 4$, $2 \to 2$, $3 \to 5$, $4 \to 1$, $5 \to 7$, $6 \to 6$, $7 \to 3$. If $\alpha$, $\beta \in \Omega$ and $g: \alpha \to \beta$, we write $g(\alpha) = \beta$; so in our case $g(1) = 4$, $g(2) = 2$, etc.

Let $G$ be a permutation group on $\Omega$, $\alpha$, $\beta \in \Omega$. If there is a $g \in G$ with $g(\alpha) = \beta$, write $\alpha \sim \beta$. Then $\sim$ is an equivalence relation on $\Omega$, and the equivalence classes are called *orbits*. $G$ is *transitive* if $\Omega$ is itself an orbit; that is, $G$ is transitive if for any $\alpha$, $\beta \in \Omega$ there is a $g \in G$ with $g(\alpha) = \beta$. For any $\alpha \in \Omega$,

$$G_\alpha = \{g \in G | g(\alpha) = \alpha\}$$

is a subgroup of $G$, the *stabilizer of $\alpha$*.

**Lemma 11.1**    *Let $G$ be a transitive permutation group on $\Omega$, $|\Omega| = n$. Then all subgroups $G_\alpha$, $\alpha \in \Omega$, are conjugate in $G$ and satisfy $|G| = n|G_\alpha|$.*

*Proof*  If $\alpha$, $\beta \in \Omega$, choose $g \in G$ with $\beta = g(\alpha)$. We claim $G_\beta = (G_\alpha)^{g^{-1}}$. For if $x \in G_\beta$, then $g^{-1}xg(\alpha) = g^{-1}x(\beta) = g^{-1}(\beta) = \alpha$, so $g^{-1}xg \in G_\alpha$, $x \in gG_\alpha g^{-1} = G_\alpha^{g^{-1}}$. And if $x \in G_\alpha^{g^{-1}}$, say $x = gyg^{-1}$, $y \in G_\alpha$, then $x(\beta) = gyg^{-1}(\beta) = gy(\alpha) = g(\alpha) = \beta$, so $x \in G_\beta$.

Let $\Omega = \{\alpha = \alpha_1, \alpha_2, \ldots, \alpha_n\}$, and choose $1 = g_1, g_2, \ldots, g_n \in G$ with $g_i(\alpha) = \alpha_i$. We will be done if we show that $g_1G_\alpha, \ldots, g_nG_\alpha$ are the left cosets of $G_\alpha$ in $G$. If $i \neq j$, $x, y \in G_\alpha$, then $g_ix(\alpha) = g_i(\alpha) = \alpha_i$ and $g_jy(\alpha) = g_j(\alpha) = \alpha_j$, so $g_iG_\alpha \cap g_jG_\alpha$ is empty. If $x \in G$ then $x(\alpha) = \alpha_i$ for some $i$, and we see $g_i^{-1}x(\alpha) = g_i^{-1}(\alpha_i) = \alpha$ so $x = g_i(g_i^{-1}x) \in g_iG_\alpha$.

56

**Definition** A transitive permutation group $G$ on $\Omega$ is *regular* if each $G_\alpha = 1$. (Then $|G| = |\Omega|$.)

**Definition** $G$ is called *k-fold transitive* (or *doubly transitive* for $k = 2$) if, for any two ordered $k$-tuples $(\alpha_1, \ldots, \alpha_k)$ and $(\beta_1, \ldots, \beta_k)$ of distinct elements of $\Omega$, there is a $g \in G$ with $g(\alpha_i) = \beta_i$ all $i$. (Distinct means $\alpha_i \neq \alpha_j$, $\beta_i \neq \beta_j$ for $i \neq j$; $\alpha_i = \beta_j$ is allowed.)

**Lemma 11.2** *Let $G$ be transitive on $\Omega$, $k > 1$. Then*:
*(1) If $G$ is k-fold transitive on $\Omega$, any $G_\alpha$ is $(k-1)$-fold transitive on $\Omega - \{\alpha\}$.*
*(2) If some $G_\alpha$ is $(k-1)$-fold transitive on $\Omega - \{\alpha\}$, then $G$ is k-fold transitive on $\Omega$.*

*Proof* (1) is trivial, (2) an easy exercise.

**Definition** Let $G$ be a permutation group on $\Omega = \{1, \ldots, n\}$, $k$ a field, $V$ a $k$-vector space with basis $\{v_1, \ldots, v_n\}$. $V$ becomes a $kG$-module by setting

$$gv_i = v_{g(i)}, \quad \text{all} \quad g \in G, \quad i \in \Omega.$$

In this representation, we see that each matrix of each element of $G$ is a permutation matrix. The character $\theta$ of this representation is the *permutation character of $G$*; if char $k = 0$, then

$$\theta(g) = \text{number of fixed points of } g \text{ on } \Omega.$$

**Theorem 11.3** *Let $G$ be a permutation group on $\Omega = \{1, \ldots, n\}$, $k$ a splitting field for $G$ of characteristic zero, $\theta$ the permutation character of $G$. Then*:
*(a) If $G$ has $t$ orbits on $\Omega$, then*

$$\sum_{g \in G} \theta(g) = t|G|; \quad \text{i.e.,} \quad (\theta, 1_G)_G = t.$$

*In other words, $\theta = t1_G + $ some other irreducible characters of $G$, so $\theta$ is never irreducible unless $G = 1$.*
*(b) $G$ is transitive on $\Omega$ if and only if $(\theta, 1_G)_G = 1$.*
*(c) If $G$ is transitive on $\Omega$ and $G_1$ has $s$ orbits on $\Omega$ (including the orbit $\{1\}$), then*

$$\sum_{g \in G} \theta(g)^2 = (\theta, \theta)_G|G| = s|G|.$$

(*d*)   *Assume G transitive on* $\Omega$. *Then G is doubly transitive if and only if*

$$\theta = 1_G + \chi,$$

$\chi$ *an irreducible character of G.*

*Proof*   (a) In the case $t = 1$, $G$ is transitive and

$$\sum_{g \in G} \theta(g) = \sum_{i=1}^{n} |\{g \in G | g(i) = i\}|$$

$$= \sum_{i=1}^{n} |G_i| = n|G_1| = |G|.$$

In the general case, let

$$\Omega = \Omega_1 \cup \cdots \cup \Omega_t, \qquad \text{(disjoint union)},$$

where the $\Omega_i$ are the orbits of $G$, and denote by $\theta_i(g)$ the number of fixed points of $g \in G$ on $\Omega_i$. Then

$$\sum_{g \in G} \theta(g) = \sum_{g \in G} \sum_{i=1}^{t} \theta_i(g)$$

$$= \sum_{i=1}^{t} \sum_{g \in G} \theta_i(g) = \sum_{i=1}^{t} |G| = t|G|.$$

Note that

$$\sum_{g \in G} \theta(g) = \sum_{g \in G} \theta(g) \, 1_G(g^{-1}) = |G|(\theta, 1_G)_G.$$

Part (b) is the case $t = 1$ of part (a).

(c) All $G_i$ are conjugate, so each has exactly $s$ orbits on $\Omega$. Each $g \in G$ is in $G_i$ for $\theta(g)$ values of $i$. Hence

$$\sum_{g \in G} \theta(g)^2 = \sum_{i=1}^{n} \sum_{g \in G_i} \theta(g) = \sum_{i=1}^{n} s|G_i|$$

$$= sn|G_1| = s|G|.$$

(The second equality is an application of part (a) to the permutation group $G_i$ on $\Omega$.)

(d) Let $\theta = 1_G + \sum_{j=1}^{m} a_j \chi_j$, the $\chi_j$ irreducible characters not $1_G$, by part (b). Then $(\theta, \theta)_G = 1 + \sum_{j=1}^{m} a_j^2$. Hence $\theta = 1_G +$ an irreducible character if and only if $(\theta, \theta)_G = 2$. By (c), this is equivalent to $s = 2$, i.e., that $G_1$ is transitive on $\Omega - \{1\}$. By Lemma 11.2, this is equivalent to having $G$ doubly transitive on $\Omega$.

*Example* Consider the symmetric group $S_4$ on the set $\{1, 2, 3, 4\}$. It is certainly doubly transitive, with permutation character $\theta$ satisfying $\theta(1) = 4$, $\theta((12)) = 2$, $\theta((123)) = 1$, $\theta((1234)) = \theta((12)(34)) = 0$. Hence by Theorem 11.3(d), $S_4$ has an irreducible character $\chi$ of degree 3 with values $\chi((12)) = 1$, $\chi((123)) = 0$, $\chi((1234)) = \chi((12)(34)) = -1$. $S_4$ has a normal subgroup $N$, $|N| = 4$, with $S_4/N \cong S_3$ of order 6. Over $\mathbf{C}$, for example, we found the three irreducible representations of $S_3$. Using these three characters, the character $\chi$ above, and the orthogonality relations it is easy to finish the character table of $S_4$.

*Remarks* The books of Wielandt [2], Passman [4], and Huppert [6] contain many recent applications of representation theory to the study of finite permutation groups. The paper [2] of D. G. Higman has been important in recent study of finite simple groups. Representation theory of the symmetric group is extensively studied in the book of Robinson [3]; recent papers on this theory include Bayar [1] and Gündüzalp [1].

## EXERCISES

**1**   Complete the character table of $S_4$.

**2**   $A_5$ is a simple group of order 60, the group of all even permutations of the set $\{1, 2, 3, 4, 5\}$. We use cycle notation for the elements of $A_5$.

(a)   Using Sylow's theorem, verify that $A_5$ has five conjugacy classes $\mathscr{C}_1, \mathscr{C}_2, \mathscr{C}_3, \mathscr{C}_4, \mathscr{C}_5$, where $\mathscr{C}_1 = \{1\}$, $(123) \in \mathscr{C}_2$, $(12)(34) \in \mathscr{C}_3$, $(12345) \in \mathscr{C}_4$, $(13524) \in \mathscr{C}_5$, $|\mathscr{C}_1| = 1$, $|\mathscr{C}_2| = 20$, $|\mathscr{C}_3| = 15$, $|\mathscr{C}_4| = |\mathscr{C}_5| = 12$.

(b)   Since $A_5$ is doubly transitive on $\{1, 2, 3, 4, 5\}$, write down an irreducible character of $A_5$ of degree 4, with all of its values.

(c)   Show that $A_5$ permutes its six Sylow 5-subgroups in a doubly transitive manner by conjugation. Then write down an irreducible character of $A_5$ of degree 5, with all of its values.

(d)   Find the degrees of the other two nonprincipal irreducible characters of $A_5$.

(e)   Using the orthogonality relations and the Exercise in §5, complete the character table of $A_5$.

# §12

## T. I. Sets and Exceptional Characters

**Definition**    Let $S$ be any subset of the group $G$. The *normalizer* of $S$ in $G$ is $N_G(S) = \{g \in G \mid S^g = S\}$, and the *centralizer* of $S$ in $G$ is $C_G(S) = \{g \in G \mid gx = xg,\ \text{all } x \in S\}$. Clearly $C_G(S)$ and $N_G(S)$ are subgroups of $G$, and it is easy to show $C_G(S) \lhd N_G(S)$.

**Definition**    A *T. I. set* (*trivial intersection set*) in a group $G$ is a subset $S$ of $G$ such that
   (a)   $S \subseteq N_G(S)$;
   (b)   for all $g \in G$, either $S^g = S$ or $S^g \cap S \subseteq \{1\}$.

**Theorem 12.1** (Brauer-Suzuki)    *Let $G$ be a finite group, $k = \mathbf{C}$, and let $S$ be a T. I. set in $G$. Denote $N = N_G(S)$. Let $\chi$, $\theta$ be generalized characters of $N$, and assume that $\chi$ and $\theta$ are 0 on $N - S$. Then:*

   (1)   *for any $x \in S - \{1\}$, $\chi^G(x) = \chi(x)$.*
   (2)   *if also $\chi(1) = 0$, then $(\chi, \theta)_N = (\chi^G, \theta^G)_G$.*

*Proof*   (1)   Assume $x \in S - \{1\}$, and define $\dot{\chi}$ on $G$ to be $\chi$ on $N$, 0 off $N$. If $y \in G - N$, then $y^{-1}xy \notin S$, so

$$\chi^G(x) = \frac{1}{|N|} \sum_{y \in G} \dot{\chi}(y^{-1}xy) = \frac{1}{|N|} \sum_{y \in N} \chi(y^{-1}xy)$$

$$= \frac{1}{|N|} |N| \chi(x) = \chi(x).$$

   (2)   $S$ has $|G : N|$ conjugates. $\chi^G(1) = |G : N| \chi(1) = 0$. Hence

$$(\chi^G, \theta^G)_G = \frac{1}{|G|} \sum_{x \in G} \chi^G(x) \, \overline{\theta^G(x)}$$

$$= \frac{1}{|G|} |G : N| \sum_{x \in S} \chi^G(x) \, \overline{\theta^G(x)}$$

$$= \frac{1}{|N|} \sum_{x \in S} \chi(x) \, \overline{\theta(x)} = \frac{1}{|N|} \sum_{x \in N} \chi(x) \, \overline{\theta(x)} = (\chi, \theta)_N.$$

In this equation, the third equality comes from part (1), and the fourth from the hypothesis $\chi = 0$ on $N - S$.

**Corollary 12.2**     *Let $S$ be a T. I. set in the finite group $G$, with $N = N_G(S)$. Assume that $\chi_1, \ldots, \chi_n$ are distinct irreducible complex characters of $N$ ($n \geq 2$), such that all $\chi_i$ vanish outside $S$ and $\chi_1(1) = \cdots = \chi_n(1)$. Then there exists a sign $\varepsilon = \pm 1$, and distinct irreducible complex characters $\theta_1, \ldots, \theta_n$ of $G$, such that*

$$\chi_i^G - \chi_j^G = \varepsilon(\theta_i - \theta_j), \qquad all \;\; 1 \leq i, \;\; j \leq n.$$

$\theta_1, \ldots, \theta_n$ *are called the* exceptional characters *associated with* $\chi_1, \ldots, \chi_n$.

*Proof*  Set $\alpha = \chi_1 - \chi_2$. By Theorem 12.1(2), $(\alpha^G, \alpha^G)_G = (\alpha, \alpha)_N = 2$. Hence

$$\chi_1^G - \chi_2^G = \alpha^G = \theta_1 - \theta_2,$$

for some irreducible characters $\theta_1$, $\theta_2$ of $G$. ($\alpha^G(1) = 0$, so $\alpha^G$ cannot be $\theta_1 + \theta_2$ or $-\theta_1 - \theta_2$.) We are done, with $\varepsilon = +1$, if $n = 2$.

If $n \geq 3$, we use Theorem 12.1(2) again to see that

$$1 = (\chi_1 - \chi_3, \chi_1 - \chi_2) = (\chi_1^G - \chi_3^G, \chi_1^G - \chi_2^G)$$

$$= (\chi_1^G - \chi_3^G, \theta_1 - \theta_2),$$

and

$$2 = (\chi_1 - \chi_3, \chi_1 - \chi_3) = (\chi_1^G - \chi_3^G, \chi_1^G - \chi_3^G).$$

Hence either $\chi_1^G - \chi_3^G = \theta_1 - \theta_3$ or $\chi_1^G - \chi_3^G = \theta_3 - \theta_2$, some irreducible character $\theta_3$ not $\theta_1$ or $\theta_2$.

In the former case,

$$\chi_2^G - \chi_3^G = -(\chi_1^G - \chi_2^G) + (\chi_1^G - \chi_3^G) = \theta_2 - \theta_3$$

so we are done with $\varepsilon = +1$ if $n = 3$. If $n > 3$, we repeat the arguments to finish the proof with $\varepsilon = +1$.

In the latter case, relabel $\theta_2$ as $\theta_1$ and $\theta_1$ as $\theta_2$, so we get

$$\chi_1^G - \chi_2^G = -(\theta_1 - \theta_2), \qquad \chi_1^G - \chi_3^G = -(\theta_1 - \theta_3).$$

We also find $\chi_2^G - \chi_3^G = -(\theta_2 - \theta_3)$, done if $n = 3$. If $n > 3$, we continue, eventually finishing the proof with $\varepsilon = -1$.

*Remark*   Exceptional characters first appeared in Suzuki [12], and have been important in most applications of characters since; we will see several. The theory has been generalized by Feit [3, 5, 6], Dade [2], Leonard and McKelvey [1], and Reynolds [2]. A recent paper by Hamel [1] relates exceptional characters to group algebra structure.

EXERCISE

Let $G$ be a finite group, $1 \neq x \in G$, $H = C_G(x)$, and define $S = \{y \in G \mid x \text{ is a power of } y\}$, so $S$ is a subset of $H$.

(a)   Assume

$$(*) \quad \begin{cases} \text{If } y_1, y_2 \in S \text{ are conjugate in } G, \\ \text{then } y_1, y_2 \text{ are conjugate in } H. \end{cases}$$

Prove that $S$ is a T. I. set in $G$ with normalizer $H$.

(b)   If $x$ has order 2, prove that (*) always holds.

# §13

## Frobenius Groups

**Definition** Let $G$ be a transitive permutation group on $\Omega$, $|\Omega| = n$, $\alpha \in \Omega$, $H = G_\alpha$. Assume that $H \neq 1$, but that only the identity element of $G$ fixes more than one element of $\Omega$. Then $G$ is a *Frobenius group*, and $H$ is a *Frobenius complement* in $G$.

**Lemma 13.1** *Let $G$ be a finite group, $1 \neq H \neq G$ a subgroup of $G$. Then the following are equivalent:*
  *(a)  $G$ is a Frobenius group with complement $H$.*
  *(b)  $H$ is a T. I. set and $H = N_G(H)$.*

*Proof* $(a) \Rightarrow (b)$ By the definition of Frobenius group, $\alpha \neq \beta$ implies $G_\alpha \cap G_\beta = \{1\}$; all conjugates of $H$ have the form $G_\beta$, so $H$ is a T. I. set. If $G_\alpha^g = G_\alpha$, also $G_\alpha^g = G_{g^{-1}(\alpha)}$ so $g^{-1}(\alpha) = \alpha$, $g \in G_\alpha = H$.

*Proof* $(b) \Rightarrow (a)$ $G$ is a permutation group on the left cosets of $H$ by $g: xH \to gxH$, and is transitive. $H = \{g \in G | g: H \to H\}$ is the subgroup fixing the coset $H$. If we show that any $h \in H - \{1\}$ does not fix any $xH \neq H$, we will be done. But $hxH = xH$ would mean $x^{-1}hx \in H$, $h \in H^{x^{-1}} \cap H = \{1\}$, a contradiction.

**Theorem 13.2** (Frobenius) *Let $G$ be a Frobenius group on $\Omega$ with complement $H$. Set*

$$K = \{g \in G | g \text{ fixes no element of } \Omega\} \cup \{1\}.$$

*Then $K$ is a normal subgroup of $G$, and satisfies $K \cap H = 1$, $KH = G$, $|K| = |G:H|$. $K$ is called the* Frobenius kernel *of $G$.*

*Remark* This theorem has never been proved without group representation theory.

*Proof* We will use the T. I. set Theorem 12.1 with $H = S = N$. Let $1_H = \chi_1, \ldots, \chi_t$ be the irreducible characters of $H$, say $d_i = \chi_i(1) = \deg \chi_i$. Set

$$\alpha_i = d_i\chi_1 - \chi_i, \qquad \text{all } 2 \leq i \leq t, \quad \text{so } \alpha_i(1) = 0.$$

By Theorem 12.1(2),

$$(\alpha_i^G, \alpha_i^G)_G = (\alpha_i, \alpha_i)_H = d_i^2 + 1.$$

By Frobenius reciprocity,

$$(\alpha_i^G, 1_G)_G = (\alpha_i, 1_H)_H = d_i.$$

The last two equations and the fact $\alpha_i^G(1) = 0$ imply

$$\alpha_i^G = d_i 1_G - \theta_i,$$

where $\theta_i$ is an irreducible character of $G$ of degree $d_i$. We note that $(\alpha_i, \theta_{iH})_H = (\alpha_i^G, \theta_i)_G = -1$, so $\theta_{iH}$ must involve $\chi_i$; $\chi_i(1) = \theta_i(1)$, so $\theta_{iH} = \chi_i$, proving in particular that the $\theta_i$ are distinct.

Define $\theta = 1_G + \sum_{i=2}^t d_i\theta_i$, so

$$\theta(1) = 1 + \sum_{i=2}^t d_i^2 = |H|.$$

If $x \in K - \{1\}$, then

$$\alpha_i^G(x) = \frac{1}{|H|} \sum_{y \in G} \dot\alpha_i(y^{-1}xy) = 0$$

since no conjugate of $x$ is in $H$. Hence

$$\theta_i(x) = d_i 1_G(x) - \alpha_i^G(x) = d_i$$

proving that

$$\theta(x) = 1 + \sum_{i=2}^t d_i^2 = |H| = \theta(1).$$

We have proved that $K \subseteq \ker \theta$.

If $x \in H - \{1\}$, we use the fact $\theta_{iH} = \chi_i$ to see that

$$\theta(x) = 1_H(x) + \sum_{i=2}^t d_i\chi_i(x) = \rho_H(x) = 0,$$

$\rho_H$ the regular character of $H$. Hence $x \notin \ker \theta$. Any element not in $K$ is in a conjugate of $H$, so it is not in $\ker \theta$, and we have proved $K = \ker \theta \lhd G$, $H \cap K = 1$. Then $KH$ is a subgroup of $G$ of order $|H||K|$. But

$$|K| = |G| - |G:H|(|H| - 1)$$
$$= |G| - |G| + |G:H| = |G:H|,$$

so $KH = G$.

*Remark*   Frobenius groups are very important in the study of finite groups. We shall discuss their structure and representations.

*Notation*   For any subset $S$ of a group $G$, write $S^\# = S - \{1\}$.

**Theorem 13.3**      *Let $G$ be a Frobenius group with kernel $K$ and complement $H$. Then:*

(1)   $|H| \mid (|K| - 1)$.
(2)   *If $h \in H^\#$, then $k \to h^{-1}kh$ is an automorphism of $K$ fixing only the element $1 \in K$.*
(3)   *Any subgroup of $H$ of order $p^2$ or $pq$ is cyclic.*
(4)   *If $|H|$ is even, $K$ is abelian.*
(5)   *In any case, $K$ is nilpotent.*

*Remark*   We will only prove (1)–(4). Using (3), the possibilities for the group $H$ can be determined. This work, due to Zassenhaus, and a proof of (5), due to Thompson [1, 2], can be found in the recent book *Permutation Groups*, by D. S. Passman [4].

*Proof of Theorem 13.3 (1) and (2)*   If $k = h^{-1}kh$, then $hk = kh$ or $k^{-1}hk = h$, $h \in H \cap H^k = \{1\}$ when $k \in K^\#$; so (2) holds. By (2), $H$ acts on $K^\#$ as a permutation group with orbits of length $|H|$, so $|H| \mid |K^\#|$, (1) holds.

**Definition**      If $\alpha$ is an automorphism of the group $H$, and $h^\alpha = h$ implies $h = 1$, then $\alpha$ is called *fixed-point-free*. If $A$ is an automorphism group of $H$ and all elements of $A$ except 1 are fixed-point-free, then $A$ is a *regular group of automorphisms of $H$*. (2) says that the Frobenius complement is always a regular group of automorphisms of the Frobenius kernel.

**Lemma 13.4**     *If $\alpha$ is a fixed-point-free automorphism of a finite group $G$, then every $g \in G$ has form $g = x^{-1}x^\alpha$, some $x \in G$.*

*Proof*   If $x$, $y \in G$, then $x^{-1}x^\alpha = y^{-1}y^\alpha$ implies $yx^{-1} = y^\alpha(x^\alpha)^{-1} = y^\alpha(x^{-1})^\alpha = (yx^{-1})^\alpha$, so $yx^{-1} = 1$, $y = x$. As $x$ ranges through all elements of $G$, so does $x^{-1}x^\alpha$.

*Proof of Theorem 13.3(4)*   Choose $h \in H$ of order 2. For any $k \in K$, we can by Lemma 13.4 find $x \in K$, $k = x^{-1}x^h$. So $kk^h = x^{-1}x^h(x^{-1})^hx^{h^2} = x^{-1}x = 1$, proving that, for all $k \in K$, $k^h = k^{-1}$. For any $x$, $y \in K$, then, $x^{-1}y^{-1} = x^hy^h = (xy)^h = (xy)^{-1} = y^{-1}x^{-1}$, implying $K$ abelian.

**Lemma 13.5**     *Let $\alpha$ be a fixed-point-free automorphism of $G$, $p$ a prime dividing $|G|$. Then there is a unique Sylow $p$-subgroup $P$ of $G$ with $P^\alpha = P$.*

*Proof of Existence*   Let $Q$ be any Sylow $p$-subgroup of $G$, and choose $g \in G$ with $Q^\alpha = Q^g$. Let $g = x^{-1}x^\alpha$, $x \in G$, by Lemma 13.4. Then $(xQx^{-1})^\alpha = x^\alpha Q^\alpha(x^\alpha)^{-1} = x^\alpha Q^g(x^\alpha)^{-1} = x^\alpha g^{-1}Qg(x^\alpha)^{-1} = x^\alpha(x^\alpha)^{-1}xQx^{-1}x^\alpha(x^\alpha)^{-1} = xQx^{-1}$, so take $P = xQx^{-1}$.

*Proof of Uniqueness*   Assume also $P_1^\alpha = P_1$, and set $P_1 = P^g$, so $P^g = (P^g)^\alpha = (g^{-1}Pg)^\alpha = (g^\alpha)^{-1}P^\alpha g^\alpha = (g^\alpha)^{-1}Pg^\alpha = P^{g^\alpha}$, so $P^{g(g^\alpha)^{-1}} = P$, $g(g^\alpha)^{-1} \in N_G(P)$. $P^\alpha = P$ implies $N_G(P)^\alpha = N_G(P)$, and $\alpha$ is fixed-point-free on $N_G(P)$, so by Lemma 13.4 there is $n \in N_G(P)$ with $g(g^\alpha)^{-1} = n^{-1}n^\alpha$. Hence $ng = (ng)^\alpha$, $g = n^{-1} \in N_G(P)$, $P_1 = P^g = P$, done.

**Lemma 13.6**     *Let $G$ be a finite regular group of automorphisms of a $k$-vector space $V$. If $|G| = p^2$ or $pq$, $p$ and $q$ primes, then $G$ is cyclic.*

*Proof*   Assume $G$ not cyclic, $p > q$. If $|G| = p^2$, then $G$ is abelian and contains exactly $p + 1$ subgroups of order $p$. If $|G| = pq$, then $G$ contains exactly $p + 1$ proper subgroups, one of order $p$ and $p$ of order $q$, by Sylow's theorem. In either case, let $H_0$, $H_1$, ... ,$H_p$ be these subgroups, and choose some $0 \neq v \in V$. If $g \in G^\#$, then

$$g(\sum_{x \in G} xv) = \sum_{x \in G} gxv = \sum_{x \in G} xv,$$

so by regularity $\sum_{x \in G} xv = 0$. Similarly, we find that $\sum_{x \in H_i} xv = 0$, any $i$. Therefore

$$0 = \sum_{x \in G} xv = -pv + pv + \sum_{x \in G} xv$$

$$= -pv + \sum_{i=0}^{p} \sum_{x \in H_i} xv = -pv,$$

proving that char $k = p$.

Choose $x \in G$ of order $p$. Then $x^p = 1$, so $0 = (x^p - 1)v = (x - 1)^p v$. Choose $0 \le i < p$ such that $(x - 1)^i v \ne 0$ but $(x - 1)^{i+1} v = 0$, and set $w = (x - 1)^i v \ne 0$. We see that $0 = (x - 1)w$ so $xw = w$, contradicting regularity, the final contradiction.

*Proof of Theorem 13.3(3)* Let $H_0$ be a subgroup of $H$ of order $p^2$ or $pq$, and assume $H_0$ not cyclic. Let $r$ be a prime dividing $|K|$, so $r \ne p$, $r \ne q$; we shall find an $r$-Sylow subgroup $R$ of $K$ with $R^{H_0} = R$.

If $|H_0| = p^2$, so $H_0$ is abelian and generated by elements $\alpha$, $\beta$ of order $p$, choose by Lemma 13.5 the unique $r$-Sylow subgroup $R$ of $K$ with $R^\alpha = R$. Then $(R^\beta)^\alpha = \alpha(\beta(R)) = \beta(\alpha(R)) = \beta(R) = R^\beta$, so by the uniqueness $R^\beta = R$, $R^{H_0} = R$.

If $|H_0| = pq$, then $H_0$ is generated by elements $\alpha$ of order $p$, $\beta$ of order $q$. The subgroup $\langle \alpha \rangle$ generated by $\alpha$ is normal in $H_0$, say $\beta^{-1} \alpha \beta = \alpha^t$, some $t$. Choose $R$ by Lemma 13.5 to be the unique $r$-Sylow subgroup of $K$ with $R^\alpha = R$. Then $(R^\beta)^\alpha = \alpha\beta(R) = \beta\beta^{-1}\alpha\beta(R) = \beta\alpha^t(R) = \beta(R) = R^\beta$, so by the uniqueness $R = R^\beta$, $R^{H_0} = R$.

In either case, let $V$ be the subgroup of the center $Z(R)$ of $R$ consisting of elements of order $\le r$. Then $V \ne 1$ and $V^{H_0} = V$, $H_0$ regular on $V$, $V$ a vector space over the field $GF(r)$ of $r$ elements. By Lemma 13.6, $H_0$ is cyclic, the final contradiction.

*Remark* The following lemma, useful in several situations, will be used here to determine the irreducible characters of Frobenius groups.

**Lemma 13.7** (Brauer's Lemma on Character Tables)   *Let $G$ be a group, $k$ a splitting field for $G$ of characteristic zero. Let $\mathscr{C}_1 = \{1\}$, $\mathscr{C}_2, \ldots, \mathscr{C}_h$ be the conjugacy classes of $G$, $1_G = \chi_1, \chi_2, \ldots, \chi_h$ the irreducible characters of $G$. Let $A$ be a group which acts as permutations on the two sets $\{\chi_1, \ldots, \chi_h\}$ and $G$, and assume*

   *(a)   for each $\alpha \in A$ and each $i$, $\mathscr{C}_i^\alpha$ is some $\mathscr{C}_j$;*

   *(b)   $\chi_i^\alpha(g) = \chi_i(g^\alpha)$ for all $g \in G$, $\alpha \in A$, all $i$.*
*Then:*

   *(1)   For any $\alpha \in A$, the number of conjugacy classes fixed by $\alpha$ equals the number of irreducible characters fixed by $\alpha$.*

(2)  *The number of orbits of $A$ in $\{\mathscr{C}_1,\ldots,\mathscr{C}_h\}$ equals the number of orbits of $A$ in $\{\chi_1,\ldots,\chi_h\}$.*

*Example*  Let $\alpha$ be an automorphism of $G$, and define $\chi_i^\alpha$ by $\chi_i^\alpha(g) = \chi_i(g^\alpha)$, all $g \in G$. Let $\chi_i$ be afforded by the representation $T_i$. $\chi_i^\alpha$ is the character of the representation $T_i^\alpha$ defined by $T_i^\alpha(g) = T_i(g^\alpha)$ (a representation since $T_i(g^\alpha) T_i(h^\alpha) = T_i(g^\alpha h^\alpha) = T_i((gh)^\alpha))$, so $\chi_i^\alpha$ is some $\chi_j$. The hypotheses, and hence the conclusion, of our lemma now hold with $A = \langle\alpha\rangle$.

*Proof of Lemma 13.7*  Choose representatives $x_i \in \mathscr{C}_i$, and denote $X = (\chi_i(x_j)) = $ character table of $G$, so

$$X^\alpha = (\chi_i^\alpha(x_j)) = (\chi_i(x_j^\alpha)).$$

Permuting the rows of a matrix is accomplished by premultiplication by a permutation matrix, in this case $P(\alpha)X = X^\alpha$. Permuting the columns is done by postmultiplication, so $X^\alpha = XQ(\alpha)$ also.

The fact $(X^\alpha)^\beta = X^{\beta\alpha}$ implies (since $X$ is nonsingular) that $P(\alpha) P(\beta) = P(\alpha\beta)$ and $Q(\alpha) Q(\beta) = Q(\beta\alpha)$. If we define $\varphi(\alpha) = P(\alpha)$ and $\psi(\alpha) = {}^tQ(\alpha)$, then $\varphi$ and $\psi$ are permutation representations of $A$.

Now $P(\alpha)X = X^\alpha = XQ(\alpha)$, so $Q(\alpha) = X^{-1}P(\alpha)X$, and we see:

number of fixed points of $\alpha$ on rows

$$= \mathrm{tr}\, P(\alpha) = \mathrm{tr}(X^{-1}P(\alpha)X) = \mathrm{tr}\, Q(\alpha)$$

$$= \text{number of fixed points of } \alpha \text{ on columns},$$

proving (1).

Let $\varphi$ afford the character $\theta_1$, $\psi$ the character $\theta_2$. By Theorem 11.3(a),

(number of orbits of $A$ on $\{\chi_1,\ldots,\chi_h\})|A|$

$$= \sum_{\alpha\in A} \theta_1(\alpha) = \sum_{\alpha\in A} \mathrm{tr}\, P(\alpha)$$

$$= \sum_{\alpha\in A} \mathrm{tr}\, Q(\alpha) = \sum_{\alpha\in A} \mathrm{tr}\, {}^tQ(\alpha) = \sum_{\alpha\in A} \theta_2(\alpha)$$

$$= \text{(number of orbits of } A \text{ on } \{\mathscr{C}_1,\ldots,\mathscr{C}_h\})|A|, \text{ done.}$$

**Theorem 13.8**     *Let $G$ be a Frobenius group with kernel $K$ and complement $H$, and let $k$ be a field of characteristic zero which is a splitting field for $G$ and its subgroups. Assume $H$ has $h(H)$ conjugacy classes, $K$*

has $h(K)$ conjugacy classes. Then $G$ has exactly $h(H) + [h(K) - 1]/|H|$ conjugacy classes, and the following irreducible characters:

(a) $h(H)$ irreducible characters $\chi_1, \ldots, \chi_{h(H)}$ with $K$ in their kernel; if $\mu_1, \ldots, \mu_{h(H)}$ are the irreducible characters of $H$, then these satisfy

$$\chi_i(hk) = \mu_i(h), \qquad all \quad h \in H, \quad k \in K.$$

(b) Whenever $\tau \neq 1_K$ is an irreducible character of $K$, then $\tau^G$ is an irreducible character of $G$. This provides $[h(K) - 1]/|H|$ irreducible characters of $G$ with $K$ not in their kernel. Each such $\tau^G$ satisfies

$$\tau^G|_H = \tau(1) \rho_H.$$

*Proof* We first show that if $k \in K^\#$ and $h \in H^\#$, then $k$ is not conjugate to $k^h$ in $K$. If on the contrary $k^h = k^x$ for $x \in K$, use Lemma 13.4 to write $x = y^{-1}y^h$, $y \in K$. Then

$$k^h = k^x = x^{-1}kx = (y^h)^{-1}yky^{-1}y^h,$$

or $y^h k^h (y^h)^{-1} = yky^{-1}$, $(yky^{-1})^h = yky^{-1}$, $yky^{-1} = 1$, the desired contradiction. This implies that, on nontrivial conjugacy classes of $K$, $H$ has orbits of length $|H|$, so $|H|$ divides $h(K) - 1$. Also, any $h \in H^\#$ fixes only the conjugacy class $\{1\}$, so by Lemma 13.7(1) fixes only the character $1_K$ of $K$.

Let $\tau$ be an irreducible character of $K$, $\tau \neq 1_K$. Then

$$(\tau^G, \tau^G)_G = \frac{1}{|G|} \sum_{g \in G} \tau^G(g)\tau^G(g^{-1})$$

$$= \frac{1}{|G|} \sum_{k \in K} \tau^G(k)\tau^G(k^{-1}) = \frac{|K|}{|G|} (\tau^G|_K, \tau^G|_K)_K.$$

But for any $k \in K$, we use the fact $G = HK$, $H \cap K = 1$ to see

$$\tau^G(k) = \frac{1}{|K|} \sum_{g \in G} \tau(g^{-1}kg) = \frac{1}{|K|} \sum_{y \in H} \sum_{x \in K} \tau(x^{-1}y^{-1}kyx)$$

$$= \frac{1}{|K|} \sum_{y \in H} \sum_{x \in K} \tau(y^{-1}ky) = \sum_{y \in H} \tau(y^{-1}ky) = \sum_{y \in H} \tau^y(k),$$

so $\tau^G|_K = \sum_{y \in H} \tau^y$, a sum of $|H|$ distinct irreducible characters of $K$. Therefore

$$(\tau^G|_K, \tau^G|_K)_K = |H| \qquad and \qquad (\tau^G, \tau^G)_G = \frac{|K|}{|G|} \cdot |H| = 1,$$

proving that $\tau^G$ is irreducible. Note that

$$(\tau^G, (\tau^y)^G)_G = (\tau^G|_K, \tau^y)_K = 1,$$

so $(\tau^y)^G = \tau^G$, all $y \in H$. This procedure must yield $[h(K) - 1]/|H|$ different irreducible characters of $G$.

Clearly $\tau^G(1) = |G:K|\tau(1) = |H|\tau(1)$. If $y \in H^\#$ then $\tau^G(y) = 0$, so $\tau^G|_H = \tau(1)\rho_H$ by Lemma 7.6.

We have seen that $K^\#$ contains $[h(K) - 1]/|H|$ conjugacy classes of $G$. Elements in $G - K^\#$ are conjugate to elements of $H$, so $G$ has at most $h(H)$ more conjugate classes. But (a) and (b) provide $h(H) + [h(K) - 1]/|H|$ irreducible characters of $G$, so these are all and we are done.

## EXERCISES

**1**  If $m > 1$ is odd and $D_{2m}$ is the dihedral group defined in Exercise 2, page 52, show that $D_{2m}$ is a Frobenius group.

**2**  Show that if $G$ is a Frobenius group with kernel $K$ and complement $H$, then any subgroup $1 \neq H_0$ of $H$ satisfies $N_G(H_0) \subseteq H$.

**3**  (a)  Let $P$ be the group of order $7^3$ and exponent 7 generated by elements $a, b, c$ with relations $a^7 = b^7 = c^7 = 1$, $[a, b] = c$, $c \in Z(P)$. Show that

$$\alpha: a \to a^2c, \qquad b \to b^2, \quad c \to c^4$$

defines an automorphism $\alpha$ of $P$ of order 3.

(To show $\alpha$ is an automorphism, it is enough to show that $\alpha(a)$, $\alpha(b)$, $\alpha(c)$ generate $P$ and satisfy the same relations as do $a, b, c$.)

(b)  Show that $\alpha$ is fixed-point-free.

(c)  Show that if $H = \langle \alpha \rangle$, then the group $G$ with generators $a, b, c, \alpha$ and relations

$$a^7 = b^7 = c^7 = 1, \quad [a, b] = c, \quad ca = ac, \quad cb = bc,$$

$$\alpha^3 = 1, \quad a^\alpha = a^2c, \quad b^\alpha = b^2, \quad c^\alpha = c^4$$

is a Frobenius group with kernel $P = \langle a, b, c \rangle$ and complement $H$.

(This shows that some Frobenius groups have nonabelian kernels.)

**4**  We shall prove the following theorem of Burnside [1]. Let $G$ be a transitive permutation group on the set $\Omega$, where $|\Omega| = p$ is a prime. Then one of the following holds:

(i)  $G$ is doubly transitive.

(ii)  Either $|G| = p$, or $G$ is a Frobenius group with kernel of order $p$; hence $G$ is solvable.

To do this, we assume (i) does not hold, and will prove (ii). Prove the following statements, assuming (i) false.

(a)  If $P$ is a Sylow $p$-subgroup of $G$, then $|P| = p$.

(b)  If $\theta$ is the permutation character of $G$, then $\theta = 1_G + \sum_{i=1}^{t} \chi_i$, where $t \geq 2$ and the $\chi_i$ are irreducible complex characters of $G$ different from $1_G$.

(c)  $\theta|_P$ is the regular character of $P$. The $\chi_i$ are all different.

(d)  Let $\varepsilon$ be a primitive $|G|$th root of 1. The linear characters of $P$ other than $1_P$ are algebraically conjugate under the Galois group $H = \mathrm{Gal}(\mathbf{Q}(\varepsilon)/\mathbf{Q}(\varepsilon^p))$. Hence $\chi_1, \ldots, \chi_t$ are conjugate under $H$.

(e)  If $g \in G$ does not have order $p$, then $p \nmid |\langle g \rangle|$ and $\chi_1(g) = \cdots = \chi_t(g)$. $\chi_1(g)$ is a nonnegative integer.

(f)  If $0 < i \leq \chi_1(1)$, let $n_i = |\{g \in G | \chi_1(g) = i\}|$. Let $N$ be the number of Sylow $p$-subgroups of $G$. Show that the relations $(\chi_1, 1_G)_G = 0$ and $(\chi_1, \chi_2)_G = 0$ imply

$$\sum_{i=1}^{\chi_1(1)} i n_i = N\chi_1(1), \qquad \sum_{i=1}^{\chi_1(1)} i^2 n_i = N\chi_1(1)^2.$$

(g)  Show that the equations of (f) imply $n_1 = n_2 = \cdots = n_{\chi_1(1)-1} = 0$. Conclude that (ii) holds.

# §14

## Clifford's Theorem

**Theorem 14.1** (Clifford's theorem) (Clifford [1]).    *Let $k$ be any field, $V$ an irreducible $kG$-module, $N$ a normal subgroup of $G$. Of course for $g \in G$, $W$ a $kN$-submodule of $V$, we denote $gW = \{gw|w \in W\} \subseteq V$.*

*(1)   If $0 \neq W$ is a $kN$-submodule of $V$, then $V = \sum_{g \in G} gW$. If $W$ is irreducible, then so is every $gW$, proving that $V_N$ is a completely reducible $kN$-module.*

*(2)   Let $W_1, \ldots, W_m$ be representatives of the isomorphism classes of irreducible $kN$-submodules of $V$, and denote by $V_i$ the sum of all $kN$-submodules of $V$ which are isomorphic to $W_i$. Then $V = V_1 \oplus \cdots \oplus V_m$. (The $V_i$ are the* homogeneous components *of $V$.)*

*(3)   If $g \in G$, then each $gV_i$ is some $V_j$. $G$ is a transitive permutation group on $\{V_1, \ldots, V_m\}$.*

*(4)   If $H_1 = \{g \in G|gV_1 = V_1\}$, then $V_1$ is irreducible as $kH_1$-module, and $V \cong V_1^G = kG \otimes_{kH_1} V_1$.*

*(5)   For some integer $e$, $V_N \cong e(W_1 \oplus \cdots \oplus W_m)$ (that is, the direct sum of $e$ copies of $W_1 \oplus \cdots \oplus W_m$).*

*(6)   Let $V$ afford the character $\theta$ of $G$, $W_i$ the character $\chi_i$ of $N$. Then*

$$\theta_N = e(\chi_1 + \cdots + \chi_m),$$

*where $W_i \cong g_i W_1$, some $g_i \in G$, and $\chi_i = \chi_1^{g_i}$; $\chi_1^{g_i}$ is defined by $\chi_1^{g_i}(x) = \chi_1(x^{g_i})$, all $x \in N$.*

*Proof of (1)*    Irreducibility of $V$ as $kG$-module implies $V = \sum_{g \in G} gW$. If $n \in N$ and $gw \in gW$, then $ngw = gg^{-1}ngw = gn^g w \in gW$, so each $gW$ is a $kN$-module. If $gW$ had a proper $kN$-submodule $W_0$, then $g^{-1}W_0$ would be a proper submodule of $W$, contradicting its irreducibility. So all $gW$ are irreducible, and $V_N$ is completely reducible.

*Proof of (2)*    Let $L$ be an irreducible $kN$-submodule of $V_i$. Let

72

$\varphi: V_i \to W_i \oplus \cdots \oplus W_i$ be an isomorphism, $\pi_j$ the projection onto the $j$th summand. Some $\pi_j \circ \varphi(L) \neq 0$, so $\pi_j \circ \varphi: L \to W_i$ is an isomorphism, $L \cong W_i$. For all $i$, we have seen that all composition factors of $V_i$ are isomorphic to $W_i$, so $(V_1 + \cdots + V_{i-1}) \cap V_i = (0)$, implying $V = V_1 \oplus \cdots \oplus V_m$.

*Proof of* (3)   *Claim 1*: $V_i = \Sigma\{xW_1 | x \in G, xW_1 \cong W_i\}$. For if $U_i = \Sigma\{xW_1 | x \in G, xW_1 \cong W_i\}$, then $U_i \subseteq V_i$, and $V = U_1 + \cdots + U_m$ by (1), implying all $U_i = V_i$ by (2).

*Claim 2*: $xW_1 \cong yW_1$ implies $gxW_1 \cong gyW_1$. For if $\varphi: xW_1 \cong yW_1$ then $g\varphi g^{-1}: gxW_1 \to gyW_1$ is a $k$-isomorphism. And for $n \in N$, $(g\varphi g^{-1})[n(gxw_1)] = g\varphi[(g^{-1}ng)xw_1] = g(g^{-1}ng)\varphi(xw_1) = n(g\varphi g^{-1})(gxw_1)$.

Claims 1 and 2 imply that $gV_i \subseteq V_j$, some $V_j$. If $gV_i \neq V_j$, then $\dim V_i < \dim V_j$. But $gV_i \subseteq V_j$ implies $V_i \subseteq g^{-1}V_j$. Since $g^{-1}V_j \subseteq$ some $V_l$ this means $g^{-1}V_j \subseteq V_i$, $\dim V_j \leq \dim V_i$, a contradiction if $gV_i \neq V_j$. So $gV_i =$ some $V_j$. Irreducibility of $V$ as $kG$-module forces $G$ to be transitive on $\{V_1, \ldots, V_m\}$.

*Proof of* (4)   By (3), $m = |G: H_1|$; if we choose $x_1 = 1, \ldots, x_m \in G$ with $x_j V_1 = V_j$, then $x_1 H_1, \ldots, x_m H_1$ are the left cosets of $H_1$ in $G$. Recall that $V_1^G = kG \otimes_{kH_1} V_1 = (x_1 \otimes V_1) \oplus \cdots \oplus (x_m \otimes V_1)$. We define

$$\varphi: V_1 \oplus \cdots \oplus V_m \to V_1^G$$

by

$$\varphi: \sum_{i=1}^{m} x_i u_i \to \sum_{i=1}^{m} x_i \otimes u_i, \quad \text{all} \quad u_1, \ldots, u_m \in V_1.$$

For any $g \in G$, if $gx_i = x_j h$, $h \in H_1$, then we have

$$gx_i u_i = x_j \cdot h u_i \quad \text{and} \quad g(x_i \otimes u_i) = x_j \otimes h u_i,$$

proving that $\varphi: V \to V_1^G$ is a $kG$-isomorphism. $V_1$ is irreducible as $kH_1$-module, for if it had a proper submodule $V_1'$, this construction would make $(V_1')^G$ a proper $kG$-submodule of $V_1^G \cong V$, a contradiction.

*Proof of* (5)   Each $W_i \cong x_i W_1$, so $\dim W_i = \dim W_1$. $V_i = x_i V_1$, so $\dim V_i = \dim V_1$. Set $e = \dim V_1 / \dim W_1$.

*Proof of* (6)   By (5), $\theta_N = e(\chi_1 + \cdots + \chi_m)$. We have to show

$x_l W_1$ affords the character $\chi_1^{x_l}$. Let $\{u_1, \ldots, u_t\}$ be a $k$-basis of $W_1$. Then $\{x_l u_1, \ldots, x_l u_t\}$ is a basis of $x_l W_1$. If $x \in N$, we can write

$$(x_l^{-1} x x_l) u_j = \sum_{i=1}^{t} \alpha_{ij} u_i, \qquad \text{so} \qquad \chi_1(x_l^{-1} x x_l) = \sum_{i=1}^{t} \alpha_{ii}.$$

We see that

$$x \cdot x_l u_j = x_l(x_l^{-1} x x_l u_j) = x_l \left( \sum_{i=1}^{t} \alpha_{ij} u_i \right) = \sum_{i=1}^{t} \alpha_{ij} x_l u_i,$$

so the character $\chi_l$ afforded by $x_l W_1$ satisfies

$$\chi_l(x) = \sum_{i=1}^{t} \alpha_{ii} = \chi_1(x_l^{-1} x x_l) = \chi_1(x^{x_l}) = \chi_1^{x_l}(x).$$

*Remark*   Clifford's paper [1] has been the starting point for a large amount of recent research, especially by Dade; see Glauberman [3], Isaacs [4], Dade [3, 7, 8], and Cline [1]. This research is summarized in the first few pages of Dade's preprint [3].


EXERCISE


    Let $\chi$ be an irreducible complex character of $G$, $H \lhd G$. Prove the following are equivalent:
    (a)   $\chi = \theta^G$, some irreducible character $\theta$ of $H$.
    (b)   $\chi = 0$ on $G - H$, and $\chi|_H$ is a sum of distinct irreducible characters of $H$.

# §15

## *M-Groups*

**Definition**    A representation of $G$ is called *monomial*, if it is a direct sum of representations induced from one-dimensional representations of subgroups of $G$.

*Remark*   In our matrix interpretation of induced representations (Lemma 9.1), we see that the matrices of a monomial representation in a suitable basis are actually monomial; that is, each matrix has only one nonzero entry in each row and each column.

**Definition**    A finite group $G$ is an *M-group* if every irreducible representation of $G$ over $\mathbf{C}$ is monomial.

**Lemma 15.1**    *Let $\mathcal{M}$ be a set of finite groups with the following properties:*
  *(1)   If $G \in \mathcal{M}$, then all subgroups and homomorphic images of $G$ are in $\mathcal{M}$.*
  *(2)   If $G \in \mathcal{M}$ is not abelian, then there is an abelian $A \lhd G$ such that $A \nsubseteq Z(G)$.*
  *Then every group in $\mathcal{M}$ is an M-group.*

*Proof*   By induction on $|G|$; let

$$T: G \to GL(V), \qquad \dim_{\mathbf{C}} V = n,$$

be an irreducible representation of $G$. If $T$ is not faithful, then $T$ is a representation of $G/\ker T$ and $G/\ker T \in \mathcal{M}$ by (1), so $T$ is monomial by induction. Hence we may assume $T$ faithful.

We may also assume $G$ nonabelian, since $G$ abelian implies $\deg T = 1$, done. By (2) there is an abelian $A \lhd G$, $A \nsubseteq Z(G)$. By Clifford's theorem, $V_A = V_1 \oplus \cdots \oplus V_m$, the $V_i$ homogeneous components, each $V_i$ a direct sum of $n/m$ isomorphic one-dimensional $A$-modules. Let $V_A = W_1 \oplus \cdots \oplus W_n$, $\dim_\mathbf{C} W_i = 1$, say $W_i = \mathbf{C}w_i$.

Suppose all the $W_i$ are $\mathbf{C}A$-isomorphic; pick $a \in A - Z(G)$, and set $aw_1 = \lambda w_1$, $\lambda \in \mathbf{C}$. Let $\varphi_i \colon W_1 \to W_i$ be a $\mathbf{C}A$-isomorphism, say $\varphi_i(w_1) = \gamma_i w_i$. Then

$$aw_i = a(\gamma_i^{-1} \varphi_i(w_1)) = \gamma_i^{-1} a\varphi_i(w_1) = \gamma_i^{-1} \varphi_i(aw_1) = \gamma_i^{-1} \varphi_i(\lambda w_1)$$
$$= \gamma_i^{-1} \lambda\varphi_i(w_1) = \gamma_i^{-1} \lambda\gamma_i w_i = \lambda w_i.$$

We have shown that $av = \lambda v$ for all $v \in V$, implying $a \in Z(G)$, a contradiction.

So not all $W_i$ are $\mathbf{C}A$-isomorphic, proving $m > 1$. By Theorem 14.1(4), $V \cong W^G$, where $W$ is a $\mathbf{C}H$-module, $H$ a proper subgroup of $G$. $H \in \mathcal{M}$ by (1), so by induction $W \cong U^H$, $U$ a one-dimensional $\mathbf{C}H_0$-module, $H_0$ a subgroup of $H$. By transitivity of induction (Theorem 9.4(a)), $V \cong U^G$, done.

**Definitions**     A finite group $G$ is *nilpotent* if it is the direct product of its Sylow subgroups. It is *supersolvable* if there is a chain

$$1 = G_0 \subset G_1 \subset \cdots \subset G_r = G,$$

where all $G_i \lhd G$ and all $|G_{i+1}/G_i|$ are primes.

**Lemma 15.2**     *Nilpotent groups are supersolvable. Supersolvable groups are solvable. Subgroups and factor groups of nilpotent groups are nilpotent. Subgroups and factor groups of supersolvable groups are supersolvable.*

*Proof*   An easy exercise.

**Lemma 15.3**     *If $G \neq 1$ is solvable, then $G' \neq G$.*

*Proof*   Solvable means

$$1 = G_0 \subset G_1 \subset \cdots \subset G_s = G,$$

all $G_{i+1}/G_i$ abelian. So $G_{s-1} \neq G$, $G/G_{s-1}$ abelian, and by Lemma 8.3, $G' \subseteq G_{s-1}$.

**Lemma 15.4**     *If $G' \subseteq Z(G)$, then $G$ is nilpotent.*

*Proof* Let $P$ be a Sylow $p$-subgroup of $G$, $N = N_G(P)$. $P$ is the only Sylow $p$-subgroup in $N$, so $N^g = N$ would imply $P^g = P$, $g \in N$, $N_G(N) = N$. If $z \in Z(G)$, then $N^z = N$ so $z \in N$, $N \supseteq Z(G) \supseteq G'$. $G/G'$ is abelian so $N/G' \lhd G/G'$, $N \lhd G$, $N = N_G(N) = G$. So all Sylow $p$-subgroups are normal in $G$. If $Q$ is a Sylow $q$-subgroup, $x \in P$, $y \in Q$, then

$$[x, y] = x^{-1}y^{-1}xy \in P \cap Q = 1$$

so $xy = yx$. This is enough to show $G$ is nilpotent.

**Theorem 15.5** (*Huppert* [1]).   *If $G$ has a normal subgroup $N$ such that $G/N$ is supersolvable and $N$ is solvable with all Sylow subgroups of $N$ abelian, then $G$ is an M-group.*

*Proof* Let $\mathscr{M}$ be the set of all groups satisfying the hypothesis. We will use Lemma 15.1; (1) of that lemma is clear, so we will prove (2), if $G \in \mathscr{M}$ nonabelian.

*Case 1* If $N$ is not abelian, let $A$ be largest possible among the abelian normal subgroups of $G$ lying in $N$. If $A \nsubseteq Z(G)$, (2) holds. If $A \subseteq Z(G)$, then $A \subseteq Z(N)$; we shall obtain a contradiction. Choose $B/A$ minimal among the nontrivial normal subgroups of $G/A$ lying in $N/A$. By Lemma 15.3, $(B/A)' \neq B/A$, so $(B/A)'$ is trivial, $B/A$ is abelian. Hence $B' \subseteq A \subseteq Z(B)$. By Lemma 15.4, $B$ is nilpotent, and $B \subseteq N$. Sylow subgroups of $N$ are abelian so $B$ is abelian, contradicting the maximality of $A$.

*Case 2* If $N$ is abelian and $N \nsubseteq Z(G)$, we are done. If $N \subseteq Z(G)$, use supersolvability of $G/N$ to write

$$N = G_0 \subset G_1 \subset \cdots \subset G_r = G,$$

all $G_i \lhd G$, all $|G_{i+1}/G_i|$ primes. $G$ is not abelian, so we can choose $i$ with $G_i \subseteq Z(G)$, $G_{i+1} \nsubseteq Z(G)$. If $x \in G_{i+1} - G_i$, we see that $A = G_{i+1}$ is generated by $x$ and $G_i \subseteq Z(G)$, so $A$ is abelian and (2) holds.

By Lemma 15.1, all $G \in \mathscr{M}$ are M-groups.

**Corollary 15.6**   *Nilpotent groups are M-groups. Supersolvable groups are M-groups. If $G$ has a normal subgroup $N$ with $N$ and $G/N$ abelian (such a $G$ is called* metabelian), *then $G$ is an M-group.*

**Theorem 15.7** (Taketa [1]).     *Every M-group is solvable.*

*Proof*  If not, let $G$ be a nonsolvable $M$-group of minimal order. Proper homomorphic images of $G$ are $M$-groups, and hence solvable.

Suppose $A$ and $B$ are distinct minimal normal subgroups of $G$. Then $G/A \times G/B$ is solvable, and

$$g \rightarrow (gA, gB)$$

is a homomorphism of $G$ into $G/A \times G/B$ with kernel $A \cap B = 1$, forcing $G$ solvable, a contradiction.

Therefore $G$ has a unique minimal normal subgroup $N \neq 1$; choose $x \in N^{\#}$. Let $T$ be an irreducible representation of $G$ with $T(x) \neq I$; $\ker T \not\supseteq N$ so $\ker T = 1$, and $T$ is faithful.

Let $\varphi$ be a faithful irreducible matrix representation of $G$ over $\mathbf{C}$ of minimal degree. $G$ is an $M$-group, so we may take all $\varphi(g)$ as monomial matrices. We form a new permutation matrix representation $\varphi_0$ of $G$ by changing all nonzero entries of matrices $\varphi(g)$ to 1. The kernel $K$ of $\varphi_0$ is

$$\{g \in G | \varphi(g) \text{ is diagonal}\} \cong \{\varphi(g) | \varphi(g) \text{ is diagonal}\},$$

so $K$ is abelian.

$\varphi_0$ is reducible, by Theorem 11.3(a). If $\varphi_0$ were faithful, we could find an irreducible component $\psi$ of $\varphi_0$, $\deg \psi < \deg \varphi_0 = \deg \varphi$, $\psi(x) \neq I$; hence $\ker \psi \not\supseteq N$, $\ker \psi = 1$, $\psi$ is faithful, contradicting minimality of $\deg \varphi$. So $\varphi_0$ is not faithful.

Therefore $1 \neq K \lhd G$, where $K$ is abelian. $G/K$ is solvable, so $G$ is solvable, the final contradiction.

*Remark*  Dade (see Huppert [6; Satz V. 18.11]) has shown that every finite solvable group is a subgroup of an $M$-group. For other results on $M$-groups, see Itô [4], Berman [4, 5], Huppert [5], Dornhoff [1], Basmaji [1], Ward [1], Seitz [2], Seitz and Wright [1], and Kerber [4].

EXERCISES

**1**  Let $Q$ be the quaternion group of order 8, with generators $c, d$ and relations

$$c^4 = d^4 = 1, \qquad c^2 = d^2, \qquad d^{-1}cd = c^{-1}.$$

(a)  Show that $Q$ has an automorphism $\beta$ of order 3 defined by $c^\beta = d$, $d^\beta = cd$.

(b)  Let $G$ be the group generated by elements $c, d, \beta$ with relations $c^4 = d^4 = 1$, $c^2 = d^2$, $c^d = c^{-1}$, $\beta^3 = 1$, $c^\beta = d$, $d^\beta = cd$, so that $|G| = 24$ and $Q = \langle c, d \rangle \lhd G$. Show that $G' = Q$.

(c)  Show that $Z(G) = \langle c^2 \rangle$ and $G/Z(G) \cong A_4$.

(d)  Show that $G$ has an irreducible complex representation of degree 2.

(e)  Show that $G$ has no subgroup of index 2.

(f)  Conclude that $G$ is not an $M$-group.

**2**  Let $P$ be the group of order 32 with defining relations

$$a^4 = 1, \quad b^{-1}ab = a^{-1}, \quad a^2 = b^2 = c^2 = d^2, \quad c^4 = 1,$$

$$d^{-1}cd = c^{-1}, \quad ac = ca, \quad ad = da, \quad bc = cb, \quad bd = db.$$

Denote $z = a^2 = b^2 = c^2 = d^2$. We see that $Q_1 = \langle a, b \rangle$ and $Q_2 = \langle c, d \rangle$ are quaternion subgroups of $P$ which centralize each other and have the same center $\langle z \rangle$. ($P$ is called the *central product* of $Q_1$ and $Q_2$.)

(a)  Show that $P$ has an automorphism $\alpha$ of order 3 satisfying $a^\alpha = b$, $b^\alpha = ab$, $c^\alpha = d$, $d^\alpha = cd$; $\alpha$ fixes $Q_1$ and $Q_2$.

(b)  Let $G$ be the group $P\langle \alpha \rangle$ of order 96, $P \lhd G$, with generators $a, b, c, d, \alpha$ satisfying all the relations above. Show that $G/\langle z \rangle$ is a Frobenius group of order 48, and hence an $M$-group.

(c)  Verify that $A = \langle ac, bd \rangle$ is an elementary abelian subgroup of order 4 fixed by $\alpha$.

(d)  Let $H = A\langle \alpha \rangle \times \langle z \rangle$, a subgroup of $G$ of order 24, and let $\lambda$ be the linear character of $H$ with kernel $A\langle \alpha \rangle$. Show that $\lambda^G = \chi$ is irreducible.

(e)  Let $1_G$, $\mu_2$, $\mu_3$ be the three linear characters of $G$, and show that $\chi$, $\chi\mu_2$, $\chi\mu_3$ are distinct irreducible characters of $G$ with kernel not containing $\langle z \rangle$. Conclude that $G$ is an $M$-group.

(f)  Using Exercise 1, show that subgroups of $M$-groups need not be $M$-groups.

A sol gp can be embedded in an M-gp.

# §16

## Brauer's Characterization of Characters

**Definition**     Let $p$ be a prime. An element of finite order of a group is a *p-element* if its order is a power of $p$, and a *p'-element* if its order is relatively prime to $p$.

**Lemma 16.1**     *Let $G$ be any group, $p$ a prime, and assume $g \in G$ has finite order $p^a m$, $p \nmid m$. Then there are unique elements $g_1, g_2 \in G$ such that $g = g_1 g_2 = g_2 g_1$, $g_1$ of order $p^a$, $g_2$ of order $m$. $g_1$ is called the* p-part *of $g$, $g_2$ the* p'-part *of $g$.*

    *Proof of Existence*    Choose integers $s$, $t$ with $sp^a + tm = 1$, and set $g_1 = g^{tm}$, $g_2 = g^{sp^a}$. Then $g = g_1 g_2 = g_2 g_1$, where $g_1^{p^a} = 1$, $g_2^m = 1$.

    *Proof of Uniqueness*    If also $g = g_3 g_4 = g_4 g_3$, $g_3$ a $p$-element, $g_4$ a $p'$-element, then $g_3$ and $g_4$ have relatively prime orders so the order of their product $g$ is the product of their orders. So $g_3^{p^a} = 1$, $g_4^m = 1$, $g_3 = g_3^{sp^a + tm} = g_3^{tm} = (gg_4^{-1})^{tm} = g^{tm} = g_1$. By cancellation, $g_4 = g_2$.

**Definition**     A finite group $G$ is *p-elementary* for the prime $p$, if it is the direct product of a cyclic group and a $p$-group. $G$ is *elementary*, if it is $p$-elementary for some $p$. Note that elementary groups are nilpotent, and hence $M$-groups.

**Theorem 16.2** (Brauer's Characterization of Characters) (Brauer [2, 3, 4]) *Let $G$ be a finite group, $k = \mathbf{C}$. Then:*
    *(1)*   *If $\chi$ is a character of $G$, then*

$$\chi = \sum_i a_i \lambda_i^G, \quad \text{all} \quad a_i \in \mathbf{Z},$$

*where the $\lambda_i$ are linear characters of suitable elementary subgroups of $G$.*

80

(2) *Let* $\theta: G \to \mathbf{C}$ *be a class function. Then* $\theta$ *is a generalized character of* $G$ *if and only if, for every elementary subgroup* $E$ *of* $G$, $\theta|_E$ *is a generalized character of* $E$.

(3) *Let* $\theta: G \to \mathbf{C}$ *be a class function. Then* $\theta$ *is an irreducible character of* $G$ *if and only if the following hold:*
  (a)  *for every elementary subgroup* $E$ *of* $G$, $\theta|_E$ *is a generalized character of* $E$.
  (b)  $(\theta, \theta)_G = 1.$
  (c)  $\theta(1) > 0.$

*Proof, Step 1, Notation* Let $R$ be any subring of $\mathbf{C}$ containing $1 \in \mathbf{C}$, and let $\chi_1, \ldots, \chi_h$ be the irreducible characters of $G$ over $\mathbf{C}$. Denote

$$X_R(G) = \{ \sum_{i=1}^{h} r_i \chi_i | \text{all } r_i \in R \},$$

so $X_R(G)$ is a ring; in the case $R = \mathbf{Z}$, $X_{\mathbf{Z}}(G)$ is the ring of generalized characters of $G$. Denote

$V_R(G) = \{ \sum_i r_i \psi_i^G | r_i \in R, \ \psi_i \text{ an irreducible character of an elementary subgroup } E_i \text{ of } G \},$

$U_R(G) = \{ \chi: G \to \mathbf{C} | \chi \text{ is a class function, and for every elementary subgroup } E \text{ of } G, \ \chi|_E \in X_R(E) \}.$

Then $U_R(G)$ is a ring.

Let $m$ be the exponent of $G$, let $\varepsilon \in \mathbf{C}$ be a primitive $m$th root of 1, and denote $S = \mathbf{Z}[\varepsilon]$ = the subring of $\mathbf{C}$ generated by $\mathbf{Z}$ and $\varepsilon$.

*Step 2*  $V_R(G) \subseteq X_R(G) \subseteq U_R(G)$, and $V_R(G)$ is an ideal in $U_R(G)$.

*Proof*  The inclusions are trivial. If $\sum_i r_i \psi_i^G \in V_R(G)$ and $\varphi \in U_R(G)$, $\psi_i$ an irreducible character of the elementary group $E_i$, then $\psi_i^G \varphi = (\psi_i \varphi_{E_i})^G$ by Theorem 9.4(b). Therefore $(\sum r_i \psi_i^G)\varphi = \sum r_i (\psi_i \varphi_{E_i})^G \in V_R(G)$, proving that $V_R(G)$ is an ideal.

*Step 3*  Let $E = A \times B$ be an elementary subgroup of $G$ such that $A$ is abelian and $(|A|, |B|) = 1$. Fix $a \in A$. Then there is a $\Phi = \Phi_a \in V_S(G)$, such that
  (i)  $\Phi(g) \in \mathbf{Z}$, all $g \in G$;
  (ii)  If $g \in G$ is not conjugate to an element of $aB$, then $\Phi(g) = 0$;
  (iii)  $\Phi(a) = |C_G(a): B|.$

*Proof*  Let $\tau_1, \ldots, \tau_{|A|}$ be the irreducible complex characters of $A$; each $\tau_i$ has its values in $S$. We define irreducible characters $\mu_1, \ldots \mu_{|A|}$ of $E$ by setting $\mu_i(a_1 b) = \tau_i(a_1)$, all $a_1 \in A$, $b \in B$.

Define

$$\psi = \sum_{i=1}^{|A|} \overline{\mu_i(a)}\, \mu_i \in X_S(E).$$

For any $x \in A$, $y \in B$, we have

$$\psi(xy) = \sum_{i=1}^{|A|} \overline{\mu_i(a)}\, \mu_i(xy)$$

$$= \sum_{i=1}^{|A|} \overline{\tau_i(a)}\, \tau_i(x) = |A| \quad \text{if} \quad x = a,$$

$$= 0 \quad \text{if} \quad x \neq a,$$

using the orthogonality relations in $A$. We define

$$\Phi = \psi^G = \sum_{i=1}^{|A|} \overline{\mu_i(a)}\, \mu_i^G \in V_S(G).$$

Also, for any $g \in G$, denote $T(g) = \{t \in G | t^{-1}gt \in aB\}$. Using the values of $\psi$ on $E = A \times B$ above, we have

$$\Phi(g) = \frac{1}{|E|} \sum_{t \in G} \dot{\psi}(t^{-1}gt) = \frac{1}{|E|} \sum_{\substack{t \in G, \\ t^{-1}gt \in E}} \psi(t^{-1}gt)$$

$$= \frac{1}{|E|} \sum_{\substack{t \in G, \\ t^{-1}gt \in aB}} \psi(t^{-1}gt) = \frac{1}{|E|} \sum_{t \in T(g)} |A| = \frac{|A||T(g)|}{|E|}.$$

If $t \in T(g)$, $b \in B$, then $b^{-1}t^{-1}gtb \in aB$. Hence $tb \in T(g)$, and $T(g)$ consists of left cosets of $B$, say $|T(g)| = n_g|B|$. Therefore $\Phi(g) = |A|\,|B|n_g/|E|$ $= n_g \in \mathbf{Z}$, proving (i). If the hypothesis of (ii) holds, then $|T(g)| = 0$ so $\Phi(g) = 0$, proving (ii).

If $t \in T(a)$, then $t^{-1}at \in aB$; $t^{-1}at$ has the same order as $a$ and $(|A|,|B|) = 1$, so $t^{-1}at = a$, $t \in C_G(a)$. Hence

$$\Phi(a) = \frac{|A||T(a)|}{|E|} = \frac{|C_G(a)|}{|B|} = |C_G(a) : B|,$$

proving (iii).

*Step 4*  Let $p$ be a prime, and let $\mathscr{C}_1 = \{1\}, \ldots, \mathscr{C}_k$ be those conjugacy classes of $G$ which consist of $p'$-elements. Then for each $1 \leq i \leq k$, there is a $\tau_i \in V_S(G)$ such that:

    (1)  $\tau_i(g) \in \mathbf{Z}$, all $g \in G$.

    (2)  $\tau_i(g) = 0$ if the $p'$-part of $g$ is not in $\mathscr{C}_i$.

    (3)  $\tau_i(g) \equiv 1 \pmod{p}$ if the $p'$-part of $g$ is in $\mathscr{C}_i$.

*Proof* Let $a_i \in \mathscr{C}_i$, and let $B_i$ be a Sylow $p$-subgroup of $C_G(a_i)$. Then $E_i = \langle a_i \rangle \times B_i$ is $p$-elementary, and by Step 3 there is $\Phi_i \in V_S(G)$ such that:

    (i)  $\Phi_i(g) \in \mathbf{Z}$, all $g \in G$.

    (ii)  $\Phi_i(g) = 0$ if $g$ is not conjugate to an element of $a_i B_i$.

    (iii)  $\Phi_i(a_i) = |C_G(a_i) : B_i| \not\equiv 0 \pmod{p}$.

Choose $m_i \in \mathbf{Z}$ with $m_i \Phi_i(a_i) \equiv 1 \pmod{p}$, and set $\tau_i = m_i \Phi_i$. By (i), (1) holds.

    Set $g = g_1 g_2$, $g_1$ the $p$-part of $g$, $g_2$ the $p'$-part. If $g$ is conjugate to an element of $a_i B_i$, say $t^{-1} g t = a_i b_i$, then $t^{-1} g_1 t \cdot t^{-1} g_2 t = a_i b_i$ gives two expressions of the same element $a_i b_i$ as a product of commuting $p$- and $p'$-parts. By the uniqueness in Lemma 16.1, then, $t^{-1} g_2 t = a_i$, $g_2 \in \mathscr{C}_i$. Hence $g_2 \notin \mathscr{C}_i$ would by (ii) imply $\Phi_i(g) = 0$; this proves (2).

    Now suppose the $p'$-part of $g$ is in $\mathscr{C}_i$, so that for some $t \in G$, $t^{-1} g t = a_i b$, $b$ a $p$-element in $C_G(a_i)$. We know $\tau_i(a_i) \equiv 1 \pmod{p}$ and $\tau_i(g) = \tau_i(a_i b)$, so for (3) it is enough to show $\tau_i(a_i) \equiv \tau_i(a_i b) \pmod{p}$.

    Let $H = \langle a_i b \rangle$. $\tau_i \in V_S(G) \subseteq U_S(G)$, so $\tau_i|_H \in X_S(H)$. Let $\{\mu_1, \ldots, \mu_l\}$ be the linear characters of $H$, so

$$\tau_i|_H = \sum_{j=1}^{l} s_j \mu_j, \qquad \text{some} \quad s_j \in S.$$

Let $b^{p^d} = 1$. Taking congruences modulo the ideal $pS$ of $S$, we have

$$\tau_i(a_i b) \equiv \tau_i(a_i b)^{p^d} = \left( \sum_{j=1}^{l} s_j \mu_j(a_i b) \right)^{p^d} \equiv \sum_{j=1}^{l} s_j^{p^d} \mu_j((a_i b)^{p^d})$$

$$= \sum_{j=1}^{l} s_j^{p^d} \mu_j(a_i^{p^d}) \equiv \left( \sum_{j=1}^{l} s_j \mu_j(a_i) \right)^{p^d} = \tau_i(a_i)^{p^d} \equiv \tau_i(a_i),$$

and we are done.

    *Step 5*  If $p$ is a prime, then there is a $\Psi \in V_S(G)$ such that for any $g \in G$, $\Psi(g) \in \mathbf{Z}$ and

$$\Psi(g) \equiv 1 \pmod{p}.$$

*Proof* In Step 4, let $\Psi = \tau_1 + \cdots + \tau_k$. The $p'$-part of any $g \in G$ is in exactly one $\mathscr{C}_i$, so

$$\Psi(g) \equiv 1 \pmod{p}, \qquad \text{all} \quad g \in G.$$

*Step 6* If $\alpha \colon G \to \mathbf{C}$ is a class function and $\alpha(g) \in |G|S$ for all $g \in G$, then $\alpha \in V_S(G)$.

*Proof* Let $g_1, \ldots, g_h$ be representatives of the conjugacy classes of $G$. By Step 3 with $A = \langle g_i \rangle$, $B = 1$, there are $\Phi_i \in V_S(G)$ such that

$$\Phi_i(g_j) = 0 \qquad \text{if} \quad i \neq j$$

$$= |C_G(g_i)| \quad \text{if} \quad i = j.$$

Let $\alpha(g_i) = |G|s_i$, $s_i \in S$. Then

$$\alpha = \sum_{i=1}^{h} \frac{|G|}{|C_G(g_i)|} s_i \Phi_i \in V_S(G).$$

*Step 7* Let $p$ be a prime, $|G| = p^n g_0$, $p \nmid g_0$. Then the constant function $g_0 \in V_S(G)$.

*Proof* We use the $\Psi$ of Step 5. $\Psi(g)^p \equiv \Psi(g) \equiv 1 \pmod{p}$, and by induction on $l$ one can show that, for all $l$,

$$\Psi(g)^{p^l} \equiv 1 \pmod{p^l}.$$

$\Psi \in V_S(G)$ and $V_S(G)$ is an ideal, so $\Psi^{p^n} \in V_S(G)$, $\Psi(g)^{p^n} \equiv 1 \pmod{p^n}$. Define the class function $\beta$ on $G$ by $\beta = \Psi^{p^n} - 1$. Then $g_0\beta(g) \in |G|S$ for all $g \in G$, so by Step 6, $g_0\beta \in V_S(G)$. Finally, $g_0 = g_0\Psi^{p^n} - g_0\beta \in V_S(G)$.

*Step 8* $1 \in V_S(G)$.

*Proof* Let

$$|G| = \prod_{i=1}^{t} p_i^{b_i}, \qquad n_j = \prod_{i \neq j} p_i^{b_i}.$$

The G.C.D. of $\{n_1, \ldots, n_t\}$ is 1, so we can find integers $a_1, \ldots, a_t$ such that $1 = \sum_{j=1}^{t} a_j n_j$. By Step 7, all $n_j \in V_S(G)$, so $1 \in V_S(G)$.

*Step 9* $1 \in V_{\mathbf{Z}}(G)$.

*Proof* If $r = \varphi(m)$ (Euler $\varphi$-function), then $1, \varepsilon, \ldots, \varepsilon^{r-1}$ are a **Z**-basis of $S = \mathbf{Z}[\varepsilon]$. Each $\varepsilon^i = \varepsilon^i 1_G$ is a constant class function in $X_S(G)$. We first show that $1, \varepsilon, \ldots, \varepsilon^{r-1}$ are linearly independent over $X_{\mathbf{Z}}(G) \subseteq X_S(G)$. For suppose we had

$$\sum_{i=0}^{r-1} \varepsilon^i \sum_{j=1}^{h} a_{ij} \chi_j = 0, \qquad \text{the} \quad a_{ij} \in \mathbf{Z}.$$

$\chi_1, \ldots, \chi_h$ are linearly independent over **C**, so for all $j$ we have

$$\sum_{i=0}^{r-1} \varepsilon^i a_{ij} = 0.$$

$\varepsilon^0, \ldots, \varepsilon^{r-1}$ are linearly independent over **Z**, so all $a_{ij} = 0$.

By Step 8, $1 \in V_S(G)$, so $1 = \sum_{i=1}^{k} \psi_i^G$, where $E_1, \ldots, E_k$ are elementary subgroups of $G$ and $\psi_i \in X_S(E_i)$. Hence we can write

$$\psi_i = \sum_{j=0}^{r-1} \varepsilon^j \psi_{ij}, \quad \psi_{ij} \in X_{\mathbf{Z}}(E_i), \quad 1 = \sum_{j=0}^{r-1} \varepsilon^j \sum_{i=1}^{k} \psi_{ij}^G.$$

Since $\varepsilon^0, \ldots, \varepsilon^{r-1}$ are linearly independent over $X_{\mathbf{Z}}(G)$ and $\varepsilon^0 = 1$, we may equate coefficients of 1 on both sides and get

$$1 = \sum_{i=1}^{k} \psi_{i0}^G \in V_{\mathbf{Z}}(G).$$

*Step 10, Conclusion* $1 \in V_{\mathbf{Z}}(G) \subseteq V_R(G)$ for any subring $R$ of **C**, so by Step 2, $V_R(G) = X_R(G) = U_R(G)$.

Since elementary groups are $M$-groups, the fact that every character is in $X_{\mathbf{Z}}(G)$ and so in $V_{\mathbf{Z}}(G)$ proves (1). The equality $X_{\mathbf{Z}}(G) = U_{\mathbf{Z}}(G)$ proves (2). (3) comes from (2) and known properties of generalized characters.

*Remarks* The proof we have given is the one given by Brauer and Tate [1] and independently by Asano [1]. Other proofs are given in Springer [1], Osima [2], and Tachikawa [1]. The theorem was generalized to arbitrary fields of characteristic zero independently by Berman [1, 2] and Witt [1]. Related results are given in Green [1] and Fischer [1].

## EXERCISE

Let $|G| = p^n m$, $p \nmid m$, $p$ a prime. Let $\chi$ be an irreducible complex character of $G$, and assume that $p^n | \chi(1)$. We will prove

(*)   If $x \in G$ has order a multiple of $p$, then $\chi(x) = 0$.
      Define the function $\theta$ on $G$ by

$$\theta(y) = \chi(y) \qquad \text{if} \quad (p, |\langle y \rangle|) = 1$$

$$= 0 \qquad \text{otherwise.}$$

(a)   If $E$ is an elementary subgroup of $G$ and $p \nmid |E|$, show that $\theta|_E$ is a generalized character of $E$.

(b)   If $E$ is an elementary subgroup of $G$ and $p \mid |E|$, show that $E = P \times Q$, $P$ a $p$-group, $p \nmid |Q|$.

(c)   With $E = P \times Q$ as in (b), show that if $\xi$ is an irreducible character of $E$ then

$$(\theta|_E, \xi)_E = \frac{1}{|Q|} \sum_{y \in Q} \frac{\chi(y)}{|P|} \overline{\xi(y)}.$$

(d)   Using Lemma 5.9, show that all quotients $\chi(y)/|P|$ in (c) are algebraic integers.

(e)   In (c), show that $(\theta|_E, \xi)_E$ is an ordinary integer. Conclude that $\theta|_E$ is a generalized character of $E$.

(f)   Show that $\theta$ is a generalized character of $G$, and prove (*).

(*)   will be an easy consequence of modular representation theory in §62, Part B.

# §17

## Brauer's Theorem on Splitting Fields

**Lemma 17.1**    *Let $F$ be a subfield of $\mathbf{C}$, $G$ a finite group with $h$ conjugacy classes. If $FG$ has $h$ pairwise nonisomorphic irreducible modules with degrees $n_1, \ldots, n_h$ satisfying $\sum_{i=1}^{h} n_i^2 = |G|$, then $F$ is a splitting field for $G$.*

*Proof*   By Wedderburn's Theorem 2.17 we know that $FG \cong \mathrm{Mat}_{m_1}(D_1)$ $\oplus \cdots \oplus \mathrm{Mat}_{m_k}(D_k)$, all the $D_i$ division rings with $F$ in their center, where $FG$ has exactly $k$ irreducible modules. Hence $k \geq h$; but $h = \dim_F Z(FG) \geq k$, so $h = k$. We may arrange the irreducible modules $V_i$ of dimensions $n_i$ over $F$, so that $n_i = m_i(\dim_F D_i)$. Hence $\dim_F FG = |G| = \sum_{i=1}^{k} m_i^2(\dim_F D_i)$. But $|G| = \sum_i n_i^2 = \sum_i m_i^2(\dim_F D_i)^2$, so all $\dim_F D_i = 1$, all $D_i = F$. This implies that $F$ is a splitting field for $G$.

**Theorem 17.2** (Brauer [1, 2]).    *Let $G$ be a finite group of exponent $m$, $\varepsilon \in \mathbf{C}$ a primitive $m$th root of $1$. Then the algebraic number field $\mathbf{Q}(\varepsilon)$ is a splitting field for $G$.*

*Proof (Feit)*   Let $F = \mathbf{Q}(\varepsilon)$; by Lemma 17.1 it is enough to show that if $\theta$ is any irreducible $\mathbf{C}G$-character, then $\theta$ is an $FG$-character (i.e., a character of an $FG$-module). By Theorem 16.2(1), we write

$$\theta = \sum_i a_i \lambda_i^G, \qquad a_i \in \mathbf{Z},$$

where the $\lambda_i$ are linear characters of elementary subgroups of $G$. Clearly the $\lambda_i^G$ are characters of $FG$-modules. Let

$$\tau_1 = \sum_{a_i > 0} a_i \lambda_i^G, \qquad \tau_2 = -\sum_{a_i < 0} a_i \lambda_i^G,$$

so $\tau_1$ and $\tau_2$ are characters, and let $W_1$, $W_2$ be $FG$-modules affording

87

$\tau_1$ and $\tau_2$, respectively. Let $V$ be the $\mathbf{C}G$-module affording $\theta$, and denote $W_i^{\mathbf{C}} = \mathbf{C} \otimes_F W_i$, so $W_i^{\mathbf{C}}$ is a $\mathbf{C}G$-module with character $\tau_i$. Since $\theta = \tau_1 - \tau_2$ or $\tau_1 = \theta + \tau_2$, Lemma 7.2 implies that

$$W_1^{\mathbf{C}} \cong W_2^{\mathbf{C}} \oplus V.$$

Among all pairs $(V_1, V_2)$ of $FG$-modules satisfying $V_1^{\mathbf{C}} \cong V_2^{\mathbf{C}} \oplus V$, choose a pair with $\dim_F V_2$ minimal. We are done if $V_2 = (0)$, so assume $V_2 \neq (0)$. Let $U$ be an irreducible $FG$-submodule of $V_2$, and let $U$ afford the character $\chi$, $V_1$ and $V_2$ the characters $\theta_1$ and $\theta_2$. Then $\theta_1 = \theta_2 + \theta$, and $(\chi, \theta_2)_G > 0$ implies $(\chi, \theta_1)_G > 0$. By Corollary 7.3, some composition factor of $V_1$ is isomorphic to $U$. By Theorem 3.1, $V_1 \cong U \oplus V_1'$ and $V_2 \cong U \oplus V_2'$ for modules $V_1'$ and $V_2'$. Therefore,

$$U^{\mathbf{C}} \oplus (V_1')^{\mathbf{C}} \cong U^{\mathbf{C}} \oplus (V_2')^{\mathbf{C}} \oplus V.$$

$(V_1')^{\mathbf{C}}$ has character $\theta_1 - \chi$, $(V_2')^{\mathbf{C}} \oplus V$ has character $(\theta_2 - \chi) + \theta = \theta_1 - \chi$, so by Lemma 7.2,

$$(V_1')^{\mathbf{C}} \cong (V_2')^{\mathbf{C}} \oplus V.$$

$\dim_F V_2' < \dim_F V_2$ so this contradicts the minimality of $\dim_F V_2$, and we are done.

*Remark*   Related results appear in Berman [7] and Solomon [2].

# §18

## Normal p-Complements and the Transfer

**Lemma 18.1**  *Let $G$ be a group with subgroup $H$, and let $T$ be a cross section of $H$ in $G$, so $G = \bigcup_{t \in T} tH$, $t_1 H = t_2 H \Rightarrow t_1 = t_2$. For any $g \in G$, let $\bar{g}$ denote the unique element of $gH \cap T$. If $g \in G$ and $x \in G$, then $\overline{xg} = [x\bar{g}]^-$;\* and $\bar{g}^{-1} g \in H$.*

*Proof* $gH = \bar{g}H$, so $xgH = x\bar{g}H$ and $\overline{xg} = [x\bar{g}]^-$. The equation $gH = \bar{g}H$ implies $\bar{g}^{-1} g \in H$.

**Definition**  Let $G$ be a group with a subgroup $H$ of finite index, and assume that $K$ is a normal subgroup of $H$ with $H/K$ abelian. Let $T$ be a cross section of $H$ in $G$. We define the function $V = V_T \colon G \to H/K$ by

$$V(x) = \prod_{t \in T} \overline{xt}^{-1} xt K, \quad \text{all} \quad x \in G.$$

Since $H/K$ is abelian, the order of the factors in the product is immaterial. $V = V_T$ is the *transfer of $G$ into $H/K$*.

**Lemma 18.2**  $V = V_T \colon G \to H/K$ is a homomorphism.

*Proof*  If $x \in G$, then $x \colon tH \to xtH$ permutes the cosets of $H$, so as $t$ ranges over $T$ so does $\overline{xt}$. Since $H/K$ is abelian, we can write

$$V(xy) = \prod_{t \in T} (\overline{xyt})^{-1} xytK = \prod_{t \in T} ([x\overline{yt}]^-)^{-1} x\overline{yt}\,\overline{yt}^{-1} ytK$$

$$= \prod_{t \in T} ([x\overline{yt}]^-)^{-1} x\overline{yt}K \cdot \prod_{t \in T} \overline{yt}^{-1} ytK = V(x)V(y),$$

using Lemma 18.1 for the second equality.

---

\* Note that [expression]$^-$ means the same as $\overline{\text{expression}}$; similarity for $\sim$ and $\wedge$ in place of $-$. [Expression]$^{-,\wedge}$ means that both $\sim$ and $\wedge$ are applied to [expression].

**Lemma 18.3**      *The transfer* $V = V_T \colon G \to H/K$ *is independent of the choice of cross section* $T$.

*Proof*  Let $S$ be a second cross section, and for any $g \in G$ let $\tilde{g}$ be the unique element of $gH \cap S$, so $V_S(x) = \prod_{s \in S} ([xs]\tilde{\ })^{-1} xsK$, any $x \in G$. Enumerate $T = \{t_1, t_2, \ldots\}$, $S = \{s_1, s_2, \ldots\}$, where $t_i H = s_i H$, say $s_i = t_i h_i$, $h_i \in H$. For a fixed $x \in G$, let $\sigma$ be the permutation of $\{1, 2, \ldots\}$ defined by $[xs_i]\tilde{\ } = s_{\sigma(i)}$.

$$xs_i H = s_{\sigma(i)} H = t_{\sigma(i)} H = xt_i H,$$

so also $\overline{xt_i} = t_{\sigma(i)}$. Hence

$$h_{\sigma(i)}^{-1} t_{\sigma(i)}^{-1} xt_i h_i = h_{\sigma(i)}^{-1} \overline{xt_i}^{-1} xt_i h_i,$$

and we see that

$$
\begin{aligned}
V_S(x) &= \prod_i ([xs_i]\tilde{\ })^{-1} xs_i K = \prod_i s_{\sigma(i)}^{-1} xt_i h_i K \\
&= \prod_i h_{\sigma(i)}^{-1} t_{\sigma(i)}^{-1} xt_i h_i K = \prod_i h_{\sigma(i)}^{-1} \overline{xt_i}^{-1} xt_i h_i K \\
&= (\prod_i h_{\sigma(i)}^{-1})(\prod_i h_i) \prod_i \overline{xt_i}^{-1} xt_i K = \prod_i \overline{xt_i}^{-1} xt_i K = V_T(x).
\end{aligned}
$$

**Definition**      For any group $G$ and prime $p$, $G'(p)$ denotes the intersection of all normal subgroups $N$ of $G$ with $G/N$ an abelian $p$-group. By Lemma 8.3, $G'(p) \supseteq G'$, so $G/G'(p)$ is an abelian $p$-group. Also, $G' = \bigcap_p G'(p)$, the intersection taken over all primes $p$.

*Remark*  Our objective is to determine $G/G'(p)$. If $S$ is a Sylow $p$-subgroup of $G$, then $G = SG'(p)$, so $G/G'(p) \cong S/S \cap G'(p)$.

**Definition**      If $S$ is a Sylow $p$-subgroup of $G$, we define $S^* = S_G^*$, the *focal subgroup of $S$ in $G$*, to be the subgroup of $S$ generated by all $x^{-1}y$, where $x, y \in S$ are conjugate in $G$. Clearly $S^* \supseteq S'$, so $S/S^*$ is abelian.

**Theorem 18.4**      *Let $S$ be a Sylow subgroup of $G$, $S^*$ the focal subgroup of $S$ in $G$. Then*

$$S^* = S \cap G'(p) = S \cap G', \qquad so \quad G/G'(p) \cong S/S^*.$$

*Proof*  If $x, y \in S$ are conjugate in $G$, say $y = g^{-1}xg$, then $x^{-1}y = x^{-1}g^{-1}xg \in G'$, so $S^* \subseteq S \cap G' \subseteq S \cap G'(p)$. We must show that $S \cap G'(p) \subseteq S^*$. We consider the transfer $V \colon G \to S/S^*$, and evaluate it on any $x \in S$. $x$ acts as a permutation on the cosets of $S$ in $G$; let its orbit decomposition be

$$(S)(x_2 S \, xx_2 S \cdots x^{e_2-1}x_2 S)(x_3 S \, xx_3 S \cdots x^{e_3-1}x_3 S) \cdots .$$

For a cross section $T$ of $S$ in $G$ we choose $x_1 = 1, x_2, xx_2, \ldots, x^{e_2-1}x_2,$ $x_3, \ldots, x^{e_3-1} x_3, \ldots$, and denote $e_1 = 1$ so $|G:S| = \Sigma_i e_i$.

Of course $V(x) = \prod_{y \in T} \overline{xy^{-1}xy} S^*$, where $\bar{g}$ is the unique element of $gS \cap T$. We compute the terms $\overline{xy^{-1}xy}$. $\overline{xx_1^{-1}xx_1} = \bar{x}^{-1}x = 1x = x$ since $x \in S$. $\overline{xx_2^{-1}xx_2} = (xx_2)^{-1}xx_2 = 1$, if $e_2 \geq 2$. If $e_2 > 2$ we get

$$\overline{x^2 x_2^{-1} x^2 x_2} = (x^2 x_2)^{-1}x^2 x_2 = 1.$$

We keep getting 1 until the term

$$\overline{x \cdot x^{e_2-1}x_2^{-1}x \cdot x^{e_2-1}x_2} = \overline{x^{e_2}x_2}^{-1} \cdot x^{e_2}x_2 = x_2^{-1}x^{e_2}x_2.$$

We conclude that

$$V(x) = x \cdot x_2^{-1}x^{e_2}x_2 \cdot x_3^{-1}x^{e_3}x_3 \cdots S^*.$$

By definition of $S^*$, $x^{-e_i}x_i^{-1}x^{e_i}x_i \in S^*$, so $x_i^{-1}x^{e_i}x_i S^* = x^{e_i}S^*$, and

$$V(x) = xx^{e_2}x^{e_3} \cdots S^* = x^{\Sigma e_i}S^*,$$

proving

$$V(x) = x^{|G:S|}S^*.$$

$(p, |G:S|) = 1$ and $S/S^*$ is a $p$-group, so the map $V: G \to S/S^*$ is *onto*. Hence $G/\ker V \cong S/S^*$ is an abelian $p$-group. So $G'(p) \subseteq \ker V$, $|G:G'(p)| \geq |S:S^*|$. But $S^* \subseteq S \cap G'(p)$, so

$$|S:S^*| \geq |S: S \cap G'(p)| = |SG'(p): G'(p)| = |G:G'(p)|.$$

Therefore equality holds everywhere, $S^* = S \cap G'(p)$, done.

**Lemma 18.5**   *Let $G$ be a group, $S$ its Sylow $p$-subgroup. Let $H$ and $K$ be two normal subsets of $S$ which are conjugate in $G$. Then $H$ and $K$ are conjugate in $N_G(S)$.*

*Proof*  Let $K = H^g$, $g \in G$. Then $N_G(K) = N_G(H^g) = N_G(H)^g$. $S^g \subset N_G(H)^g = N_G(K)$, so $S$ and $S^g$ are Sylow $p$-subgroups of $N_G(K)$. Choose $n \in N_G(K)$ with $S = S^{gn}$. Then $gn \in N_G(S)$, and $H^{gn} = K^n = K$.

**Corollary 18.6**     *Suppose the Sylow $p$-subgroup $S$ of $G$ is abelian. Then the focal subgroup of $S$ in $G$ equals the focal subgroup of $S$ in $H = N_G(S)$. Hence $G/G'(p) \cong H/H'(p)$.*

*Proof* The first part is immediate from Lemma 18.5 and the definition of focal subgroup; elements of $S$ are conjugate in $G$ if and only if they are conjugate in $N_G(S)$. The last part follows from Theorem 18.4.

**Theorem 18.7** (Burnside)     *Suppose that the Sylow $p$-subgroup $S$ of $G$ is in the center of $N_G(S)$. Then $G$ has a normal subgroup $N$ with $G = NS$, $S \cap N = 1$, $|N| = |G:S|$.*

*Proof* The focal subgroup of $S$ in $H = N_G(S)$ is trivial, so $H/H'(p) \cong S$. By Corollary 18.6, $G/G'(p) \cong H/H'(p) \cong S$ so take $N = G'(p)$.

**Definition**   A *normal $p$-complement* in a group $G$ is a normal subgroup $N$ such that, if $S$ is a Sylow $p$-subgroup of $G$, then $G = SN$, $S \cap N = 1$, $|S| = |G:N|$, $|N| = |G:S|$. Theorem 18.7 says that if $S$ is in the center of $N_G(S)$, then $G$ has a normal $p$-complement.

**Corollary 18.8**     *If $G$ has a cyclic Sylow 2-subgroup $S$, then $G$ has a normal 2-complement.*

*Proof* Let $|S| = 2^n$. $S$ has $2^n - 2^{n-1} = 2^{n-1}$ generators, and an automorphism of $S$ is determined by the image of a generator, so the automorphism group of $S$ has order $2^{n-1}$. $S \subseteq C_G(S)$, so $N_G(S)/C_G(S)$ has odd order, and any $x \in N_G(S)$ performs an automorphism of odd order on $S$. This automorphism is trivial, so $x \in C_G(S)$, $N_G(S) = C_G(S)$, Theorem 18.7 applies.

**Definition**     A subgroup $K$ of a group $G$ is *characteristic* in $G$ if for all automorphisms $\alpha$ of $G$, $K^\alpha = K$. If $K$ is characteristic in $G$, we write $K$ char $G$.

**Lemma 18.9**     *If $K$ char $N$ and $N \lhd G$, then $K \lhd G$.*

*Proof* If $g \in G$, then $x \to g^{-1}xg$ is an automorphism of $N$, so $x \in K$ implies $g^{-1}xg \in K$, $K \lhd G$.

**Lemma 18.10**    *Any normal p-complement $N$ in $G$ is characteristic in $G$; in fact, $N = \{x \in G | x \text{ has order relatively prime to } p\}$.*

*Proof*   If $x \in G$, of course $x^{|G:N|} \in N$. If $x$ has order $m$ and $(m, p) = 1$, then $(m, |G: N|) = 1$, so we choose integers $s, t$ with $sm + t|G: N| = 1$. Therefore

$$x = x^{sm + t|G:N|} = x^{t|G:N|} = (x^{|G:N|})^t \in N.$$

*Remarks*   We have not used representation theory in this section. This approach to the transfer stems from D. G. Higman [1]. The papers of Thompson [1, 2] contain very important criteria for normal p-complements, obtained without representation theory. Recent papers relating normal p-complements and character theory include Gallagher [2], Sah [1], Suzuki [9], Brauer [9], Passman [3], and Thompson [5]. Normal p-complements will be related to modular representation theory in §65 of Part B.

## EXERCISE

Prove Theorem 13.2 without using characters in the special case when $H$ is an abelian p-group.

# §19

## Generalized Quaternion Sylow 2-Subgroups

**Definition**    An *involution* in a group $G$ is an element of order 2.

*Notation*    For any element or subset $S$ of $G$, $\langle S \rangle$ is the subgroup of $G$ generated by $S$.

**Lemma 19.1**    *If* $x, y \in G$ *are involutions and* $w = xy$, *then* $w^x = w^{-1}$, $w^y = w^{-1}$.

*Proof*  $y = xxy = xw$, so $1 = xwxw$, $w^{-1} = xwx = x^{-1}wx$. $x = xyy = wy$, so $1 = wywy$, $w^{-1} = ywy = y^{-1}wy$.

**Lemma 19.2**    *Let $G$ have conjugacy classes $\mathscr{C}_1, \ldots, \mathscr{C}_h$, $x_i \in \mathscr{C}_i$, and let $\chi_1, \ldots, \chi_h$ be the irreducible complex characters of $G$. Let $c_{ijk}$ denote the number of ordered pairs $(x, y)$ such that $x \in \mathscr{C}_i$, $y \in \mathscr{C}_j$, and $xy = x_k$. Then*

$$c_{ijk} = \frac{|\mathscr{C}_i|\,|\mathscr{C}_j|}{|G|} \sum_{t=1}^{h} \frac{\chi_t(x_i)\chi_t(x_j)\overline{\chi_t(x_k)}}{\chi_t(1)}.$$

*Proof*  Denote $C_i = \sum_{x \in \mathscr{C}_i} x \in CG$, $h_i = |\mathscr{C}_i|$, $n_i = \chi_i(1)$; then the $c_{ijk}$ clearly satisfy

$$C_i C_j = \sum_{k=1}^{h} c_{ijk} C_k.$$

In the proof of Lemma 5.9, we saw that

$$\text{if} \quad \omega_{ij} = \frac{h_i \chi_j(x_i)}{n_j}, \quad \text{then} \quad \omega_{il}\omega_{jl} = \sum c_{ijk}\omega_{kl}.$$

Substituting the values of the $\omega_{kl}$, this becomes

$$h_i h_j \chi_l(x_i)\chi_l(x_j) = n_l \sum_{k=1}^{h} c_{ijk} h_k \chi_l(x_k).$$

Multiplying by $\overline{\chi_l(x_s)}/n_l$ and summing over $l$,

$$h_i h_j \sum_{l=1}^{h} \frac{\chi_l(x_i)\chi_l(x_j)\overline{\chi_l(x_s)}}{n_l} = \sum_{l=1}^{h}\sum_{k=1}^{h} c_{ijk} h_k \chi_l(x_k)\overline{\chi_l(x_s)}.$$

By the orthogonality relations, this equals

$$\sum_{k=1}^{h} c_{ijk} h_k \delta_{ks} \frac{|G|}{h_s} = c_{ijs}|G|,$$

proving the result.

**Definition**     A *generalized quaternion group* of order $2^{n+1}$ is a group with generators $a, b$ and relations $a^{2^n} = 1$, $b^2 = a^{2^{n-1}}$, $b^{-1}ab = a^{-1}$. If $n = 2$, this group is the *quaternion group* of order 8.

**Lemma 19.3**     *A generalized quaternion group contains only one involution.*

*Proof*  Clearly $\langle a \rangle$ has index 2, so any element has form $a^i$ or $a^i b$, some $i$; $a^{2^{n-1}}$ is the only element $a^i$ of order 2. $(a^i b)^2 = a^i b a^i b = a^i b^2 b^{-1} a^i b = a^i a^{2^{n-1}} a^{-i} = a^{2^{n-1}} \neq 1$, so no element $a^i b$ is an involution.

**Lemma 19.4**     *If $G$ is a group with $G' \subseteq Z(G)$ and $x, y \in G$, then for any positive integer $n$,*

$$(xy)^n = x^n y^n [y, x]^{n(n-1)/2}.$$

*Proof*  By induction on $n$, true for $n = 1$. If true for $n$, then

$$\begin{aligned}
(xy)^{n+1} &= (xy)^n xy = x^n y^n [y, x]^{n(n-1)/2} xy \\
&= [y, x]^{n(n-1)/2} x^n y^{n-1} \cdot yx \cdot y \\
&= [y, x]^{n(n-1)/2} x^n y^{n-1} \cdot xy[y, x] \cdot y \\
&= \cdots = [y, x]^{n(n-1)/2 + n} x^{n+1} y^{n+1} \\
&= x^{n+1} y^{n+1} [y, x]^{n(n+1)/2}, \qquad \text{done.}
\end{aligned}$$

We now generalize Lemma 6.1.

**Lemma 19.5**     *Let $P$ be a finite $p$-group, $1 \neq N$ a normal subgroup of $P$. Then $N \cap Z(P) \neq 1$.*

*Proof*  Let $\mathscr{C}_1 = \{1\}$, $\mathscr{C}_2, \ldots, \mathscr{C}_k$ be the conjugacy classes of $P$ contained in $N$. Then $|N| = \sum_{i=1}^{k} |\mathscr{C}_i| = 1 + \sum_{i=2}^{k} |\mathscr{C}_i|$ is a power of $p$, so for some $i > 1$, $p \nmid |\mathscr{C}_i|$. But $|\mathscr{C}_i|$ is a power of $p$, so $|\mathscr{C}_i| = 1$, and if $\mathscr{C}_i = \{x\}$ then $x \in N \cap Z(P)$.

**Theorem 19.6**     *Let $P$ be a finite 2-group containing only one involution. Then $P$ is cyclic or generalized quaternion.*

*Proof*  Assume $P$ not cyclic, and let $A$ be a maximal abelian normal subgroup of $P$. Then $A$ is cyclic; we claim $A = C_P(A)$. If $A \neq C_P(A)$, then $C_P(A)/A \cap Z(P/A) \neq 1$, so there is $N \lhd P$, $|N:A| = 2$, $N \subseteq C_P(A)$. $N$ is then abelian, contradicting maximality of $A$; so $A = C_P(A)$. If $g \in P$, then $g_A: a \to gag^{-1}$ is an automorphism of $A$, and $g \to g_A$ is a homomorphism $P \to \operatorname{Aut}(A)$ with kernel $A$. Hence $P/A$ is isomorphic to a group of automorphisms of $A$.

Let $A = \langle a \rangle$, and consider any $b \in P - A$ with $b^2 \in A$. $\langle b^2 \rangle \neq A$, so there is a subgroup $A_1$ of $A$, $|A_1 : \langle b^2 \rangle| = 2$. Then $Q = A_1 \langle b \rangle$ has two cyclic subgroups $U_1 = A_1$ and $U_2 = \langle b \rangle$ of index 2. Let $D = U_1 \cap U_2$: then $Q' \subseteq D \subseteq Z(Q)$. By Lemma 19.4, if $x, y \in Q$ then $(xy)^4 = [y,x]^6 x^4 y^4$. $[y,x]^2 = y^{-1}x^{-1}yx[y,x] = y^{-1}[y.x]x^{-1}yx = y^{-2}x^{-1}y^2 x = [y^2, x] = 1$, so $(xy)^4 = x^4 y^4$, and $\psi: x \to x^4$ is a homomorphism in $Q$. If $D = 1$, then $U_1 \neq U_2$ are two subgroups of $P$ of order 2, a contradiction; hence $D \neq 1$, and since $Q/D$ is not cyclic im $\psi$ is a proper subgroup of $D$. Hence $|\ker \psi| \geq 8$, and $Q$ has two cyclic subgroups $V_1 \neq V_2$ of order 4 (in $\ker \psi$). $|U_1| \geq 4$, so we may assume $V_1 \subseteq U_1$. If $V_1 \neq U_1$, then $V_1 \subseteq D \subseteq Z(Q)$, so $V_1 V_2$ is an abelian noncyclic subgroup of $P$, a contradiction; hence $V_1 = U_1$ and $|Q| = 8$. Let $U_1 = \langle c \rangle$. $Q$ has only one subgroup of order 2, so $c^2 = b^2$; $(cb)^2 = c^2 b^2 [b, c] = [b, c]$, so $1 \neq [b, c]$ has order 2, $[b, c] = c^2 = b^2$, $c^b = c[c, b] = c^3 = c^{-1}$, $Q = $ quaternion group of order 8.

If $|\langle a \rangle| = 2^m \geq 4$, we have seen $b^2 = c^2 = a^{2^{m-1}}$. $ab \in P - A$ and $|\langle b, A \rangle : A| = 2$, so $(ab)^2 \in A$. Hence, by replacing $b$ by $ab$ in the previous paragraph, $(ab)^2 = a^{2^{m-1}}$. So

$$abab = a^{2^{m-1}}, \quad bab = a^{2^{m-1}-1}, \quad bab \cdot a^{2^{m-1}} = a^{2^{m-1}} = a^{-1}, \quad bab^{-1} = a^{-1}.$$

We have proved that $\langle a, b \rangle$ is generalized quaternion.

We must show $P = \langle a, b \rangle$. If there exists some $g \in P - \langle a, b \rangle$, then we may assume $g^2 \in \langle a, b \rangle - \langle a \rangle$; for if $g^2 \in A$ then the preceding paragraphs with $g$ in place of $b$ would imply $g_A: a \to a^{-1}$, impossible since $b_A: a \to a^{-1}$ and $P/A \cong$ a subgroup of Aut($A$). Hence $g_A^2: a \to a^{-1}$. Let $g_A: a \to a^k$, $k = 2l + 1$ odd. Then $g_A^2: a \to a^{k^2}$, so modulo $2^m$, $-1 \equiv k^2 \equiv (2l + 1)^2 \equiv 4l^2 + 4l + 1$, and $0 \equiv 4l^2 + 4l + 2$, a contradiction.

**Theorem 19.7** (*Brauer-Suzuki* [*1*]).   *Let G be a group with a generalized quaternion Sylow 2-subgroup S of order at least 16, and let K be the largest normal subgroup of G of odd order. Then G/K has a center of order 2.*

**Corollary 19.8**   *A group with a generalized quaternion Sylow 2-subgroup S of order at least 16 is not simple.*

*Remarks*   Theorems 19.7 and 19.8 also hold when $|S| = 8$, but all known proofs require modular representation theory. A proof will appear in §66 of Part B.

Feit and Thompson [2, 4] have proved that noncyclic finite simple groups have even order. Because of Theorem 19.6, we see that any non-cyclic simple group $G$ must have a Sylow 2-subgroup $S$ containing more than one involution; $S$ must have an involution in its center, so in fact it must contain an elementary abelian group of order 4.

*Proof of Theorem 19.7, Step 1, Notation*   Let $S$ be generated by $x$ and $y$ with $x^{2^n} = 1$, $y^2 = x^{2^{n-1}}$, $y^{-1}xy = x^{-1}$; by our hypothesis, $n \geq 3$. Denote $X = \langle x \rangle$, $T = \langle x^2 \rangle$, $R = \langle x^4 \rangle$, $N = N_G(T)$, $C = C_G(T)$.

*Step 2*   $N$ has a normal subgroup $H$ such that $|H|$ is odd, $N = SH$, and $C = XH$.

*Proof*   Clearly $T^x = T$, $T^y = T$ so $T \lhd S$, $S \subseteq N$. Also, $(x^2)^y = x^{-2} \neq x^2$ since $n \geq 3$, $y \notin C = C_G(T)$, $C \cap S = X$. We claim $X$ is a Sylow 2-subgroup of $C$. We have $C \lhd N$ and $N/C \supseteq SC/C \cong S/C \cap S = S/X$, of order 2, so $2 \mid |N: C|$; but $|S: X| = 2$, so $X$ is a Sylow 2-subgroup of $C$.

$X$ is cyclic, so by Corollary 18.8, $C$ has a normal subgroup $H$ of odd order, $C = XH$. $H$ char $C \lhd N$, so $H \lhd N$ by Lemmas 18.9 and 18.10. Any element $n$ of $N$ of odd order performs an automorphism on $T$ of odd order. As in the proof of Corollary 18.8, this means $n \in C$. By Lemma 18.10, then, $n \in H$. We conclude that $H$ contains all Sylow $p$-subgroups of $N$ for odd primes $p$, and $N = SH$.

*Step 3*   The set $A = C - RH$ is a T. I. set in $G$ with normalizer $N$.

*Proof*   Clearly $XH \supset TH \supset RH$, $XH = C$. If $z \in XH - TH$, then $\langle z^{2|H|} \rangle = T$; if $z \in TH - RH$, then $\langle z^{|H|} \rangle = T$. Suppose now $z \in A \cap A^u$ for some $u \in G$; then we have $\langle z^m \rangle = T$ and $\langle z^m \rangle = T^u$ for some $m \in \{2|H|, |H|\}$, so $T = T^u$ and $u \in N_G(T) = N$. But $RH = \{t \in C | t^{2^{n-2}|H|} = 1\}$ is characteristic in $C$. Since $N$ normalizes $C$ it also normalizes $RH$, and hence normalizes $A$.

*Step 4*   $N = SH$ has an irreducible complex character $\psi$ of degree 2 with kernel $RH$.

*Proof*   $C/RH = XH/RH$ is cyclic of order 4 with generator $xH$. Hence we get a linear character $\lambda$ of $C$ with kernel $RH$ by defining $\lambda(H) = 1$, $\lambda(x) = i$. (Here $i^2 = -1$, $i^4 = 1$.) Set $\psi = \lambda^N$. $N = C \cup yC$, so $\lambda^N(x) = \lambda(x) + \lambda(y^{-1}xy) = \lambda(x) + \lambda(x^{-1}) = i + i^{2^{n-1}} = i + i^{-1} = i - i = 0$. By Frobenius reciprocity, we see

$$1 \le (\lambda^N, \lambda^N)_N = (\lambda^N|_C, \lambda)_C.$$

$\lambda^N|_C$ has degree 2, so $(\lambda^N|_C, \lambda)_C > 1$ would mean $\lambda^N|_C = 2\lambda$; but $\lambda^N(x) = 0 \ne 2\lambda(x) = 2i$. We conclude $1 = (\lambda^N, \lambda^N)_N$, so $\psi = \lambda^N$ is irreducible. $RH \lhd N$ so if $z \in RH$, then

$$\lambda^N(z) = \lambda(z) + \lambda(y^{-1}zy) = 1 + 1 = 2 = \lambda^N(1),$$

proving $RH \subseteq \ker \lambda^N = \ker \psi$. But $|N/RH| = 8$. 2-groups of order less than 8 are abelian. Therefore their complex irreducible representations have degree 1; this implies that $RH = \ker \psi$.

*Step 5*   Let $\theta = 1_C^N - \psi$. Then $(\theta, \theta)_N = 3$, $\theta(1) = 0$, and $\theta(y) = 0$ for all $y \in N - A$.

*Proof*   $1_C^N$ has degree $|N: C| = 2$. $(1_C^N, 1_N) = (1_C, 1_C)_C = 1$, so $1_C^N = 1_N + \mu$, $\mu \ne 1_N$ a linear character of $N$. Therefore $\theta = 1_N + \mu - \psi$, $(\theta, \theta)_N = 3$, and $\theta(1) = 1_N(1) + \mu(1) - \psi(1) = 1 + 1 - 2 = 0$.

$RH \lhd N$, so since $1_C$ has $RH$ in its kernel, $1_C^N$ has $RH$ in its kernel (just as does $\lambda^N = \psi$). Therefore $\theta = 1_C^N - \psi$ is 0 on $RH$. Also, $\theta = (1_C - \lambda)^N$ is 0 on $N - C = N - XH$ since $C \lhd N$. Hence $\theta$ is 0 on $N - A$.

*Step 6*  There exist distinct nonprincipal irreducible characters $\chi_1, \chi$ of $G$ such that $\theta^G = 1_G + \chi_1 - \chi$ and $\chi(1) = 1 + \chi_1(1)$.

*Proof*  By Theorem 12.1, $(\theta^G, \theta^G)_G = 3$. But by Frobenius reciprocity, $(\theta^G, 1_G)_G = (\theta, 1_N)_N = 1$.

*Step 7*  If $y \in G$ is an involution or has odd order, then $\theta^G(y) = 0$, so $\chi(y) = 1 + \chi_1(y)$.

*Proof*  If $g \in G$ is not conjugate to an element of $A$, then

$$\theta^G(g) = \frac{1}{|N|} \sum_{y \in G} \dot\theta(y^{-1}gy) = \frac{1}{|N|} \sum_{\substack{y \in G, \\ y^{-1}gy \in A}} \theta(y^{-1}gy) = 0.$$

$A$ contains no involutions or elements of odd order.

*Step 8*  If $u$ and $v$ are involutions in $G$, then $w = uv$ has odd order.

*Proof*  If $w$ has even order $2s$, then Lemma 19.1 shows $u$ and $v$ centralize the involution $y = w^s$. Hence $\langle u, y \rangle$ is a 2-group, contained in a Sylow 2-subgroup $S_0$; but $S_0$ contains only one involution by Lemma 19.3, so $u = y$. Similarly $v = y$ so $uv = uu = 1$, a contradiction.

*Step 9*  For any involution $u$ of $G$,

$$1 + \frac{\chi_1(u)^2}{\chi_1(1)} - \frac{\chi(u)^2}{\chi(1)} = 0.$$

*Proof*  Since $S$ has only one involution by Lemma 19.3, $G$ has only one conjugacy class $\mathscr{C}_2$ of involutions. Let $\mathscr{C}_1 = \{1\}, \mathscr{C}_2, \ldots, \mathscr{C}_h$ be the conjugacy classes of $G$, $1_G = \zeta_1, \zeta_2, \ldots, \zeta_h$ the irreducible characters. For any $y \in \mathscr{C}_j$, we apply Lemma 19.2; $c_{22j}$ is the number of ordered pairs $(u, v)$ of involutions of $G$ with $uv = y$, and

$$\beta(y) = c_{22j} = \frac{|G|}{|C_G(u)|^2} \sum_{i=1}^{h} \frac{\zeta_i(u)^2}{\zeta_i(1)} \overline{\zeta_i(y)}.$$

Here $\beta(y)$ is a class function on $G$, and

$$(\bar{\beta}, \zeta_i)_G = \frac{|G|}{|C_G(u)|^2} \frac{\zeta_i(u)^2}{\zeta_i(1)}.$$

But

$$(\bar{\beta}, \theta^G)_G = \frac{1}{|G|} \sum_{y \in G} \bar{\beta}(y)\overline{\theta^G(y)}.$$

If $y$ has even order then $\beta(y) = 0$ by Step 8, and if $y$ has odd order then $\theta^G(y) = 0$ by Step 7. We see that

$$0 = (\bar{\beta}, \theta^G)_G = (\bar{\beta}, 1_G + \chi_1 - \chi)_G$$

$$= \frac{|G|}{|C_G(u)|^2}\left(1 + \frac{\chi_1(u)^2}{\chi_1(1)} - \frac{\chi(u)^2}{\chi(1)}\right).$$

*Step 10*   ker $\chi$ contains every involution of $G$.

*Proof*  By Steps 6 and 7, $\chi_1(1) = \chi(1) - 1$ and $\chi_1(u) = \chi(u) - 1$. Substituting in Step 9, we see that

$$1 + \frac{(\chi(u) - 1)^2}{\chi(1) - 1} - \frac{\chi(u)^2}{\chi(1)} = 0,$$

$$\chi(1)^2 - 2\chi(1)\chi(u) + \chi(u)^2 = 0,$$

$$(\chi(1) - \chi(u))^2 = 0, \qquad \chi(1) = \chi(u).$$

By Lemma 7.8, $u \in \ker \chi$.

*Step 11*   If $M$ is the normal subgroup of $G$ generated by all of its involutions, then $Q = S \cap M$ is a cyclic Sylow 2-subgroup of $M$.

*Proof*  $|M: S \cap M| = |SM: S| \,\big|\, |G: S|$ is not divisible by 2, so $Q$ is a Sylow 2-subgroup of $M$.

If $Q$ is not cyclic, then for some $i$, $yx^i \in Q$. $Q = S \cap M$ is normal in $S$, so also $(yx^i)^x = x^{-1}yxx^i = yy^{-1}x^{-1}yxx^i = yxxx^i = yx^{i+2} \in Q$, $x^2 \in Q$. $x^2 \in Q$ and $Q$ not cyclic implies $|S: Q| \leq 2$, and $S/Q$ is abelian.

The group $G_1 = SM$ has a generalized quaternion Sylow 2-subgroup $S$, so by Step 10, $G_1$ has a nonlinear irreducible character $\phi$ whose kernel $K_1$ contains all involutions of $G_1$. $M \subseteq K_1$, so $G_1/K_1$ is a homomorphic image of $G_1/M = SM/M \cong S/S \cap M = S/Q$. Therefore $G_1/K_1$ is abelian and $\phi$ is linear, a contradiction.

*Step 12, Conclusion* Using Step 11 and Corollary 18.8, $M$ has a normal 2-complement $L$, $M = QL$. $L$ char $M \lhd G$, so $L \lhd G$ and $L \subseteq K$. $KQ = KM$, so $KQ \lhd G$. Setting $\bar{G} = G/K$, we see that $\bar{Q} = KQ/K \lhd \bar{G}$. $\bar{Q}$ is cyclic, so its element of order 2 is in the center of $G/K$. $\bar{G}$ has no normal subgroups of odd order, so $Z(\bar{G})$ is a 2-group. But Sylow 2-subgroups of $\bar{G}$ are isomorphic to $S$ and have centers of order 2, so $|Z(\bar{G})| = 2$.

EXERCISES

**1** Let $P$ be a finite $p$-group. Prove that $P$ has a faithful irreducible complex character if and only if $Z(P)$ is cyclic.

**2** (Hoheisel [1]) Let $G$ be a finite group, and let the integers $c_{ijk}$ be as in Lemma 19.2. By the Lemma, if we know the character table of $G$ then we know all $c_{ijk}$. Conversely, suppose we know all $c_{ijk}$.
   (a)   Show that we then know all the $\omega_{ij}$ of Lemma 5.9.
   (b)   Show that we know $|\mathscr{C}_j|$ for all conjugacy classes $\mathscr{C}_j$ of $G$.
   (c)   Show that we can determine the entire character table of $G$.

# §20

## A Theorem of Tate

**Definitions** (*not standard*)    Let $G$ be a finite group, $p$ a fixed prime, $P$ a Sylow $p$-subgroup of $G$. We denote by $M(G)$ the smallest normal subgroup of $G$ with $G/M(G)$ a $p$-group, by $A(G)$ the smallest normal subgroup of $G$ with $G/A(G)$ an abelian $p$-group, and by $E(G)$ the smallest normal subgroup of $G$ with $G/E(G)$ an elementary abelian $p$-group. It is clear that $G \supseteq E(G) \supseteq A(G) \supseteq M(G)$ and $G = PM(G) = PA(G) = PE(G)$.

Let $H$ be a subgroup of $G$ containing $P$, so $p \nmid |G:H|$. Then $H/H \cap M(G) \cong G/M(G)$ is a $p$-group, so $M(H) \subseteq H \cap M(G)$. Similarly we see that $A(H) \subseteq H \cap A(G)$ and $E(H) \subseteq H \cap E(G)$. We say that $H$ *controls* $M(G)$, if $M(H) = H \cap M(G)$. (*Reason*: then $H/M(H) \cong G/M(G)$.) Similarly, we say $H$ *controls* $A(G)$ if $A(H) = H \cap A(G)$, and $H$ *controls* $E(G)$ if $E(H) = H \cap E(G)$.

*Example*    Corollary 18.6 states that if the Sylow $p$-subgroup $P$ of $G$ is abelian, then $N_G(P)$ controls $A(G)$. (In this case $A(G) = M(G)$.)

**Lemma 20.1**    *Let $P$ be a Sylow $p$-subgroup of $G$, $H$ a subgroup of $G$ with $P \subseteq H$.*

    (*i*)    *If $H$ controls $M(G)$, then $H$ controls $A(G)$.*

    (*ii*)    *If $H$ controls $A(G)$, then $H$ controls $E(G)$.*

*Proof*    (i)    We have $G/M(G) \cong H/M(H)$.    $A(G) \supseteq M(G)$, and $A(G)/M(G) = A(G/M(G)) \cong A(H/M(H)) = A(H)/M(H)$.    Therefore $|H : H \cap A(G)| = |G : A(G)| = |H : A(H)|$, forcing $A(H) = H \cap A(G)$.

    (ii)    Similar.

**Theorem 20.2** (*Tate* [1]).    *Let $P$ be a Sylow $p$-subgroup of $G$, $H$ a subgroup of $G$ with $P \subseteq H$. If $H$ controls $E(G)$, then $H$ controls $M(G)$.*

**Corollary 20.3**    *If the Sylow p-subgroup P of G satisfies $E(P) = P \cap E(G)$, then G has a normal p-complement.*

*Proof*  $P$ controls $E(G)$, so by Tate's theorem $P$ controls $M(G)$. $1 = M(P) = P \cap M(G)$, so $M(G)$ is a normal $p$-complement.

Our purpose is to give a proof of Theorem 20.2 using characters, due to Thompson and communicated to us by D. S. Passman. Related results appear in Thompson [5] and Glauberman [4]. Theorem 20.2 was originally proved using cohomology of groups.

**Corollary 20.4**    *If G has no normal p-complement and P is a Sylow p-subgroup of G, then $E(P) \neq P \cap E(G)$.*

Note that for a $p$-group $P$, $E(P)$ is often denoted $\Phi(P)$, the *Frattini subgroup* of $P$.

**Definition**    The linear complex characters of a finite group $G$ form an abelian group under multiplication. The *order* of a linear character is its order in that group.

**Definition**    Let $\theta$ be a complex character of a finite group $G$, afforded by the matrix representation $x \to M(x)$, all $x \in G$. *det $\theta$* is the linear character det $\theta : x \to \det M(x)$, all $x \in G$. Since similar matrices have the same determinant, det $\theta$ depends only on $\theta$, not on the choice of $M$.

**Lemma 20.5**    *Let $K \lhd G$, $G/K$ a p-group. Let $\theta$ be an irreducible character of K such that $\theta$ is fixed by G, $\theta(1)$ is prime to p, and det $\theta$ has order prime to p. Then there is a unique irreducible character $\chi$ of G with $\chi_K = \theta$, det $\chi$ of order prime to p.*

*Proof*  By "$\theta$ is fixed by $G$" we mean $\theta = \theta^g$ for all $g \in G$, $\theta^g$ as defined in Clifford's Theorem 14.1. The proof is by induction on $|G:K|$, true if $G = K$.

Choose a subgroup $G_0 \lhd G$, $K \lhd G_0$, $|G_0 : K| = p$. Let $\beta$ be an irreducible character of $G_0$ with $(\theta^{G_0}, \beta) = (\theta, \beta_K) > 0$. By Clifford's theorem, $\beta_K = n\theta$, some $n$. Let $1_{G_0} = \mu_0, \mu_1, \ldots, \mu_{p-1}$ be the linear characters of $G_0/K$, with $\mu_i = \mu_1^i$. $\beta = \mu_0\beta, \mu_1\beta, \ldots, \mu_{p-1}\beta$ are all irreducible characters of $G_0$ with $(\mu_i\beta)_K = \beta_K = n\theta$. $\mu_i\beta = \mu_j\beta$ for $i \neq j$ would mean $\beta|_{G_0 - K} = 0$, so

$$1 = (\beta, \beta) = \frac{1}{|G_0|} \sum_{x \in G_0} |\beta(x)|^2 = \frac{n^2}{p|K|} \sum_{x \in K} |\theta(x)|^2$$

$$= \frac{n^2}{p} (\theta, \theta) = \frac{n^2}{p},$$

a contradiction. Therefore the $\mu_i \beta$ are all different, forcing $n = 1$,

$$\theta^{G_0} = \beta + \mu_1 \beta + \cdots + \mu_{p-1} \beta, \qquad \beta_K = \theta.$$

Then

$$(\det \mu_i \beta)(x) = \det \mu_i M_0(x) = \mu_i^{\theta(1)}(x) \det M_0(x)$$

$$= \mu_i^{\theta(1)} \det \beta(x) \qquad \text{for all} \quad x \in G,$$

where $M_0$ is a matrix representation affording $\beta$. Let $\det \theta$ have order $m$, $(m, p) = 1$. Then $(\det \theta)^m = 1_K$, so $\mu_i^{m\theta(1)}(\det \beta)^m$ is a linear character of $G_0/K$.

If $x \in G_0 - K$, then $(\det \beta)^m(x)$ and $\mu_1(x)$ are $p$th roots of 1, say $(\det \beta)^m(x) = \mu_1(x)^t$, $t \in \mathbf{Z}$; we have

$$\mu_i^{m\theta(1)}(\det \beta)^m = 1 \qquad \text{iff} \quad \mu_i(x)^{m\theta(1)}(\det \beta)^m(x) = 1$$

$$\text{iff} \quad \mu_1(x)^{im\theta(1)+t} = 1$$

$$\text{iff} \quad p | (im\theta(1) + t),$$

a result true for a unique choice $i_0$ of $i$.

Choose $\chi_0 = \mu_{i_0} \beta$. We claim $\chi_0$ is fixed by $G$. If $i \neq i_0$, we saw that $\det \mu_i \beta$ has order divisible by $p$. If $g \in G$, then $\chi_0|_K = \theta$ so $\chi_0^g|_K = \theta$. $\det \chi_0$ has order prime to $p$, so $\det \chi_0^g$ has order prime to $p$. $\chi_0^g|_K = \theta$ now forces $\chi_0^g = \chi_0$.

By induction, there is a unique irreducible character $\chi$ of $G$ with $\chi|_{G_0} = \chi_0$, $\det \chi$ of order prime to $p$. $\chi$ must be the character we seek.

*Proof of Theorem 20.2, Notation*     Let $G$ be a counterexample, so $M(H) \neq H \cap M(G)$, $H$ a subgroup of $G$, $H \supseteq P$, $P$ a Sylow $p$-subgroup of $G$, $E(H) = H \cap E(G)$. Let $|G| = p^n m$, $p \nmid m$. Let $\varepsilon$ be a primitive $|G|$th root of 1, $\delta$ a primitive $pm$th root of 1, say $\delta = \varepsilon^{p^{n-1}}$. Let $\varphi$ be the Euler function $\varphi(x) =$ number of integers $\leq x$ and relatively prime to $x$. Then

$$\dim_{\mathbf{Q}} \mathbf{Q}(\varepsilon) = \varphi(p^n m) = \varphi(p^n) \varphi(m) = p^{n-1}(p-1)\varphi(m),$$

$$\dim_{\mathbf{Q}} \mathbf{Q}(\delta) = \varphi(pm) = \varphi(p)\varphi(m) = (p-1)\varphi(m),$$

so $\dim_{Q(\delta)} Q(\varepsilon) = p^{n-1}$. Let $\mathscr{Q}$ be the Galois group of $Q(\varepsilon)$ over $Q(\delta)$. $|\mathscr{Q}| = p^{n-1}$, so $\mathscr{Q}$ is a $p$-group.

*Step 1*   $H \cap M(G)$ has a linear character $\lambda$ of order $p$ with $M(H) \subseteq$ ker $\lambda$. $P$ and $\mathscr{Q}$ fix $\lambda$.

*Proof*   $(H \cap M(G))/M(H)$ is a nontrivial $p$-group normalized by the $p$-group $P$, so there is a subgroup $K$ normalized by $P$, $M(H) \subseteq K \neq H \cap M(G)$, $|H \cap M(G) : K| = p$.

Let $\lambda$ be a faithful linear character of $(H \cap M(G))/K$. There are $p-1$ such and $P$ is a $p$-group, so $P$ fixes $\lambda$. $\mathscr{Q}$ fixes $p$th roots of 1, so $\mathscr{Q}$ fixes $\lambda$.

*Step 2*   There is an irreducible constituent $\theta$ of $\lambda^{M(G)}$ with $\theta$ fixed by $P \times \mathscr{Q}$, $\theta(1)$ prime to $p$, $(\theta, \lambda^{M(G)})$ prime to $p$.

*Proof*   Since $\lambda$ is invariant under $P \times \mathscr{Q}$, so is $\lambda^{M(G)}$. $\lambda^{M(G)}$ has degree $|M(G) : H \cap M(G)| = |G : H|$ prime to $p$. Absence of a constituent $\theta$ of $\lambda^{M(G)}$ of the type required would force $p | \lambda^{M(G)}(1)$, a contradiction.

*Step 3*   There is an irreducible character $\chi$ of $G$ such that $\chi|_{M(G)} = \theta$, det $\chi$ has order prime to $p$, and $\mathscr{Q}$ fixes $\chi$.

*Proof*   $G = PM(G)$, so $\theta$ is fixed by $G$. det $\theta$ is a character of $M(G)/M(G)'$, a $p'$-group, so det $\theta$ has order prime to $p$. By Lemma 20.5 there is a unique $\chi$ satisfying the first two conditions. Any $\mathscr{Q}$-conjugate of $\chi$ satisfies the same conditions and must be $\chi$.

*Step 4*   There is an irreducible constituent $\psi$ of $\chi_H$ such that $\psi$ is fixed by $\mathscr{Q}$ and $\psi(1) = (\psi_{H \cap M(G)}, \lambda)$ is prime to $p$.

*Proof*   $H \cap M(G) \triangleleft H$ and $H = P(H \cap M(G))$, so since $\lambda$ is fixed by $P$, $\lambda$ is fixed by $H$. Define $\mathscr{S}$ to be the set of all irreducible constituents $\psi$ of $\chi_H$ such that $\psi$ is fixed by $\mathscr{Q}$ and $(\psi_{H \cap M(G)}, \lambda) > 0$. By Clifford's theorem, $\psi \in \mathscr{S}$ implies $\psi_{H \cap M(G)} = \psi(1)\lambda$. We have

$$\sum_{\psi \in \mathscr{S}} (\psi, \chi_H)\psi(1) = \sum_{\psi \in \mathscr{S}} (\psi, \chi_H)(\psi_{H \cap M(G)}, \lambda)$$

$$\equiv (\chi_{H \cap M(G)}, \lambda) = (\theta_{H \cap M(G)}, \lambda)$$

$$= (\theta, \lambda^{M(G)}) \not\equiv 0 \ (\mathrm{mod}\ p).$$

Choosing $\psi \in \mathscr{S}$ with $\psi(1) = (\psi_{H \cap M(G)}, \lambda)$ prime to $p$, we have Step 4.

*Step 5, Conclusion*   Since $\psi_{H \cap M(G)} = \psi(1)\lambda$, and $M(H) \subseteq \ker \lambda$, we have $M(H) \subseteq \ker \psi$. Therefore $\psi$ is a character of the $p$-group $H/M(H)$. $(\psi(1), p) = 1$, so $\psi$ is linear and $\psi_{H \cap M(G)} = \lambda$. $\psi$ is fixed by $\mathscr{D}$, so since $\mathscr{D}$ moves all $p^2$th roots of 1, $E(H) \subseteq \ker \psi$. $E(H) = H \cap E(G)$ by hypothesis. Therefore $\ker \psi \supseteq H \cap E(G) \supseteq H \cap M(G)$, a contradiction to the fact $\psi_{H \cap M(G)} = \lambda \neq 1$.

# §21

*Mackey Decomposition*

**Lemma 21.1**  *Let $A$ be a ring, $V$ a completely reducible $A$-module, so $V = \oplus \sum_{i \in I} V_i$, the $V_i$ irreducible $A$-modules. Let $W$ be an irreducible submodule of $V$, and let $J = \{i \in I \,|\, V_i \cong W\}$. Then $W \subseteq \oplus \sum_{j \in J} V_j$.*

*Proof*  Choose $0 \neq w \in W$, and let $w = \sum_i v_i$, $v_i \in V_i$, where only finitely many $v_i \neq 0$. If $v_j \neq 0$, then $\varphi \colon aw \to av_j$ is a nonzero homomorphism from $Aw = W$ to $Av_j = V_j$. $W$ and $V_j$ are irreducible, so $W \cong V_j$ and $j \in J$, done.

**Lemma 21.2**  *Let $A$ be a ring, $V$ a completely reducible $A$-module, and suppose $V = \oplus \sum_{i \in I} V_i = \oplus \sum_{j \in J} W_j$, the $V_i$ and $W_j$ irreducible. If $W$ is any irreducible $A$-module, let*

$$I_W = \{i \in I \,|\, V_i \cong W\}, \qquad J_W = \{j \in J \,|\, W_j \cong W\}.$$

*Then*

$$\oplus \sum_{i \in I_W} V_i = \oplus \sum_{j \in J_W} W_j.$$

*Proof*  By Lemma 21.1, each side is contained in the other.

The main results of this section come from Mackey [1]. $G$ denotes a finite group.

**Theorem 21.3** (Mackey Decomposition Theorem)  *Let $R$ be a commutative ring, $H$ and $A$ subgroups of $G$, $W$ an $R$-free $RH$-module. Let $\{x_i\}$ be a cross section of $H$ in $G$, so $G = x_1 H \cup \cdots \cup x_t H$, $t = |G\colon H|$. Then:*

*(1)  For any $x \in G$, $x \otimes W$ is an $R(xHx^{-1})$-module with a module action $xhx^{-1}(x \otimes w) = xh \otimes w = x \otimes hw$.*

107

(2)   *For any $(A, H)$-double coset $D = AxH$ of $G$, the RA-module*

$$L(D) = (x \otimes W)_{xHx^{-1} \cap A}{}^A$$

*depends only on the double coset $D$, not on the particular choice of $x \in D$.*

(3)                                            $(W^G)_A = \oplus \sum_D L(D),$

*where $D$ runs over the $(A, H)$-double cosets of $G$.*

*Proof* (1) is clear. Now assume the $x_i$ chosen so that $D = x_1 H \cup \cdots \cup x_s H$, and denote

$$M_D = \sum_{i=1}^{s} x_i \otimes W.$$

By its construction we know

$$W^G = \sum_{i=1}^{t} x_i \otimes W,$$

so $M_D$ is an $RA$-submodule of $W^G$; in fact,

$$(W^G)_A \cong \oplus \sum_D M_D$$

as $RA$-modules. $M_D$ does not depend on the choice of $\{x_i\}$, for if $x_i H = x_i' H$, then

$$x_i \otimes W = x_i \otimes HW = x_i H \otimes W = x_i' H \otimes W = x_i' \otimes W.$$

$D = AxH = x_1 H \cup \cdots \cup x_s H$, so let $x_i H = a_i x H$, $a_i \in A$. We claim that $a_1(xHx^{-1} \cap A), \ldots, a_s(xHx^{-1} \cap A)$ are the distinct cosets of $xHx^{-1} \cap A$ in $A$. They are distinct, for $a_i(xHx^{-1} \cap A) = a_j(xHx^{-1} \cap A)$ iff $a_j^{-1}a_i \in xHx^{-1} \cap A$ iff $a_j^{-1}a_i \in xHx^{-1}$ iff $x^{-1}a_j^{-1}a_i x \in H$ iff $a_j xH = a_i xH$ iff $i = j$. They exhaust $A$, for if $a \in A$ then $ax \in D$, $ax = a_i xh$ some $i$ and some $h \in H$, $a \in a_i xHx^{-1}$, $a \in a_i xHx^{-1} \cap a_i A = a_i(xHx^{-1} \cap A)$.

Now

$$M_D = \sum_{i=1}^{s} x_i H \otimes W = \sum_{i=1}^{s} a_i xH \otimes W = \sum_{i=1}^{s} a_i(x \otimes W).$$

But

$$(x \otimes W)_{xHx^{-1} \cap A}{}^A = \sum_{i=1}^{s} a_i(x \otimes W),$$

where $A$ acts the same as it does on $M_D$, so

$$M_D = (x \otimes W)_{xHx^{-1} \cap A}{}^A.$$

This proves both (2) and (3).

**Lemma 21.4**   *Let $R$ be a commutative ring, $H$ a subgroup of $G$, $W$ an R-free RH-module, and $x \in G$. Then we can define an $R(xHx^{-1})$-module $W^x$, called a conjugate module to $W$, as follows. The underlying R-free R-module is still $W$, but the module action in $W^x$ is $\cdot$ , where $xhx^{-1} \cdot w = hw$. Then:*

(*i*)   $W^x \cong x \otimes W$ *as $R(xHx^{-1})$-modules.*

(*ii*)   *If $R = k$ is a field and $\theta$ is the character afforded by $W$, then $W^x$ affords the character $\theta^x$, where $\theta^x(y) = \theta(y^x)$, all $y \in xHx^{-1}$.*

(*iii*)   *If $x,\, y \in G$ and $H \lhd G$, then $(W^x)^y \cong W^{yx}$. In particular, if $I(W) = \{g \in G \,|\, W^g \cong W\}$, then $I(W) \supseteq H$ is a subgroup of $G$.*

*Proof*   That $W^x$ is an $R(xHx^{-1})$-module is clear.

(i)   Let $W$ have $R$-basis $w_1, \dots, w_t$, so $x \otimes W$ has $R$-basis $x \otimes w_1,$ $\dots, x \otimes w_t$. If $h \in H$, let $hw_j = \sum_i \alpha_{ij} w_i$. Then $xhx^{-1} \cdot w_j = hw_j = \sum_i \alpha_{ij} w_i$ and

$$xhx^{-1}(x \otimes w_j) = xh \otimes w_j = x \otimes hw_j$$
$$= x \otimes \left( \sum_i \alpha_{ij} w_i \right) = \sum_i \alpha_{ij}(x \otimes w_i),$$

so $W^x \cong x \otimes W$.

(ii)   Let $y = xhx^{-1}$, $h \in H$. With the notation of the last paragraph, $\theta^x(y) = \sum_i \alpha_{ii}$ and $\theta(h) = \theta(y^x) = \sum_i \alpha_{ii}$, done.

(iii)   By (i), we must show $yx \otimes W \cong y \otimes (x \otimes W)$. For any $h \in H$, let $h = yxh'x^{-1}y^{-1}$, $h = yh''y^{-1}$, $h'' = xh'''x^{-1}$. Then $h''' = x^{-1}h''x = x^{-1}y^{-1}hyx = h'$, and $h(yx \otimes w) = yxh'x^{-1}y^{-1}(yx \otimes w) = yxh' \otimes w = yx \otimes h'w$ while

$$h(y \otimes (x \otimes w)) = yh''y^{-1}(y \otimes (x \otimes w))$$
$$= yh'' \otimes (x \otimes w) = y \otimes h''(x \otimes w)$$
$$= y \otimes xh'''x^{-1}(x \otimes w) = y \otimes (xh''' \otimes w)$$
$$= y \otimes (x \otimes h'''w),$$

done.

**Theorem 21.5**    *Let* $H \lhd G$, $R$ *a commutative ring*, $W$ *an R-free RH-module, and let* $I(W) = \{g \in G \mid W^g \cong W\}$. *Let* $G = y_1 I(W) \cup \cdots \cup y_t I(W)$, $y_1 = 1$. *Then*

$$(W^G)_H \cong |I(W) : H| \, (\oplus \sum_{i=1}^{t} W^{y_i}).$$

*In particular, if* $R = k$ *is a field of characteristic zero and* $W$ *affords the character* $\theta$, *then*

$$I(W) = I(\theta) = \{x \in G \mid \theta^x = \theta\}$$

*and*

$$(\theta^G)_H = |I(\theta) : H| \, (\sum_{i=1}^{t} \theta^{y_i}).$$

*Proof*    Let $I(W) = x_1 H \cup \cdots \cup x_m H$, $m = |I(W) : H|$, $x_1 = 1$, so all $W^{x_j} \cong W$ and $G = \bigcup_{i,j} y_i x_j H$. By 21.3 in the case $A = H$, we have

$$(W^G)_H = \oplus \sum_{i,j} y_i x_j \otimes W \cong \oplus \sum_{i,j} W^{y_i x_j}$$

$$\cong \oplus \sum_{i,j} (W^{x_j})^{y_i} \cong m(\oplus \sum_i W^{y_i}).$$

The result for characters is now a consequence of Lemmas 7.2 and 21.4(ii).

**Lemma 21.6**    *Let* $k$ *be a splitting field of characteristic zero for* $G$ *and its subgroups. Let* $H$ *and* $A$ *be subgroups of* $G$, $\theta$ *a character of* $H$, $\eta$ *a character of* $A$. *Let* $Ax_1 H, \ldots, Ax_m H$ *be all of the* $(A, H)$-*double cosets of* $G$, *and let* $K_i = x_i H x_i^{-1} \cap A$, $1 \le i \le m$. *Then*

$$(\theta^G, \eta^G)_G = \sum_{i=1}^{m} (\theta^{x_i}{}_{K_i}, \eta_{K_i})_{K_i}.$$

*Proof*    By Theorem 21.3,

$$(\theta^G)_A = \sum_{i=1}^{m} (\theta^{x_i}{}_{K_i})^A.$$

Using Frobenius reciprocity twice, we therefore have

$$(\theta^G, \eta^G)_G = ((\theta^G)_A, \eta)_A = \sum_{i=1}^{m} ((\theta^{x_i}{}_{K_i})^A, \eta)_A = \sum_{i=1}^{m} (\theta^{x_i}{}_{K_i}, \eta_{K_i})_{K_i}.$$

# §22

## Itô's Theorem on Character Degrees

**Theorem 22.1**    *If $\chi$ is an irreducible character of $G$ over* $\mathbf{C}$, *then* $\chi(1)$
*divides* $|G:Z(G)|$.

*Proof (Glauberman)*    Denote $Z = Z(G)$, and let $\mathscr{C}_1, \ldots, \mathscr{C}_h$ be the
conjugacy classes of $G$, with $x_i \in \mathscr{C}_i$, $|\mathscr{C}_i| = h_i$. For any $i$ and any $z \in Z$,
$z\mathscr{C}_i$ is one of the $\mathscr{C}_j$; we define an equivalence relation on $\{\mathscr{C}_1, \ldots, \mathscr{C}_h\}$
by writing $\mathscr{C}_i \sim \mathscr{C}_j$ if and only if $\mathscr{C}_i = z\mathscr{C}_j$, some $z \in Z$. If $\mathscr{C}_i \sim \mathscr{C}_j$,
then $h_i = h_j$.

We may assume the notation chosen so that $\{\mathscr{C}_1, \ldots, \mathscr{C}_{|Z|}\}, \ldots,$
$\{\mathscr{C}_{(t-1)|Z|+1}, \ldots, \mathscr{C}_{t|Z|}\}$ are all of the equivalence classes containing
$|Z|$ distinct members; the other equivalence classes are smaller.

Let $\varphi$ be the matrix representation affording $\chi$, and assume $\varphi$ is
faithful. If $j > t|Z|$, then $\mathscr{C}_j = z\mathscr{C}_j$, some $z \in Z^\#$. $\varphi(z) = \alpha_z I$ is a scalar
matrix, so $\chi(x_j) = \chi(zx_j) = \operatorname{tr} \varphi(z)\varphi(x_j) = \operatorname{tr} \alpha_z\varphi(x_j) = \alpha_z\chi(x_j)$. $\chi$ is
faithful so $\alpha_z \neq 1$, forcing $\chi(x_j) = 0$, all $j > t|Z|$.

For any $x \in G$ and any $z \in Z$, $|\chi(zx)| = |\alpha_z\chi(x)| = |\alpha_z| \, |\chi(x)| = |\chi(x)|$, so $|\chi(x)|$ is constant on equivalent conjugacy classes. By the orthogonality relations,

$$|G| = \sum_{x \in G} |\chi(x)|^2 = \sum_{i=1}^{h} h_i|\chi(x_i)|^2 = \sum_{i=1}^{t|Z|} h_i|\chi(x_i)|^2$$

$$= \sum_{i=1}^{t} |Z|h_{i|Z|}|\chi(x_{i|Z|})|^2$$

$$= \sum_{i=1}^{t} |Z|h_{i|Z|}\chi(x_{i|Z|}) \, \overline{\chi(x_{i|Z|})}.$$

By Lemma 5.9, $h_{i|Z|} \, \chi(x_{i|Z|}) = \chi(1)\omega_i$ where $\omega_i$ is an algebraic
integer. Therefore

111

$$\frac{|G:Z|}{\chi(1)} = \sum_{i=1}^{t} \omega_i \overline{\chi(x_{i|Z|})}$$

is an algebraic integer. Since it is also rational it is an ordinary integer, proving $\chi(1) \mid |G:Z|$.

If $\varphi$ is not faithful, say with kernel $K$, then our argument shows $\chi(1)$ divides $|G/K: Z(G/K)|$. Certainly $ZK/K \subseteq Z(G/K)$, so $\chi(1)$ divides $|G/K: ZK/K|$, and $|G/K: ZK/K| = |G:ZK|$ divides $|G:Z|$.

Another proof of Theorem 22.1 appears in Exercise 2. Until recently the only known proofs were more complicated; see, e.g., Exercise 33.1 of Curtis and Reiner [1] or Satz V.12.6 of Huppert [6].

**Lemma 22.2** *Let $k$ be a splitting field of characteristic zero for $G$ and its subgroups. Let $\chi$ be an irreducible character of $G$, $H \triangleleft G$, $\theta$ an irreducible character of $H$ such that $\theta \subset \chi_H$. Then there is a unique irreducible character $\zeta$ of $I(\theta)$ such that $\theta \subset \zeta_H$ and $\zeta \subset \chi_{I(\theta)}$. Furthermore, $\chi = \zeta^G$.*

*Proof* We first show existence of $\zeta$. Let $\chi_{I(\theta)} = \sum n_i \zeta_i$, the $\zeta_i$ irreducible, so $\chi_H = \sum n_i \zeta_{iH}$. We take $\zeta = \zeta_i$, some $i$ with $\theta \subset \zeta_{iH}$.

We next show $\chi = \zeta^G$; by Frobenius reciprocity $(\zeta^G, \chi)_G > 0$, so it is enough to show $\zeta^G$ is irreducible. Let $I(\theta)x_1 I(\theta), \ldots, I(\theta)x_m I(\theta)$, $x_1 = 1$ be all of the $(I(\theta), I(\theta))$-double cosets in $G$, and denote $K_i = x_i I(\theta) x_i^{-1} \cap I(\theta) \supseteq H$. By Lemma 21.6,

$$(\zeta^G, \zeta^G)_G = \sum_{i=1}^{m} (\zeta^{x_i}{}_{K_i}, \zeta_{K_i})_{K_i}.$$

$K_1 = I(\theta)$, so $(\zeta^{x_1}{}_{K_1}, \zeta_{K_1})_{K_1} = 1$.

If $i \neq 1$, then $x_i \notin I(\theta)$ so $\theta^{x_i} \neq \theta$. By 21.5 with $G = I(\theta)$, we have $\zeta_H = e\theta$, some integer $e$. Hence $\zeta^{x_i}{}_H = e\theta^{x_i}$, and $(\zeta^{x_i}{}_H, \zeta_H)_H = e^2(\theta, \theta^{x_i})_H = 0$. If $(\zeta^{x_i}{}_{K_i}, \zeta_{K_i})_{K_i} \neq 0$, then there is an irreducible character $\tau$ of $K_i$ with $\tau \subset \zeta^{x_i}{}_{K_i}$, $\tau \subset \zeta_{K_i}$. That means $\tau_H \subset \zeta^{x_i}{}_H$, $\tau_H \subset \zeta_H$ so $(\zeta^{x_i}{}_H, \zeta_H)_H \neq 0$, a contradiction. Therefore $(\zeta^{x_i}{}_{K_i}, \zeta_{K_i})_{K_i} = 0$ for all $i > 1$, proving that $(\zeta^G, \zeta^G)_G = 1$, $\zeta^G$ is irreducible, $\chi = \zeta^G$.

By Clifford's Theorem 14.1(6), $\chi_H = (\zeta^G)_H = e_0 \sum_i \theta^{y_i}$, some integer $e_0$, where $\{\theta^{y_i}\}$ is the set of conjugates of $\theta$ in $G$. Of course $|\{\theta^{y_i}\}| = |G: I(\theta)|$, so we see $\chi(1) = e_0 |G: I(\theta)| \theta(1)$; also $\chi(1) = |G: I(\theta)| \zeta(1)$, so $(\chi_H, \theta)_H = e_0 = \zeta(1)/\theta(1)$. $\zeta_H = e\theta$, some integer $e$, and $e = (\zeta_H, \theta)_H = \zeta(1)/\theta(1) = e_0$. Therefore $\chi_{I(\theta)} - \zeta$ satisfies $(\theta, (\chi_{I(\theta)} - \zeta)_H)_H = e_0 - e = 0$, $\theta \not\subset (\chi_{I(\theta)} - \zeta)_H$, and $\zeta$ is unique.

**Theorem 22.3** (*Itô[1]*).     *Let $\chi: G \to \mathbf{C}$ be an irreducible character of $G$, and let $H$ be an abelian normal subgroup of $G$. Then $\chi(1) \mid |G: H|$.*

*Proof*   By induction on $|G|$. If $\chi$ is not faithful, then $\chi$ is a character of $G/\ker \chi$, and by induction $\chi(1) \mid |G/\ker \chi: H/H \cap \ker \chi|$. So $\chi(1)$ divides

$$\frac{|G: \ker \chi|}{|H: H \cap \ker \chi|} = \frac{|G: H|}{|\ker \chi: H \cap \ker \chi|},$$

and we are done if $\chi$ is not faithful.

Assume now $\chi$ is faithful, and let $\lambda$ be an irreducible character of $H$ with $(\chi_H, \lambda)_H > 0$. Since $H$ is abelian, $\lambda$ is linear.

*Case 1*   If $I(\lambda) \neq G$, then by Lemma 22.2 there is an irreducible character $\zeta$ of $I(\lambda)$ with $\chi = \zeta^G$ and $\lambda \subset \zeta_H$. By induction $\zeta(1) \mid |I(\lambda): H|$, and $\chi(1) = |G: I(\lambda)| \zeta(1)$ so $\chi(1) \mid |G: H|$.

*Case 2*   If $I(\lambda) = G$, then by Clifford's Theorem 14.1, $\chi_H = e\lambda$ for some integer $e$. $\lambda(1) = 1$, so $e = \chi(1)$. If $h \in H$, $|\chi(h)| = e|\lambda(h)| = e = \chi(1)$, so by Lemma 5.10 and the fact $\chi$ is faithful, the matrix representation affording $\chi$ consists of scalar matrices on $H$. This means $H \subseteq Z(G)$, so the result follows from Theorem 22.1.

### EXERCISES

**1**   Let $A$ be *any* abelian subgroup of $G$, $\chi$ an irreducible complex character of $G$. Use Frobenius reciprocity to show that $\chi(1) \leq |G: A|$.

**2**   The following proof of Theorem 22.1 is due to Tate (see Serre [1]). If $\chi$ is a faithful irreducible complex character of $G$ of degree $n$, show that, for any $m$,

$$\underbrace{\chi \cdot \chi \cdots \chi}_{m} \text{ is an irreducible character of } \underbrace{G \times G \times \cdots \times G}_{m}$$

with kernel of order $|Z(G)|^{m-1}$ and degree $n^m$. By choosing $m$ sufficiently large and using Theorem 7.7 for $\chi \cdot \chi \cdots \chi$, conclude that $n$ divides $|G: Z(G)|$. Finally, use the foregoing to give a complete proof of Theorem 22.1.

**3**   A subgroup $H$ of the group $G$ is *subnormal* in $G$ if there is a series of subgroups

$$H = N_0 \subset \cdots \subset N_s = G,$$

each $N_i \lhd N_{i+1}$. Show that if $A$ is a subnormal abelian subgroup of the finite group $G$ and $\chi$ is an irreducible complex character of $G$, then $\chi(1)$ divides $|G : A|$. (*Hint*: use Exersise 2 of §25.)

# §23

## Algebraically Conjugate Characters

*Remark*   We know that a function $\chi: G \to k$ is a *character of G over k*, if $\chi$ is the character afforded by some $kG$-module. This is a stronger statement than merely saying $\chi(g) \in k$, all $g \in G$, for when $\chi$ is a character of $G$ over $k$ we know that all matrix entries of the matrix representation affording $\chi$ are in $k$.

**Theorem 23.1**   *Let $k$ be a field of characteristic $0$, $K$ a field which is both a finite Galois extension of $k$ and a splitting field for $G$.*

*(1)   If $\chi$ is a character of $G$ over $K$ and $\alpha \in \mathrm{Gal}(K/k)$, then the function $\chi^\alpha$ defined by $\chi^\alpha(g) = \chi(g)^\alpha$, all $g \in G$, is a character of $G$ over $K$. $\chi^\alpha$ is called* algebraically conjugate *to $\chi$; it is irreducible if $\chi$ is.*

*(2)   Let $\chi$ and $\tau$ be characters of $G$ over $K$, and assume that $\chi(g) \in k$, all $g \in G$. Then*

$$(\chi, \tau)_G = (\chi, \tau^\alpha)_G, \qquad \text{all} \quad \alpha \in \mathrm{Gal}(K/k).$$

*(3)   Assume that $K$ contains all $|G|$th roots of $1$, and for any integer $m$ with $(|G|, m) = 1$ let $\alpha_m \in \mathrm{Gal}(K/k)$ be such that $\alpha_m(\varepsilon) = \varepsilon^m$, any $|G|$th root $\varepsilon$ of $1$. Then*

$$\chi^{\alpha_m}(g) = \chi(g^m),$$

*for all characters $\chi$ of $G$ over $K$ and all $g \in G$.*

*Proof* (1) Let $\chi$ be afforded by the $KG$-module $V$, with $K$-basis $\{v_1, \ldots, v_n\}$, where $gv_j = \sum_{i=1}^n a_{ij}(g)v_i$. Define a new $KG$-module $V^\alpha$ with underlying vector space $V$, but module action $g \cdot v_j = \sum_{i=1}^n a_{ij}(g)^\alpha v_i$; $V^\alpha$ is a $KG$-module because if $g, h \in G$, then

$$g \cdot (h \cdot v_l) = g \cdot \sum_{j=1}^{n} a_{jl}(h)^z v_j = \sum_{i=1}^{n} \sum_{j=1}^{n} a_{ij}(g)^z a_{jl}(h)^z v_i$$

$$= \sum_{i=1}^{n} a_{il}(gh)^z v_j = gh \cdot v_l.$$

$V^z$ affords the character

$$\chi^z(g) = \sum_{i=1}^{n} a_{ii}(g)^z = (\sum_{i=1}^{n} a_{ii}(g))^z = \chi(g)^z.$$

$V$ is irreducible iff $V^z$ is, since $(\chi, \chi)_G = 1$ iff $(\chi^z, \chi^z)_G = 1$.

(2)  $(\chi, \tau)_G = (\chi, \tau)_G^z = \left( \dfrac{1}{|G|} \sum_{g \in G} \chi(g)\tau(g^{-1}) \right)^z$

$$= \frac{1}{|G|} \sum_{g \in G} \chi(g)\tau^z(g^{-1}) = (\chi, \tau^z)_G.$$

(3)  Let $\psi$ be the matrix representation affording $\chi$, say $\psi(g) = (a_{ij}(g))$. Then $\chi(g)$, the sum of the eigenvalues of $\psi(g)$, is a sum of $|G|$th roots of 1. $\chi(g^m)$ is the sum of the eigenvalues of $\psi(g^m) = \psi(g)^m$, so $\chi(g^m)$ is the sum of the $m$th powers of the eigenvalues of $\psi(g)$. This means $\chi(g^m) = \chi(g)^{z_m}$, so $\chi(g^m) = \chi^{z_m}(g)$.

**Definition**    Let $\chi$ be a complex character of $G$. $\chi$ is called *rational-valued* if all $\chi(g) \in \mathbf{Q}$ = rational numbers, and *real-valued* if all $\chi(g) \in \mathbf{R}$ = real numbers.

**Definition**    $g \in G$ is a *real element* of $G$, if $g$ and $g^{-1}$ are conjugate in $G$. The conjugacy class of $g$ in $G$ is also called *real*.

**Lemma 23.2**    $g \in G$ is a real element if and only if for all complex characters $\chi$ of $G$, $\chi(g)$ is real.

*Proof*  If $g \in G$ is real, then $\chi(g) = \chi(g^{-1}) = \overline{\chi(g)}$. Conversely, if $\chi(g)$ is real for all $\chi$, then $\chi(g) = \chi(g^{-1})$ for all class functions $\chi$ of $G$. Hence $g$ and $g^{-1}$ are conjugate.

Groups with all elements real are called *ambivalent*, and have recently been studied in Berggren [1] and Kerber [3].

**Theorem 23.3**    (a)  *The number of real-valued irreducible complex characters of $G$ equals the number of real conjugacy classes in $G$.*

(b)   *The following are equivalent*:
(i)   *All irreducible complex characters of G are rational-valued.*
(ii)   *For all integers m with $(m, |G|) = 1$ and all $g \in G$, g and $g^m$ are conjugate.*

*Proof* (a)   Represent the group $\{1, z\}$ of order 2 as a permutation group on G by $g^z = g^{-1}$, and as a permutation group on the irreducible characters $\chi$ of G by $\chi^z = \bar{\chi}$. Then $\chi^z(g) = \overline{\chi(g)} = \chi(g^{-1}) = \chi(g^z)$, so the result follows from Lemma 13.7.

(b)   Let $\varepsilon$ be a primitive $|G|$th root of 1. If $m$ is an integer with $(m, |G|) = 1$, choose $\alpha_m \in \mathrm{Gal}(\mathbf{Q}(\varepsilon)/\mathbf{Q})$ such that $\varepsilon^{\alpha_m} = \varepsilon^m$. Let $A$ be the multiplicative group of integers relatively prime to $|G|$ modulo $|G|$, and let $a_m$ be the residue class of $m$. Let $A$ act on G by $g^{a_m} = g^m$, and on the irreducible characters $\chi$ of G by $\chi^{a_m} = \chi^{\alpha_m}$. Then $\chi^{a_m}(g) = \chi(g)^{\alpha_m} = \chi(g^m) = \chi(g^{a_m})$ using Theorem 23.1(3), so the hypotheses of Lemma 13.7 are satisfied. We use Lemma 13.7(2). (i) holds iff $A$ fixes all irreducible characters, which will hold iff $A$ has $h$ orbits on characters ($h = $ number of conjugacy classes of G). By Lemma 13.7(2), this holds iff $A$ has $h$ orbits on conjugacy classes, which is equivalent to saying $A$ fixes all conjugacy classes. This means that (ii) holds.

**Corollary 23.4** (Burnside)   *Assume $|G|$ odd. Then $1_G$ is the only real-valued irreducible complex character of G.*

*Proof*   By Theorem 23.3(a), we need only show that $\{1\}$ is the only real conjugacy class. If $x^g = x^{-1}$ for some $x \neq 1$, then $x^{g^2} = (x^{-1})^g = (x^g)^{-1} = (x^{-1})^{-1} = x$, so $g^2 \in C_G(x)$. But $|G|$ is odd so $g \in \langle g^2 \rangle$, $g \in C_G(x)$, $x = x^{-1}$, $x^2 = 1$, $x$ has order 2, a contradiction.

*Remark*   Let $S_n$ denote the symmetric group on $n$ letters. Theorem 23.3 (b) can be used to show that all irreducible complex characters of $S_n$ are rational-valued, by noting that any $g$ and $g^m$ in 23.3(b)(ii) have the same cycle structure and hence are conjugate. Corollary 5.6 shows that rational-valued characters are actually integer-valued.

**Definition**   Denote by $\mathbf{Q}(1^{1/m})$ the field $\mathbf{Q}$ extended by a primitive $m$th root of 1. An algebraic number $\alpha$ is said to *require the mth roots of 1* if $\alpha \in \mathbf{Q}(1^{1/m})$, but $\alpha \notin \mathbf{Q}(1^{1/n})$ for any $n < m$.

The following theorem is due to Blichfeldt (see Brauer [8]).

**Theorem 23.5** *Let $\chi$ be an irreducible complex character of $G$. Let $p_1, \ldots, p_n$ be distinct primes. Assume there exist elements $x_1, \ldots, x_n \in G$ and positive integers $a_1, \ldots, a_n$ such that $\chi(x_i)$ requires the $(p_i^{a_i})$th roots of 1. Then $G$ contains an element $x$ of order $p_1^{a_1} \cdots p_n^{a_n}$.*

*Proof*  For $1 \leq i \leq n$, choose integers $b_i$ with $p_i^{b_i} \mid |G|$, $p_i^{b_i+1} \nmid |G|$. Then

$$\chi(x_i) \in \mathbf{Q}(1^{1/|G|}) \cap \mathbf{Q}(1^{1/p_i^{a_i}}) \subseteq \mathbf{Q}(1^{1/p_i^{b_i}}),$$

so $a_i \leq b_i$, all $i$. Denote

$$t_i = \frac{|G|}{p_i^{b_i}} p_i^{a_i-1} \quad \text{and} \quad F_i = \mathbf{Q}(1^{1/t_i}).$$

$\chi(x_i) \in F_i$ would imply

$$\chi(x_i) \in F_i \cap \mathbf{Q}(1^{1/p_i^{a_i}}) = \mathbf{Q}(1^{1/p_i^{a_i-1}}),$$

a contradiction, so $\chi(x_i) \notin F_i$, all $i$. If $i \neq j$, $\chi(x_i) \in F_j$. We now can find $\gamma_1, \ldots, \gamma_n \in \mathrm{Gal}(\mathbf{Q}(1^{1/|G|})/\mathbf{Q})$ such that for each $i$, $\gamma_i(\chi(x_i)) \neq \chi(x_i)$ but $\gamma_i$ is trivial on $F_i$; in particular, $\gamma_i(\chi(x_j)) = \chi(x_j)$ for $i \neq j$.

Assume our theorem is false. Then to each $g \in G$ must correspond some $i = i(g)$ with $g^{t_i} = 1$. Then $\chi(g) \in F_i$, so $\gamma_i(\chi(g)) = \chi(g)$. Since $\mathrm{Gal}(\mathbf{Q}(1^{1/|G|})/\mathbf{Q})$ is abelian, we conclude that for all $g \in G$,

$$\prod_{i=1}^{n} (1 - \gamma_i)(\chi(g)) = 0.$$

Expanding this product, we have

$$0 = \chi - \sum_{i=1}^{n} \gamma_i(\chi) + \sum_{1 \leq i < j \leq n} \gamma_i \gamma_j(\chi) - \cdots,$$

where all terms in the equation are irreducible characters by Theorem 23.1(1). Hence for some $0 \leq i_1 < \cdots < i_j \leq n$, $\chi = \gamma_{i_1} \cdots \gamma_{i_j}(\chi)$. This means that

$$\chi(x_{i_1}) = \gamma_{i_1} \cdots \gamma_{i_j}(\chi(x_{i_1})) = \gamma_{i_1}(\chi(x_{i_1})),$$

a contradiction.

# §24

## The Schur Index

*Remark* We now wish to tackle questions of the following type. Let $k$ be a field, $K$ an extension of $k$, $V$ a $KG$-module affording the character $\chi$. Suppose $\chi(g) \in k$, all $g \in G$. When can we then say $\chi$ is a character of $G$ *over* $k$? Also, suppose $\theta$ is an irreducible character of some $kG$-module. How does $\theta$ break up into a sum of irreducible characters over $K$?

These and similar questions will be studied after some theory of algebras over a field. This section is based on Reiner [1] and Huppert [6, §V.14].

**Definition** Let $A$ be a finite-dimensional algebra over the field $k$. $A$ is said to be *central-simple* (over $k$), if

(i) $Z(A) = k$, and

(ii) $A$ is a simple ring.

**Lemma 24.1** *Let $A$ be a finite-dimensional algebra over the field $k$, and assume that the only two-sided ideals of $A$ are $A$ and $(0)$. Then $A$ is a simple ring.*

*Proof* Let $V$ be an irreducible $A$-module which is finite-dimensional over $k$, and let $I = \{a \in A | aV = 0\}$. $I \neq A$ is a two-sided ideal of $A$, so $I = (0)$ and $A$ has a faithful irreducible module. If $D = \text{End}_A(V)$, then $D \supseteq k$ is a division ring with $\dim_D V$ finite. By Corollary 2.6, $A \cong \text{End}_D(V)$, so by Theorem 2.18, $A$ is a simple ring.

*Remark* If $A$ and $B$ are algebras over the field $k$, it is easy to see that $A \otimes_k B$ is also an algebra over $k$, with a multiplication satisfying $(a \otimes b)(a' \otimes b') = aa' \otimes bb'$.

**Lemma 24.2**    *Let $A$ and $B$ be finite-dimensional algebras over the field $k$. Then*:

(a) $Z(A \otimes_k B) = Z(A) \otimes_k Z(B)$.

(b) *If $A$ is a simple ring and $B$ is central-simple over $k$, then $A \otimes_k B$ and $B \otimes_k A$ are simple rings.*

*Proof* Let $\{a_1, \ldots, a_m\}$ be a $k$-basis of $A$, $\{b_1, \ldots, b_n\}$ a $k$-basis of $B$, so $\{a_i \otimes b_j\}$ is a $k$-basis of $A \otimes_k B$. Note that each element of $A \otimes_k B$ has a unique expression $\sum_{j=1}^{n} x_j \otimes b_j$, $x_j \in A$.

In proving (a), we assume that $\{a_1, \ldots, a_r\}$ is a $k$-basis of $Z(A)$ and $\{b_1, \ldots, b_s\}$ a $k$-basis of $Z(B)$, so $r \le m$, $s \le n$. Clearly $Z(A) \otimes_k Z(B) \subseteq Z(A \otimes_k B)$. Conversely, suppose

$$z = \sum_{i=1}^{m} \sum_{j=1}^{n} \alpha_{ij}(a_i \otimes b_j) \in Z(A \otimes_k B), \qquad \text{all} \quad \alpha_{ij} \in k.$$

For any $a \in A$, we have

$$0 = z(a \otimes 1) - (a \otimes 1)z = \sum_{i=1}^{m} \sum_{j=1}^{n} (\alpha_{ij}(a_i a - a a_i)) \otimes b_j.$$

For all $j$, then, $\sum_{i=1}^{m} \alpha_{ij}(a_i a - a a_i) = 0$, and

$$(\sum_{i=1}^{m} \alpha_{ij} a_i)a = a(\sum_{i=1}^{m} \alpha_{ij} a_i).$$

Therefore $\sum_{i=1}^{m} \alpha_{ij} a_i \in Z(A)$, and for $i > r$ we have $\alpha_{ij} = 0$. Similarly if $j > s$ we show $\alpha_{ij} = 0$, proving that $z \in Z(A) \otimes_k Z(B)$ and completing the proof of (a).

*Proof of (b)* Let $0 \ne I$ be a two-sided ideal of $A \otimes_k B$. Among all $0 \ne z \in I$, choose one such that in the expression $z = \sum_{i=1}^{m} a_i \otimes z_i$, $z_i \in B$, the number $l(z)$ of nonzero $z_i$'s is the smallest. ($\{a_1, \ldots, a_m\}$ is still our fixed basis of $A$.)

Some $z_i$, say $z_1$, is not zero, and $B$ is a simple ring so $Bz_1 B = B$. Choose $y_i, y_i' \in B$ with $\sum_{i=1}^{t} y_i z_1 y_i' = 1$, and set

$$z' = \sum_{i=1}^{t} (1 \otimes y_i) z (1 \otimes y_i') \in I.$$

Then

$$z' = \sum_{i=1}^{m} a_i \otimes (\sum_{j=1}^{t} y_j z_i y_j') = \sum_{i=1}^{m} a_i \otimes w_i,$$

where

$$w_1 = 1, \qquad w_i = \sum_{j=1}^{t} y_j z_i y_j', \qquad l(z') \le l(z).$$

For any $b \in B$, we have

$$(1 \otimes b)z' - z'(1 \otimes b) = \sum_{j=1}^{m} a_j \otimes (bw_j - w_j b) \in I.$$

But $bw_1 - w_1 b = b - b = 0$, implying

$$l((1 \otimes b)z' - z'(1 \otimes b)) < l(z') \le l(z).$$

This forces

$$(1 \otimes b)z' - z'(1 \otimes b) = 0, \qquad \text{so } bw_j = w_j b, \quad \text{all } j, \quad \text{all } b \in B.$$

$z' \ne 0$ since $w_1 = 1$. We have all $w_j \in Z(B) = k$, so

$$z' = \sum_{i=1}^{m} a_i \otimes w_i = \sum_{i=1}^{m} a_i w_i \otimes 1 = w \otimes 1, \qquad \text{some} \quad w \in A.$$

$A$ is simple, so $A = AwA$, proving that

$$I \supseteq (A \otimes_k B) z' (A \otimes_k B) \supseteq AwA \otimes_k B | B = A \otimes_k B.$$

By Lemma 24.1, $A \otimes_k B$ is a simple ring. Similarly for $B \otimes_k A$.

Recall that if $A$ is an algebra over $k$ and $K$ is an extension field of $k$, then $A^K = K \otimes_k A$ is an algebra over $K$.

**Theorem 24.3**  *Let $A$ be a central-simple algebra over $k$. Then:*
  *(a)  If $K$ is an algebraically closed extension field of $k$, then $K$ is a splitting field for $A^K$.*
  *(b)  $\dim_k A$ is a square.*

*Proof*  (a)  $K$ is a simple ring, so by Lemma 24.2 $A^K = K \otimes_k A$ is a simple ring. By Theorem 2.16, $A^K = \operatorname{End}_D(V)$, where $V$ is the unique irreducible $A^K$-module and $D = \operatorname{End}_{A^K}(V)$ is a division ring. $D$ is finite-dimensional over $K$ with $K$ in its center, so $D = K$, implying that $K$ is a splitting field.
  (b)  $\dim_k A = \dim_K A^K = m^2$, where $m = \dim_K V$.

**Lemma 24.4**  *Let $A$ be a simple finite-dimensional algebra over $k$, so*

$A = \text{Mat}_n(D)$ *for a division ring* $D$. *Let* $V$ *be an* $A$-*module, and let* $B = \text{End}_A(V)$. *Then*:

(a)   $B \cong \text{Mat}_m(D^{\text{op}})$, *some* $m$. *Hence* $B$ *is a simple ring.*

(b)   $(\dim_k V)^2 = (\dim_k A)(\dim_k B)$.

*Proof*   We know $A$ has only one irreducible module $W$, where $D^{\text{op}} = \text{End}_A(W)$ and $\dim_{D^{\text{op}}}(W) = n$. $V$ is completely reducible, so let $V \cong W \oplus \cdots \oplus W$ ($m$ copies). By Lemma 2.1,

$$B = \text{End}_A(V) \cong \text{Mat}_m(\text{End}_A(W)) = \text{Mat}_m(D^{\text{op}}).$$

Thus (a) holds. For (b), we have

$$\dim_k V = m(\dim_k W) = mn(\dim_k D),$$

$$(\dim_k A)(\dim_k B) = (n^2 \dim_k D)(m^2 \dim_k D^{\text{op}}) = m^2 n^2 (\dim_k D)^2.$$

**Theorem 24.5**   *Let* $A$ *be a central-simple algebra over* $k$, $B$ *a simple subalgebra of* $A$ *containing* $1 \in A$. *If* $C_A(B) = \{a \in A | ab = ba, \text{ all } b \in B\}$, *then*:

(1)   $C_A(B)$ *is a simple ring.*

(2)   *Let* $A^{\text{op}}$ *be the central-simple* $k$-*algebra opposite to* $A$, *so by Lemma 24.2* $B \otimes_k A^{\text{op}}$ *is a simple ring. Then* $C_A(B)$ *and* $B \otimes_k A^{\text{op}}$ *are full matrix rings over opposite division rings.*

(3)   $\dim_k A = (\dim_k B)(\dim_k C_A(B))$.

(4)   $C_A(C_A(B)) = B$.

*Proof*   $A$ is a $(B \otimes_k A^{\text{op}})$-module under an action $(b \otimes a)x = bxa$, because $(b_2 \otimes a_2)((b_1 \otimes a_1)x) = (b_2 \otimes a_2)(b_1 x a_1) = b_2 b_1 x a_1 a_2 = (b_2 b_1 \otimes a_1 a_2)x = ((b_2 \otimes a_2)(b_1 \otimes a_1))x$.

We will now show $C_A(B) \cong \text{Hom}_{B \otimes A^{\text{op}}}(A, A)$. If $\varphi \in \text{Hom}_{B \otimes A^{\text{op}}}(A, A)$ and $\varphi(1) = c$, then $\varphi(a) = \varphi((1 \otimes a)1) = (1 \otimes a)\varphi(1) = \varphi(1)a = ca$, so $\varphi$ is determined by knowing $\varphi(1)$. If $b \in B$, then $cb = \varphi(b) = \varphi((b \otimes 1)1) = (b \otimes 1)\varphi(1) = (b \otimes 1)c = bc$, so $c \in C_A(B)$. Conversely, if $c \in C_A(B)$, then we do get a $\varphi_c \in \text{Hom}_{B \otimes A^{\text{op}}}(A, A)$ by defining $\varphi_c(a) = ca$. If $c, d \in C_A(B)$, then $(\varphi_c \varphi_d)(1) = \varphi_c(\varphi_d(1)) = \varphi_c(d) = cd = \varphi_{cd}(1)$, so the correspondence $\varphi_c \leftrightarrow c$ is the desired isomorphism.

We now apply Lemma 24.4, with $B \otimes A^{\text{op}}$ in place of $A$, $A$ in place of $V$, $C_A(B)$ in place of $B$. (1) and (2) follow directly from Lemma 24.4 (a). Lemma 24.4(b) becomes $(\dim_k A)^2 = (\dim_k B)(\dim_k A)(\dim_k C_A(B))$, so (3) also holds.

Certainly $B \subseteq C_A(C_A(B))$. But by (3) applied to $C_A(B)$ in place of $B$, $\dim_k A = (\dim_k C_A(B))(\dim_k C_A(C_A(B)))$. Hence

$$\dim_k C_A(C_A(B)) = \frac{\dim_k A}{\dim_k C_A(B)} = \dim_k B,$$

so (4) must hold.

**Theorem 24.6** *Let $A$ be a central-simple algebra over $k$, so $A = \mathrm{Mat}_n(D)$, $D$ a division ring. Then:*

(a) $\dim_k D = m^2$, *some integer* $m$.

(b) *If $F$ is a maximal subfield of $D$ (so $F \supseteq k$), then $\dim_k F = m$, and $F$ is a splitting field for the algebras $D^F$ and $A^F$.*

(c) *If $K$ is any field finite-dimensional over $k$ and such that $K$ is a splitting field for $A^K$, then $m \mid \dim_k K$.*

(d) *Let $K$ be a splitting field for $A^K$. Then $A^K$ is a simple ring. Let $V$ be the unique irreducible $A^K$-module. Then the following are equivalent:*

(i) $\underbrace{V \oplus \cdots \oplus V}_{t \text{ copies}} \cong U^K$, *some $A$-module $U$.*

(ii) $m \mid t$.

*Proof* (a) $\dim_k D = (\dim_k A)/n^2$, where $\dim_k A$ is a square by Theorem 24.3.

(b) Clearly $F \subseteq C_D(F)$; if $d \in C_D(F) - F$, then $F(d)$ is a subfield larger than $F$, a contradiction, so $F = C_D(F)$. $D$ is itself central-simple over $k$, so Theorem 24.5(3) says $\dim_k D = (\dim_k F)^2$. By (a), $\dim_k F = m$.

By Theorem 24.5(2) applied to $D$, $F = C_D(F)$ and $F \otimes_k D^{\mathrm{op}}$ are full matrix rings over opposite division rings $E$ and $E^{\mathrm{op}}$. Therefore $F = E = E^{\mathrm{op}}$; let $F \otimes_k D^{\mathrm{op}} \cong \mathrm{Mat}_t(F)$. We see easily that

$$(F \otimes_k D)^{\mathrm{op}} \cong F \otimes_k D^{\mathrm{op}} \cong \mathrm{Mat}_t(F),$$

and $F \otimes_k D \cong \mathrm{Mat}_t(F)^{\mathrm{op}} \cong \mathrm{Mat}_t(F^{\mathrm{op}}) = \mathrm{Mat}_t(F)$, proving that $F$ is a splitting field for $D^F = F \otimes_k D$.

Finally, we see that

$$F \otimes_k A \cong F \otimes_k (\mathrm{Mat}_n(D)) \cong \mathrm{Mat}_n(F \otimes_k D) = \mathrm{Mat}_{nt}(F),$$

so $F$ is also a splitting field for $A^F$.

(c) By Lemma 24.2 $A^K = K \otimes_k A$ is a simple ring, so since $K$ is a splitting field we have $A^K \cong \mathrm{Mat}_i(K)$, some $i$; $\dim_K A^K = \dim_k A = n^2(\dim_k D) = n^2 m^2$, so $A^K \cong \mathrm{Mat}_{mn}(K)$. $A^K$ has only one irreducible module $V$, and $\dim_K V = mn$ so $\dim_k V = mn(\dim_k K)$. $A$ is a subring of

$A^K$ under the map $a \leftrightarrow 1 \otimes a$, so $V$ is itself an $A$-module. $A$ is a simple ring, so for some integer $t$, $V$ is a direct sum of $t$ isomorphic irreducible $A$-modules. Any irreducible $A$-module $W$ has dimension $\dim_k W = n(\dim_k D) = nm^2$, so we see $\dim_k V = tnm^2$. Equating the two expressions for $\dim_k V$, we have

$$tnm^2 = mn(\dim_k K), \qquad \dim_k K = tm, \qquad m|(\dim_k K).$$

(d)   We use again some information from (c), without assuming $\dim_k K$ finite. Namely, we saw that $A^K = \text{Mat}_{mn}(K)$, and if $W$ is the only irreducible $A$-module and $V$ the only irreducible $A^K$-module, then $\dim_k W = nm^2$, $\dim_K V = mn$.

If $U$ is an $A$-module with $U^K \cong \underbrace{V \oplus \cdots \oplus V}_{t \text{ copies}}$, let $U \cong$

$\underbrace{W \oplus \cdots \oplus W}_{s \text{ copies}}$. Then $snm^2 = \dim_k U = \dim_K U^K = tmn$, so

$$sm = t, \ m|t, \text{ proving (i)} \Rightarrow \text{(ii)}.$$

Also, $\dim_K W^K = nm^2$, so $W^K \cong \underbrace{V \oplus \cdots \oplus V}_{m \text{ copies}}$, proving (ii) $\Rightarrow$ (i).

**Lemma 24.7**    *Assume* char $k \nmid |G|$, *and let* $K \supseteq k$ *be a splitting field for* $G$. *Let* $V$ *be an irreducible* $KG$-module *affording the character* $\chi$, *and let* $kG = \oplus \sum_{i=1}^{s} A_i$, *the* $A_i$ *simple rings. Then*:
   (1)   *there is exactly one* $j \in \{1, \ldots, s\}$ *such that* $A_j V \neq 0$.
   (2)   *if* $A_j$ *is chosen as in* (1), *then* $Z(A_j) \cong k(\chi)$, *where* $k(\chi)$ *is the extension of* $k$ *by all* $\chi(x)$, $x \in G$.

*Proof*   $V$ is a $kG$-module since $kG \subseteq KG$, and $1 \in kG$ so $(kG)V \neq 0$. We have two expressions for $KG$;

$$KG = \sum_{t=1}^{h} \text{Mat}_{n_t}(K) \qquad \text{and} \qquad KG = K \otimes_k kG = \sum_{i=1}^{s} A_i^K.$$

Each $A_j^K$ is a sum of some of the $\text{Mat}_{n_t}(K)$'s, and for only one $t_0$ do we have $\text{Mat}_{n_{t_0}}(K) \cdot V \neq 0$, so there is only one $j$ with $A_j^K \cdot V \neq 0$, $A_j = 1 \otimes A_j \subseteq A_j^K$, so (1) holds.

In proving (2), denote $n = \dim_K V$. For any $a \in A_j$, $a$ acts on $V$, so we have a natural map

$$T: A_j \to \text{End}_K(V) \cong \text{Mat}_n(K),$$

the representation of $A_j$ afforded by $V$. ker $T$ is a two-sided ideal in $A_j$, so ker $T = 0$ and $A_j \cong$ im $T$. Denoting im $T = B$, we will compute $Z(B)$. For $i \neq j$ we have $A_i V = 0$, so $T$ extends to all of $kG$ and we still have im $T = B$.

Let $I: V \to V$ be the identity map. For any $g \in G$, let $C_g \in kG$ denote the sum of the conjugates of $g$ and $h_g$ the number of conjugates of $g$; $h_g = |G|/|C_G(g)|$, so $h_g \neq 0$ in $k$. $C_g \in Z(KG)$, so by Theorem 10.1 we have

$$T(C_g) \in Z(\text{End}_K(V)) = KI,$$

say

$$T(C_g) = \omega(g)I \in Z(B).$$

Taking traces in this equation, we have

$$h_g\chi(g) = n\omega(g),$$

so

$$\chi(g)I = \frac{n}{h_g}\omega(g)I \in Z(B),$$

and we have proved that

$$k(\chi)I \subseteq Z(B).$$

Conversely, suppose $a = \sum_{g \in G} a_g g \in Z(A_j)$, so that all $a_g \in k$ and $T(a) \in Z(B)$. For any $g \in G$, $T(g) \in B$, so by Theorem 10.1 certainly $Z(B) \subseteq Z(\text{End}_K(V))$, and we have

$$Z(B) \subseteq KI = \{cI \mid c \in K\}.$$

Hence

$$T(a) = \sum_{g \in G} a_g T(g) = cI, \qquad \text{some} \quad c \in K.$$

Taking traces in this equation,

$$\sum_{g \in G} a_g\chi(g) = nc.$$

But by Lemma 5.2, $n = \dim_K V \neq 0$ in $K$, so $n$ has an inverse $n^{-1}$ in $k$, and we have $c = \sum_{g \in G} n^{-1}a_g\chi(g) \in k(\chi)$. We now know $k(\chi)I = Z(B)$, so $k(\chi) \cong Z(B) \cong Z(A_j)$ and we are done.

**Definition**      Let $k$ be a field, char $k \nmid |G|$, and let $K$ be an extension field of $k$ which is a splitting field of $G$. Let $\chi$ be an irreducible $KG$-character. In Lemma 24.7, let $A_j = \mathrm{Mat}_n(D)$, $D$ a division ring, so by Lemma 24.7, $Z(A_j) \cong k(\chi)$. By Theorem 24.3, $\dim_{k(\chi)} D$ is a square, say $s^2$. $s = s_k(\chi)$ is called the *Schur index of $\chi$ over $k$.*

**Definition**      Let $k$, $K$, $G$ be as above, $\chi$ some $KG$-character afforded by a $KG$-module $V$. We say $\chi$ *is realizable over $k$* if $V \cong U^K$, some $kG$-module $U$. (Of course then $U$ affords $\chi$, so this can only happen when $k = k(\chi)$.)

**Theorem 24.8**      *Let* char $k \nmid |G|$, *$K$ an extension field of $k$ which is a splitting field for $G$. Let $V$ be an irreducible $KG$-module affording the character $\chi$. Let $A_j$ be the simple direct summand of $kG$ with $A_j V \neq 0$, and let $F$ be an extension field of $k(\chi)$. Then*:

(a)  *$\chi$ is realizable over $F$ if and only if $F$ is a splitting field for the central-simple $k(\chi)$-algebra $A_j$.*

(b)  *$\chi$ is realizable over $k(\chi)$ if and only if $s_k(\chi) = 1$.*

*Proof (a)*    Let $L$ be an extension field of both $K$ and $F$, so $L$ is a splitting field for $G$. By Lemma 24.2, $F \otimes_{k(\chi)} A_j$ is a simple ring with center $F \otimes_{k(\chi)} Z(A_j)$. By Lemma 24.7 $Z(A_j) \cong k(\chi)$, so $F \otimes_{k(\chi)} Z(A_j) \cong F$ and $F \otimes_{k(\chi)} A_j$ is central-simple over $F$. Let $F \otimes_{k(\chi)} A_j \cong \mathrm{Mat}_n(D)$, $D$ a division ring with center $F$, and set $\dim_F D = m^2$. Now

$$L \otimes_F (F \otimes_{k(\chi)} A_j) \cong (L \otimes_F F) \otimes_{k(\chi)} A_j = L \otimes_{k(\chi)} A_j$$

is by Lemma 24.2 central-simple with center $L$. $V$ is an irreducible $A_j^K = K \otimes_k A_j$-module, so $V^L$ is an irreducible $L \otimes_k A_j$-module, using the fact that $K$ is a splitting field. But $V^L$ is also naturally an $L \otimes_{k(\chi)} A_j$-module, and $L \otimes_k A_j$ and $L \otimes_{k(\chi)} A_j$ induce the same $L$-endomorphisms of $V^L$, so $V^L$ is an irreducible $L \otimes_{k(\chi)} A_j$-module. By Theorem 24.6(d), $V^L \cong U^L$ for some $F \otimes_{k(\chi)} A_j$-module $U$ if and only if $m|1$. This holds if and only if $D = F$, which is equivalent to $F$ being a splitting field for $A_j$.

*Proof (b)*    Take $F = k(\chi)$ in (a), and use the definition of $s_k(\chi)$.

**Theorem 24.9** (Wedderburn)      *Every finite division ring $D$ is a field.*

*Proof*    By induction on $|D|$. Let $k$ be the field $Z(D)$, and assume

$D \neq k$, $|k| = q$. By Theorem 24.6, $|D| = q^{m^2}$ for some integer $m > 1$, and for any maximal subfield $F$ of $D$, $|F| = q^m$. If $x \in D - k$, then $C_D(x)$ is a subfield of $D$ by induction, and must be maximal; hence $|C_D(x)| = q^m$. Let $S$ be a set of representatives of noncentral conjugacy classes in the multiplicative group $D^{\#}$. Writing $D^{\#}$ as a union of conjugacy classes, we see

$$q^{m^2} - 1 = q - 1 + |S| \frac{q^{m^2} - 1}{q^m - 1}.$$

Hence

$$|S| = q^m - 1 - \frac{(q - 1)(q^m - 1)}{q^{m^2} - 1},$$

a contradiction since this right side cannot be an integer.

**Theorem 24.10** *Assume $k$ is a finite field, char $k \nmid |G|$, and let $K$ be an extension field of $k$ which is a splitting field for $G$, $\chi$ an irreducible $KG$-character. Then $s_k(\chi) = 1$, so $\chi$ is realizable over $k(\chi)$.*

*Proof* Let $V$ be the $KG$-module affording $\chi$, $A_j$ the simple direct summand of $kG$ with $A_j V \neq 0$. Then $A_j = \mathrm{Mat}_n(D)$, $D$ a finite division ring. By Theorem 24.9, $D$ is a field, so by Lemma 24.7(2), $D \cong Z(A_j) \cong k(\chi)$. $s_k(\chi)^2 = \dim_{k(\chi)} D$, so $s_k(\chi) = 1$. By Theorem 24.8(b), $\chi$ is realizable over $k(\chi)$.

**Corollary 24.11** *If $k$ is a finite field, char $k \nmid |G|$, and $m$ is the exponent of $G$, then $k(1^{1/m})$ is a splitting field for $G$.*

*Proof* Let $K$ be the algebraic closure of $k$, and let $\chi$ be any irreducible $KG$-character, $\varphi$ the matrix representation affording $\chi$. For any $x \in G$, $\varphi(x)^m = \varphi(x^m) = I$. $X^m = 1$ factors into distinct linear factors over $K$, so the minimal polynomial of $\varphi(x)$ is a product of distinct linear factors, implying that $\varphi(x)$ is similar to a diagonal matrix. This proves that $\chi(x)$ is the sum of the eigenvalues of $\varphi(x)$, which are all $m$th roots of 1; hence $\chi(x) \in k(1^{1/m})$.

Denote $F = k(1^{1/m})$; by Theorem 24.10, $s_F(\chi) = 1$. If $G$ has $h$ conjugacy classes, then since $KG$ has $h$ irreducible modules, all with characters realizable over $F$, we conclude that $FG$ has $h$ pairwise nonisomorphic irreducible modules $V_1, \ldots, V_h$, $\dim_F V_i = n_i$, $\sum_{i=1}^h n_i^2 = |G|$. By Lemma 10.2 these are all the irreducible $FG$-modules, and by our construction all $V_i^K$ are irreducible $KG$-modules, implying $F$ is a splitting field for $G$.

**Theorem 24.12**    *Assume* char $k \nmid |G|$, *and let* $\Omega \supseteq k$ *be a splitting field for* $G$, $V$ *an irreducible* $\Omega G$-*module affording the character* $\chi$. *Then*:

(1)   *Let* $K$ *be an intermediate field* $k(\chi) \subseteq K \subseteq \Omega$ *of finite degree* $(K: k(\chi))$ *over* $k(\chi)$. *If* $\chi$ *is realizable over* $K$, *then* $s_k(\chi)|(K: k(\chi))$.

(2)   *There is an intermediate field* $K$, $k(\chi) \subseteq K \subseteq \Omega$, *such that* $\chi$ *is realizable over* $K$ *and* $(K: k(\chi)) = s_k(\chi)$.

(3)   $t\chi$ *is realizable over* $k(\chi)$ *if and only if* $s_k(\chi)|t$.

(4)   $s_k(\chi)|\dim_\Omega V$.

*Proof*   (1)   Let $A_j$ be the simple direct summand of $kG$ with $A_j V \neq 0$. By Theorem 24.8(a), $K$ is a splitting field for the central-simple $k(\chi)$-algebra $A_j$. If $A_j \cong \mathrm{Mat}_n(D)$, $D$ a division ring, then by definition of $s_k(\chi)$, $\dim_{k(\chi)} D = s_k(\chi)^2$. By Theorem 24.6(c), $s_k(\chi)|(K: k(\chi))$.

(2)   Let $K$ be a maximal subfield of $D$. By Theorem 24.6(b), $(K: k(\chi)) = s_k(\chi)$ and $K$ is a splitting field for $A_j$. By Theorem 24.8(a), $\chi$ is realizable over $K$.

(3)   is immediate from Theorem 24.6(d).

(4)   Since    $A_j \cong \mathrm{Mat}_n(D)$,    $\dim_{k(\chi)} A_j = n^2 \dim_{k(\chi)} D = n^2(s_k(\chi))^2$. By Lemma 24.2, $\Omega \otimes_{k(\chi)} A_j$ is a simple ring, so $\Omega \otimes_{k(\chi)} A_j \cong \mathrm{Mat}_m(\Omega)$, some $m$. $V$ is an irreducible $\Omega \otimes_{k(\chi)} A_j$-module, so $m = \dim_\Omega V$, $\dim_{k(\chi)} A_j = m^2$, $m^2 = n^2 s_k(\chi)^2$, proving $s_k(\chi)|m$.

**Lemma 24.13**    *Assume* $k$ *is a splitting field for* $G$ *with* char $k \nmid |G|$, *and let* $V$ *be an irreducible* $kG$-*module affording the character* $\chi$. *If* $A_j$ *is the unique simple direct summand of* $kG$ *with* $A_j V \neq 0$, *and* $e_j$ *the unit element of* $A_j$, *then*

$$e_j = \frac{\chi(1)}{|G|} \sum_{g \in G} \chi(g^{-1}) g.$$

*Proof*   Let $kG = A_1 \oplus \cdots \oplus A_h$, where $1 = e_1 + \cdots + e_h$. Let $V_1, \ldots, V_h$, $V = V_j$, be the irreducible $kG$-modules, where $V_i$ affords the representation $T_i$ and the character $\chi_i$, $\chi = \chi_j$. Since $e_r V_s = 0$ if $r \neq s$ and $e_r$ is the unit of $A_r$, $T_r(e_s) = \delta_{rs}I$.

Let $\mathscr{C}_1, \ldots, \mathscr{C}_h$ be the conjugacy classes of $G$, $h_i = |\mathscr{C}_i|$, $x_i \in \mathscr{C}_i$, $C_i = \sum_{g \in \mathscr{C}_i} g \in kG$. By Lemma 5.2 each $\chi_i(1) \neq 0$ in $k$, so as in the proof of Lemma 5.9 we have

$$T_s(C_r) = \omega_{rs}I = \frac{h_r \chi_s(x_r)}{\chi_s(1)} I.$$

$\{e_1, \ldots, e_h\}$ is a basis of $Z(kG)$, so for some $\alpha_{rs} \in k$ we have $C_r = \sum_{s=1}^{h} \alpha_{rs} e_s$. We now have

$$\omega_{rs} I = T_s(C_r) = T_s \left( \sum_{t=1}^{h} \alpha_{rt} e_t \right) = \sum_{t=1}^{h} \alpha_{rt} \delta_{st} I = \alpha_{rs} I,$$

so $\omega_{rs} = \alpha_{rs}$, $C_r = \sum_{s=1}^{h} \omega_{rs} e_s$.

Using the orthogonality relations, we now have

$$\frac{\chi_j(1)}{|G|} \sum_{g \in G} \chi_j(g^{-1}) g = \frac{\chi_j(1)}{|G|} \sum_{i=1}^{h} \chi_j(x_i^{-1}) C_i$$

$$= \frac{\chi_j(1)}{|G|} \sum_{i=1}^{h} \sum_{t=1}^{h} \chi_j(x_i^{-1}) \frac{h_i \chi_t(x_i)}{\chi_t(1)} e_t$$

$$= \chi_j(1) \sum_{t=1}^{h} \frac{\delta_{jt}}{\chi_t(1)} e_t = e_j,$$

completing the proof.

**Theorem 24.14**    *Assume* char $k \nmid |G|$, *and let $W$ be an irreducible $kG$-module. Let $K \supseteq k$ be a splitting field for $G$, and let $V$ be an irreducible submodule of the $KG$-module $W^K$. Then*

$$W^K \cong s(V \oplus V^{\alpha_2} \oplus \cdots \oplus V^{\alpha_t}),$$

*the $V^{\alpha_i}$ pairwise nonisomorphic algebraically conjugate modules, $s$ an integer. If $V$ affords the character $\chi$, then $s = s_k(\chi)$. If $W$ affords the character $\theta$, then*

$$\theta = s_k(\chi) \sum_{\alpha \in H} \chi^{\alpha},$$

*where $H$ is the Galois group of $k(\chi)$ over $k$.*

*Proof*   Let $A_j$ be the unique simple direct summand of $kG$ with $A_j W \neq 0$, and let $e_j$ be the unit element of $A_j$. Also, let $B$ be the unique simple direct summand of $KG$ with $BV \neq 0$, so $B$ is a summand of $A_j^K$. If $B$ has unit element $e_\chi$, then $B = (KG) e_\chi$ and, by Lemma 24.13,

$$e_\chi = \frac{\chi(1)}{|G|} \sum_{g \in G} \chi(g^{-1}) g.$$

$k(\chi)$ is a subfield of $k(1^{1/|G|})$ and Gal $(k(1^{1/|G|})/k)$ is abelian, so $k(\chi)$

is a Galois extension of $k$; denote $H = \mathrm{Gal}(k(\chi)/k)$, and denote $e = \sum_{\alpha \in H} e_{\chi^{\alpha}}$. Each $e_{\chi^{\alpha}}$ is central in $KG$, so $e$ is central in $KG$. But

$$e = \sum_{\alpha \in H} \frac{\chi^{\alpha}(1)}{|G|} \sum_{g \in G} \chi^{\alpha}(g^{-1})g$$

is $H$-invariant, so $e \in kG$, in fact $e \in Z(kG)$.

If $x \in A_j$ then $xe_j = x$, so for any $x \in A_j^K$, $xe_j = x$. Therefore $e_{\chi^{\alpha}}e_j = e_{\chi^{\alpha}}$ for all $\alpha \in H$, and $ee_j = e$. This implies that $(kG)e = (kG)ee_j \subseteq (kG)e_j$.

$e^2 = e$ and $e(1 - e) = 0$, so $kG = (kG)e \oplus (kG)(1 - e)$; $(kG)e \subseteq (kG)e_j$ with $(kG)e$ a direct summand of $kG$ and $(kG)e_j = A_j$ a simple direct summand of $kG$, so $(kG)e = (kG)e_j = A_j$. We also see that $A_j^K = (KG)e = \oplus \sum_{\alpha \in H} (KG)e_{\chi^{\alpha}}$. If $m = \chi(1) = \chi^{\alpha}(1)$ for all $\alpha$, then $\dim_K(KG)e_{\chi^{\alpha}} = m^2$, and as $KG$-modules we have

$$(KG)e_{\chi^{\alpha}} \cong \underbrace{V^{\alpha} \oplus \cdots \oplus V^{\alpha}}_{m \text{ copies}},$$

$V^{\alpha}$ the irreducible $KG$-module affording $\chi^{\alpha}$.

$(KG)e \cdot V = (KG)e_{\chi} \cdot V = V$, so $A_j V \neq 0$. By Lemma 24.7(2), if $A_j = \mathrm{Mat}_n(D)$ for a division ring $D$, then $Z(A_j) = Z(D) \cong k(\chi)$. By definition of $s_k(\chi)$, $s_k(\chi)^2 = \dim_{k(\chi)}D$. We see that

$$\dim_k A_j = n^2(\dim_k D) = n^2 \, s_k(\chi)^2 \, \dim_k k(\chi)$$

and

$$\dim_k A_j = \dim_K A_j^K = |H|m^2 = m^2 \, \dim_k k(\chi).$$

Equating these two expressions, we have $m = ns_k(\chi)$.

$A_j$ is the direct sum of $n$ $kG$-modules isomorphic to $W$, so $A_j^K$ is the direct sum of $n$ $KG$-modules isomorphic to $W^K$. In $A_j^K$, each $V^{\alpha}$ has multiplicity $m$, so in $W^K$, $V^{\alpha}$ has multiplicity $m/n = s_k(\chi)$.

The $V^{\alpha}$ are pairwise nonisomorphic because they have different characters. All parts of Theorem 24.14 have now been proved.

**Theorem 24.15**     Let char $k = 0$, and let $K \supseteq k$ be a splitting field for $G$, $V$ an irreducible $KG$-module affording the character $\chi$. Then:

(1)   There is a unique irreducible character $\psi$ of $G$ over $k$ such that $(\psi, \chi)_G \neq 0$. In fact,

$$\psi = s_k(\chi) \sum_{\alpha \in \mathrm{Gal}(k(\chi)/k)} \chi^{\alpha}.$$

(2)   *If $\tau$ is any character of $G$ over $k$, then $s_k(\chi)|(\tau, \chi)_G$.*

*Proof*   Let $\psi$ be any irreducible character of $G$ over $k$ which satisfies $(\psi, \chi)_G \neq 0$. By Theorem 24.14,

$$\psi = s_k(\theta) \sum_{\alpha \in \mathrm{Gal}(k(\theta)/k)} \theta^\alpha,$$

for some irreducible $KG$-character $\theta$. $(\psi, \chi)_G \neq 0$ implies that $\chi$ is one of the $\theta^\alpha$, so (1) holds. (2) follows from (1).

*Remarks*   Using the characterization of characters, Brauer [3] has shown that the Schur indices of characters of $G$ can be determined from the Schur indices of certain solvable subgroups of $G$. Other recent works on the Schur index include Witt [1], Roquette [1, 2], Berman [6], Solomon [3], Burgoyne and Fong [1], Lorenz [1], Saksonov [3], Ford [1], and Janusz [1].

# §25

## Projective Representations

**Definition**     Let $G$ be a finite group, $k$ a field, $V$ a finite-dimensional $k$-vector space. A *projective representation of $G$ on $V$* is a function $T: G \to GL(V)$ such that for all $x, y \in G$ we have

$$T(x)T(y) = \alpha(x, y)T(xy), \qquad \text{some} \quad \alpha(x, y) \in k.$$

The function $\alpha: G \times G \to k$ is the *factor set of $T$*. $T$ is called *irreducible* if $V$ has no proper subspace invariant under all $T(x)$, $x \in G$.

If $T_1: G \to GL(V_1)$, $T_2: G \to GL(V_2)$ are two projective representations of $G$, we say $T_1$ and $T_2$ are *equivalent* if there is a $k$-vector space isomorphism $f: V_1 \to V_2$ such that, for all $x \in G$,

$$T_1(x) = c(x)f^{-1}T_2(x)f, \qquad \text{some} \quad c(x) \in k.$$

**Lemma 25.1**     (a)   *If $T$ is a projective representation of $G$ with factor set $\alpha$, then*

$$\alpha(x, yz)\alpha(y, z) = \alpha(x, y)\alpha(xy, z), \qquad all \quad x, y, z \in G.$$

(b)   *If $T_1$, $T_2$ are projective representations of $G$ on vector spaces $V_1$, $V_2$ with factor sets $\alpha_1$, $\alpha_2$, and $T_1$, $T_2$ are equivalent with $f: V_1 \to V_2$ an isomorphism and $c: G \to k$ a function satisfying*

$$T_1(x) = c(x)f^{-1}T_2(x)f, \qquad all \quad x \in G,$$

*then*

$$\alpha_1(x, y) = \alpha_2(x, y)c(x)c(y)c(xy)^{-1}, \qquad all \quad x, y \in G.$$

*Proof*   (a)   We have the associative law, and

132

$$T(x)\,[T(y)T(z)] = T(x)\alpha(y, z)T(yz) = \alpha(x, yz)\alpha(y, z)T(xyz),$$

$$[T(x)T(y)]T(z) = \alpha(x, y)T(xy)T(z) = \alpha(x, y)\alpha(xy, z)T(xyz).$$

(b)  We compute $T_1(x)T_1(y)$ in two ways.

$$T_1(x)T_1(y) = \alpha_1(x, y)T_1(xy) = \alpha_1(x, y)c(xy)f^{-1}T_2(xy)f,$$

and

$$T_1(x)T_1(y) = c(x)f^{-1}T_2(x)fc(y)f^{-1}T_2(y)f$$
$$= \alpha_2(x, y)c(x)c(y)f^{-1}T_2(xy)f.$$

**Definition**     For any field $k$, we denote $k^\# = k - \{0\}$. Any function $\alpha: G \times G \to k^\#$ satisfying

$$\alpha(x, yz)\alpha(y, z) = \alpha(x, y)\alpha(xy, z), \qquad \text{all}\quad x, y, z \in G,$$

is a *factor set* of $G$. Two factor sets $\alpha$, $\beta$ of $G$ are *equivalent* if there is a function $c: G \to k^\#$ such that

$$\alpha(x, y) = \beta(x, y)c(x)c(y)c(xy)^{-1}, \qquad \text{all}\quad x, y \in G.$$

**Lemma 25.2**     *If $\alpha$ and $\beta$ are factor sets of $G$, then so are $\alpha\beta$, defined by $(\alpha\beta)(x, y) = \alpha(x, y)\beta(x, y)$, and $\alpha^{-1}$, defined by $(\alpha^{-1})(x, y) = \alpha(x, y)^{-1}$. Equivalence of factor sets is an equivalence relation $\sim$. If $\alpha \sim \alpha'$ and $\beta \sim \beta'$, then $\alpha\beta \sim \alpha'\beta'$ and $\alpha^{-1} \sim (\alpha')^{-1}$.*

*Proof*  An easy, mechanical exercise.

**Definition**     For any $k$-factor set $\alpha$ of $G$, let $\{\alpha\}$ denote the equivalence class containing $\alpha$. By Lemma 25.2, multiplication $\{\alpha\}\{\beta\} = \{\alpha\beta\}$ is well-defined; the abelian group $M$ of equivalence classes of factor sets is the *Schur multiplier of $G$* (over $k$). (We remark that $M$ is actually the second cohomology group $H^2(G, k^\#)$, $k^\#$ a trivial $G$-module.)

**Theorem 25.3**     *Let $G$ be a finite group, $k$ an algebraically closed field, $M$ the Schur multiplier of $G$ over $k$. Then:*
  (1)  *For any $\{\alpha\} \in M$, $\{\alpha\}^{|G|} = 1$.*
  (2)  *If $\{\alpha\}$ has order $e$ in $M$, then there is an $\alpha' \in \{\alpha\}$ for which all $\alpha'(x, y)$ are eth roots of 1 in $k$.*
  (3)  *$M$ is a finite group, and char $k \nmid |M|$.*

*Proof*  For any factor set $\alpha$ of $G$ and any $x \in G$, define $\rho_\alpha(x) = \prod_{z \in G} \alpha(x, z)$. Then for any $x, y, z \in G$, we have

$$\alpha(x, y) = \frac{\alpha(x, yz)\alpha(y, z)}{\alpha(xy, z)},$$

so

$$\alpha(x, y)^{|G|} = \prod_{z \in G} \frac{\alpha(x, yz)\alpha(y, z)}{\alpha(xy, z)} = \frac{\rho_\alpha(x)\rho_\alpha(y)}{\rho_\alpha(xy)}.$$

This shows that

$$\alpha^{|G|}(x, y) = 1 \cdot \rho_\alpha(x)\rho_\alpha(y)\rho_\alpha(xy)^{-1},$$

so $\{\alpha\}^{|G|} = \{1\}$, proving (1).

For any $\{\alpha\}$ of order $e$ in $G$, there is $a: G \to k^\#$ with

$$\alpha(x, y)^e = a(x)a(y)a(xy)^{-1}.$$

For any $x \in G$ we choose $b(x) \in k^\#$ with $a(x)b(x)^e = 1$. If we define $\alpha'$ by

$$\alpha'(x, y) = \alpha(x, y)b(x)b(y)b(xy)^{-1},$$

then $\alpha' \in \{\alpha\}$ and

$$\alpha'(x, y)^e = \alpha(x, y)^e \frac{b(x)^e b(y)^e}{b(xy)^e} = \frac{a(x)a(y)}{a(xy)} \frac{b(x)^e b(y)^e}{b(xy)^e} = 1.$$

We have proved (2). There are at most $|G|\,|G|$th roots of 1 in $k$, so (1) and (2) imply that $|M|$ is finite.

Finally, char $k | |M|$ would imply char $k = p$, where $\{\alpha\} \in M$ has order $p$, some $\{\alpha\}$. Then we have a function $c: G \to k^\#$, $\alpha(x, y)^p = c(x)c(y)c(xy)^{-1}$. Each $c(x)$ has a unique $p$th root in $k$, so

$$\alpha(x, y) = c(x)^{1/p}c(y)^{1/p}(c(xy)^{1/p})^{-1},$$

showing $\alpha \in \{1\}$, a contradiction.

**Definition**     Suppose $H$ is a group, $N \subseteq Z(H)$, and $G \cong H/N$. Then $H$ is a *central extension* of $G$ with *kernel* $N$.

**Lemma 25.4**     *Let $H$ be a central extension of $G$ with kernel $N$, and if $\varphi: G \to H/N$ is an isomorphism let $\varphi(x) = h_x N$, $h_x \in H$. Then:*

*(1)* $h_x h_y = h_{xy} a(x, y)$, *some* $a(x, y) \in N$.
*(2)* *If $\overline{T}$ is an ordinary irreducible representation*

$$\overline{T}: H \to GL(V),$$

*$V$ a vector space over an algebraically closed field $k$, then $\overline{T}(a(x, y)) = \alpha(x, y)I$ for some $\alpha(x, y) \in k^{\#}$, and the mapping*

$$T: G \to GL(V)$$

*defined by*

$$T(x) = \overline{T}(h_x)$$

*is a projective representation of $G$ with factor set $\alpha$.*

*Proof* $h_x N h_y N = \varphi(x)\varphi(y) = \varphi(xy) = h_{xy} N$, so (1) holds.
Each $a(x, y) \in Z(H)$, so all $\overline{T}(a(x, y)) \in \mathrm{End}_{kH}(V)$, and by Theorem 4.2,

$$\overline{T}(a(x, y)) = \alpha(x, y)I \qquad \text{for some} \quad \alpha(x, y) \in k^{\#}.$$

Finally, we have

$$T(x)T(y) = \overline{T}(h_x)\overline{T}(h_y) = \overline{T}(h_x h_y) = \overline{T}(h_{xy})\overline{T}(a(x, y))$$
$$= T(xy)\alpha(x, y)I = \alpha(x, y)T(xy).$$

**Definition**    A projective representation $T$ of a group $G$ constructed as in Lemma 25.4 from a central extension $H$ of $G$ is said to be *lifted* to the ordinary representation $\overline{T}$ of $H$.

**Theorem 25.5**    *Let $G$ be a finite group, $k$ an algebraically closed field, $M$ the Schur multiplier of $G$. Then there is a central extension $H$ of $G$ with kernel $M$, such that every projective representation of $G$ is equivalent to one lifted to $H$.*

*Proof* The proof consists of four steps.

*Step 1*   $M = \langle\{\alpha_1\}\rangle \times \cdots \times \langle\{\alpha_d\}\rangle$, where $\{\alpha_i\}$ has order $e_i$, all $\alpha_i(x, y)$ are $e_i$th roots of 1 in $k$, and $\alpha_i(1, 1) = 1$.

*Proof of Step 1* By Theorem 25.3, $M$ is a finite abelian group, and hence a direct product of cyclic groups; also, we can assume that all $\alpha_i(x, y)$ are $e_i$th roots of 1. If $\alpha_i(1, 1)$ is not 1, then $\bar{\alpha}_i$, defined by

$$\bar{\alpha}_i(x, y) = \alpha_i(x, y)\alpha_i(1,1)^{-1},$$

is a factor set. $\bar{\alpha}_i \in \{\alpha_i\}$, so we can replace $\alpha_i$ by $\bar{\alpha}_i$.

*Step 2*   Let $\gamma_i$ be a primitive $e_i$th root of 1 in $k$, and define integers $0 \le \eta_i(x, y) < e_i$ by

$$\alpha_i(x, y) = \gamma_i^{\eta_i(x,y)}, \qquad \text{all} \quad x, y \in G.$$

If we define

$$a(x, y) = \{\alpha_1\}^{\eta_1(x,y)} \cdots \{\alpha_d\}^{\eta_d(x,y)} \in M,$$

then for all $x, y, z \in G$ we have

$$a(x, yz)a(y, z) = a(x, y)a(xy, z),$$

$$a(x, 1) = a(1, z) = 1, \qquad a(x, x^{-1}) = a(x^{-1}, x).$$

*Proof of Step 2*   Since $\alpha_i$ is a factor set,

$$\gamma_i^{\eta_i(x,yz)+\eta_i(y,z)} = \alpha_i(x, yz)\alpha_i(y, z) = \alpha_i(x, y)\alpha_i(xy, z) = \gamma_i^{\eta_i(x,y)+\eta_i(xy,z)}.$$

By Theorem 25.3, char $k \nmid e_i$, so $\gamma_i$ has order exactly $e_i$ and

$$\eta_i(x, yz) + \eta_i(y, z) \equiv \eta_i(x, y) + \eta_i(xy, z) \quad (\text{mod } e_i).$$

Therefore

$$a(x, yz)a(y, z) = \prod_i \{\alpha_{ij}\}^{\eta_i(x,yz)+\eta_i(y,z)}$$

$$= \prod_i \{\alpha_{ij}\}^{\eta_i(x,y)+\eta_i(xy,z)}$$

$$= a(x, y)a(xy, z).$$

$1 = \alpha_i(1, 1) = \gamma_i^{\eta_i(1,1)}$, so all $\eta_i(1, 1) = 0$, and we have

$$a(1, 1) = \prod_i \{\alpha_{ij}\}^{\eta_i(1,1)} = 1.$$

Setting $y = z = 1$, we have

$$a(x, 1)a(1, 1) = a(x, 1)a(x, 1), \qquad \text{so} \quad a(x, 1) = a(1, 1) = 1.$$

Setting $x = y = 1$, we have

$$a(1, z)a(1, z) = a(1, 1)a(1, z), \qquad \text{so} \quad a(1, z) = a(1, 1) = 1.$$

Setting $y = x^{-1}$, $z = x$, we have

$$a(x, 1)a(x^{-1}, x) = a(x, x^{-1})a(1, x), \qquad \text{so} \quad a(x^{-1}, x) = a(x, x^{-1}).$$

*Step 3* If we define $H = \{(x, m)|x \in G, m \in M\}$ with multiplication

$$(x, m)(y, m') = (xy, a(x, y)mm'),$$

then:

(1) $H$ is a group with identity $(1, 1)$, where

$$(x, m)^{-1} = (x^{-1}, m^{-1}a(x, x^{-1})^{-1}).$$

(2) $\{(1, m)|m \in M\}$ is a subgroup of $H$ contained in $Z(H)$, and is isomorphic to $M$ via the correspondence

$$m \leftrightarrow (1, m).$$

We identify $m \in M$ with $(1, m) \in H$, so $M \subseteq Z(H)$.

(3) The coset representatives $h_x = (x, 1)$ of $M$ in $H$ satisfy $h_x h_y = h_{xy}a(x, y)$, so $\varphi: x \to h_x M$ is an isomorphism $G \cong H/M$.

*Proof of Step 3* All is trivial, using the information in Step 2.

*Step 4* Let $T_0: G \to GL(V)$ be any projective representation of $G$, with factor set $\alpha_0$.

$\{\alpha_0\} \in M$ so $\{\alpha_0\} = \{\alpha_1\}^{l_1} \cdots \{\alpha_d\}^{l_d}$, some integers $l_1, \ldots, l_d$.

There is a function $c: G \to k^\#$ with

$$\alpha_0(x, y) = \alpha_1(x, y)^{l_1} \cdots \alpha_d(x, y)^{l_d}c(x)c(y)c(xy)^{-1}, \qquad \text{all} \quad x, y \in G.$$

The equivalent projective representation $T$ defined by

$$T(x) = c(x)^{-1}T_0(x), \qquad \text{all} \quad x \in G$$

lifts to an ordinary representation $\overline{T}$ of $H$, defined by

$$\overline{T}((x, \{\alpha_1\}^{u_1} \cdots \{\alpha_d\}^{u_d})) = \gamma_1^{u_1 l_1} \cdots \gamma_d^{u_d l_d} T(x).$$

*Proof of Step 4*

$$\begin{aligned}
T(x)T(y) &= c(x)^{-1}c(y)^{-1}T_0(x)T_0(y) \\
&= c(x)^{-1}c(y)^{-1}\alpha_0(x, y)T_0(xy) \\
&= c(x)^{-1}c(y)^{-1}\alpha_1(x, y)^{l_1} \cdots \alpha_d(x, y)^{l_d}c(x)c(y)c(xy)^{-1}c(xy)T(xy) \\
&= \alpha_1(x, y)^{l_1} \cdots \alpha_d(x, y)^{l_d}T(xy),
\end{aligned}$$

so $T$ has factor set $\alpha_1^{l_1} \cdots \alpha_d^{l_d}$.

$$\overline{T}((x, \{\alpha_1\}^{u_1} \cdots \{\alpha_d\}^{u_d})) \overline{T}((y, \{\alpha_1\}^{v_1} \cdots \{\alpha_d\}^{v_d}))$$

$$= \gamma_1^{(u_1+v_1)l_1} \cdots \gamma_d^{(u_d+v_d)l_d} T(x)T(y)$$

$$= \gamma_1^{(u_1+v_1)l_1} \cdots \gamma_d^{(u_d+v_d)l_d} \alpha_1(x, y)^{l_1} \cdots \alpha_d(x, y)^{l_d} T(xy)$$

$$= \gamma_1^{(u_1+v_1+\eta_1(x,y))l_1} \cdots \gamma_d^{(u_d+v_d+\eta_d(x,y))l_d} T(xy)$$

$$= \overline{T}((xy, \{\alpha_1\}^{u_1+v_1+\eta_1(x,y)} \cdots \{\alpha_d\}^{u_d+v_d+\eta_d(x,y)}))$$

$$= \overline{T}((xy, a(x, y)\{\alpha_1\}^{u_1+v_1} \cdots \{\alpha_d\}^{u_d+v_d}))$$

$$= \overline{T}((x, \{\alpha_1\}^{u_1} \cdots \{\alpha_d\}^{u_d})(y, \{\alpha_1\}^{v_1} \cdots \{\alpha_d\}^{v_d})),$$

so $\overline{T}$ is an ordinary representation of $H$.

$$T(1)T(1) = \alpha_1(1, 1)^{l_1} \cdots \alpha_d(1, 1)^{l_d} T(1) = T(1), \qquad \text{so} \quad T(1) = I.$$

Hence

$$\overline{T}((1, a(x, y))) = \gamma_1^{\eta_1(x,y)l_1} \cdots \gamma_d^{\eta_d(x,y)l_d} T(1)$$

$$= \alpha_1(x, y)^{l_1} \cdots \alpha_d(x, y)^{l_d} I$$

and $\overline{T}((x, 1)) = T(x)$, so $T$ lifts to $\overline{T}$.

**Corollary 25.6**    *The degrees of the complex irreducible projective representations of $G$ divide $|G|$.*

    *Proof*   By Theorem 25.5, such a degree is the degree of an ordinary irreducible representation of $H$, where $G \cong H/M$ and $M \subseteq Z(H)$. Hence by Theorem 22.1, the degree divides $|H: Z(H)|$, a divisor of $|G|$.

**Lemma 25.7**    *Let $H$ be a finite group, $k$ an algebraically closed field, and let $W$ be an irreducible $kH$-module affording the representation*

$$T: H \to GL(W).$$

*Let $U$ be a finite-dimensional $k$-vector space, and assume that $f \in End_k(U \otimes_k W)$ satisfies*

$$f(1 \otimes T(h)) = (1 \otimes T(h))f, \qquad all \quad h \in H.$$

*Then $f = e \otimes 1$, some $e \in End_k(U)$.*

*Proof* Let $U$ have $k$-basis $u_1, \ldots, u_m$, so each element of $U \otimes_k W$ has a unique expression $\sum_{i=1}^{m} u_i \otimes w_i$, the $w_i \in W$. For $w \in W$, define elements $f_{ij}(w) \in W$ by

$$f(u_i \otimes w) = \sum_{j=1}^{m} u_j \otimes f_{ij}(w).$$

Clearly all $f_{ij} \in \text{End}_k(W)$. If $h \in H$, we have

$$f(1 \otimes T(h))(u_i \otimes w) = f(u_i \otimes T(h)w) = \sum_{j=1}^{m} u_j \otimes f_{ij}T(h)w,$$

$$(1 \otimes T(h))f(u_i \otimes w) = (1 \otimes T(h))(\sum_{j=1}^{m} u_j \otimes f_{ij}(w))$$

$$= \sum_{j=1}^{m} u_j \otimes T(h)f_{ij}(w).$$

By hypothesis these two expressions are equal, so

$$f_{ij}T(h) = T(h)f_{ij}, \qquad \text{all} \quad h \in H.$$

$W$ is an absolutely irreducible $kH$-module, so we have $f_{ij} \in \text{End}_{kH}(W) = k$, say

$$f_{ij} = \alpha_{ij}1_W, \qquad \alpha_{ij} \in k.$$

Define $e \in \text{End}_k(U)$ by $e(u_i) = \sum_{j=1}^{m} \alpha_{ij}u_j$. Then

$$(e \otimes 1)(u_i \otimes w) = (\sum_{j=1}^{m} \alpha_{ij}u_j) \otimes w = \sum_{j=1}^{m} u_j \otimes \alpha_{ij}w$$

$$= \sum_{j=1}^{m} u_j \otimes f_{ij}(w) = f(u_i \otimes w),$$

proving that $e \otimes 1 = f$.

**Lemma 25.8** *Let $H$ be a finite group, $k$ an algebraically closed field, and let $W$ be a $k$-vector space which is a $kH$-module under two equivalent irreducible representations $T$ and $T'$. Let $X \in GL(W)$ be such that*

$$X^{-1}T(h)X = T'(h), \qquad \text{all} \quad h \in H.$$

*Let $U$ be a finite-dimensional $k$-vector space, and assume $S \in GL(U \otimes_k W)$ is such that*

$$S^{-1}(1 \otimes T(h))S = 1 \otimes T'(h), \qquad all \quad h \in H.$$

*Then there exists* $Y \in GL(U)$ *with* $S = Y \otimes X.$

*Proof*   Clearly $(1 \otimes X)^{-1} = 1 \otimes X^{-1}$ in $GL(U \otimes_k W)$, so

$$(1 \otimes X)^{-1}(1 \otimes T(h))(1 \otimes X) = 1 \otimes T'(h), \qquad all \quad h \in H.$$

Comparing this with our hypothesis on $S$, we get

$$S(1 \otimes X)^{-1}(1 \otimes T(h))(1 \otimes X)S^{-1} = 1 \otimes T(h).$$

By Lemma 25.7,

$$S(1 \otimes X)^{-1} = Y \otimes 1, \qquad some \quad Y \in End_k(U).$$

Since $S$ and $1 \otimes X$ are invertible, so are $Y \otimes 1$ and $Y$. And $S = (Y \otimes 1)(1 \otimes X) = Y \otimes X.$

*Remark*   When $V$ is an irreducible $kG$-module and $N$ a normal subgroup of $G$, Clifford's Theorem 14.1 says that $V_N$ is a direct sum of conjugate irreducible $kN$-modules. If these are not all isomorphic, Clifford's Theorem tells us that $V$ is induced from a proper subgroup of $G$. The following theorem gives us information when the conjugate irreducible $kN$-modules are all isomorphic.

**Theorem 25.9**   *Let $k$ be an algebraically closed field, $N$ a normal subgroup of the finite group $G$. Let $V$ be an irreducible $kG$-module, $W$ an irreducible $kN$-submodule of $V_N$, and assume that $W \cong W^g$, all conjugate modules $W^g$ of $W$ for $g \in G$. Let $s = \dim_k V/\dim_k W$, and let $U$ be an s-dimensional k-vector space.*

   *Then there are irreducible projective representations*

$$Y: G \to GL(U) \qquad and \quad X: G \to GL(W)$$

*such that $S(g) = Y(g) \otimes X(g)$ defines an ordinary irreducible representation of $G$ on $U \otimes_k W$, and $U \otimes_k W \cong V$ as $kG$-modules. We have $Y(n) = 1_U$ for all $n \in N$.*

*Proof*   By Clifford's theorem, we know that for some $g_2, \ldots, g_s \in G$ we have

$$V_N = W \oplus g_2 W \oplus \cdots \oplus g_s W, \qquad all \quad g_i W \cong W^{g_i} \cong W.$$

(For the properties of the module $W^g$, $g \in G$, see Lemma 21.4.) For any $g \in G$, $n \in N$, Lemma 21.4 shows that $W^g$ affords a representation $T^g$ satisfying $T^g(gng^{-1}) = T(n)$, so $T^g(n) = T(g^{-1}ng)$; $W^g$ here has the same underlying $k$-vector space as $W$. Since $W^g \cong W$, there is an $X(g) \in GL(W)$ with

$$T^g(n) = X(g)^{-1}T(n)X(g), \qquad \text{all} \quad n \in N;$$

for $n_1 \in N$ we simply take $X(n_1) = T(n_1)$.

We shall show that $X: G \to GL(W)$ is a projective representation of $G$. We have, for $g, h \in G$, $n \in N$,

$$\begin{aligned}
X(gh)^{-1}T(n)X(gh) = T^{gh}(n) &= T(h^{-1}g^{-1}ngh) = T^h(g^{-1}ng) \\
&= X(h)^{-1}T(g^{-1}ng)X(h) = X(h)^{-1}T^g(n)X(h) \\
&= X(h)^{-1}X(g)^{-1}T(n)X(g)X(h);
\end{aligned}$$

so

$$X(g)X(h)X(gh)^{-1}T(n) = T(n)X(g)X(h)X(gh)^{-1}, \qquad \text{all} \quad n \in N.$$

$W$ is an absolutely irreducible $kN$-module, so this shows

$$X(g)X(h)X(gh)^{-1} \in \text{End}_{kN}(W) = k,$$

say

$$X(g)X(h) = \alpha(g, h)X(gh), \qquad \alpha(g, h) \in k.$$

We have proved that $X$ is a projective representation of $G$ with factor set $\alpha$.

Let $U$ have $k$-basis $u_1, \ldots, u_s$ and $W$ $k$-basis $w_1, \ldots, w_m$, so $U \otimes_k W$ has $k$-basis $\{u_i \otimes w_j\}$. For $i > 1$, we have $g_i W \cong W$. Therefore, we may choose a basis $w_{i1}, \ldots, w_{im}$ of $g_i W$ such that the map $f_i: w_j \to w_{ij}$ induces a $kN$-isomorphism $W \to g_i W$. If, for example $nw_j = \sum_{l=1}^m \gamma_{lj}w_l$, then we have

$$nw_{ij} = nf_i(w_j) = f_i(nw_j)$$
$$= f_i(\sum_l \gamma_{lj}w_l) = \sum_l \gamma_{lj}f_i(w_l) = \sum_l \gamma_{lj}w_{il}.$$

Denote $w_{1j} = w_j$, so $\{w_{ij}\}$ is a basis of $V$.

We easily make $U \otimes_k W$ a $kG$-module by defining

$$g(u_i \otimes w_j) = \sum_{r,t} \beta_{ijrt}(u_r \otimes w_t) \qquad \text{whenever} \quad gw_{ij} = \sum_{r,t} \beta_{ijrt}w_{rt}.$$

This automatically makes $w_{ij} \leftrightarrow u_i \otimes w_j$ into a $kG$-isomorphism $V \cong U \otimes_k W$. If $n \in N$, say $nw_j = \sum_l \gamma_{lj}w_l$, then $nw_{ij} = \sum_l \gamma_{lj}w_{il}$, so

$$n(u_i \otimes w_j) = \sum_l \gamma_{lj}(u_i \otimes w_l) = u_i \otimes (\sum_l \gamma_{lj}w_l) = u_i \otimes nw_j.$$

Let $S$ be the representation of $G$ afforded by $U \otimes_k W$, so $S(g)x = gx$, all $x \in U \otimes_k W$. In particular, if $n \in N$ the last equation of the previous paragraph shows

$$S(n)(u_i \otimes w_j) = u_i \otimes nw_j = u_i \otimes T(n)w_j = (1 \otimes T(n))(u_i \otimes w_j),$$

so

$$S(n) = 1 \otimes T(n), \qquad \text{all} \quad n \in N.$$

For any $g \in G$ and $n \in N$ we have

$$1 \otimes T^g(n) = 1 \otimes T(g^{-1}ng) = S(g^{-1}ng) = S(g)^{-1}S(n)S(g)$$
$$= S(g)^{-1}(1 \otimes T(n))S(g).$$

By Lemma 25.8, this means that

$$S(g) = Y(g) \otimes X(g), \qquad \text{some} \quad Y(g) \in GL(U).$$

To see that $Y: G \to GL(U)$ is a projective representation, we have, for any $g, h \in G$,

$$Y(gh) \otimes X(gh) = S(gh) = S(g)S(h) = (Y(g) \otimes X(g))(Y(h) \otimes X(h))$$
$$= Y(g)Y(h) \otimes \alpha(g, h)X(gh) = \alpha(g, h)Y(g)Y(h) \otimes X(gh).$$

This implies that

$$Y(gh) = \alpha(g, h)Y(g)Y(h) \qquad \text{or} \qquad Y(g)Y(h) = \alpha(g, h)^{-1}Y(gh),$$

so $Y$ is projective with factor set $\alpha^{-1}$.

If $Y$ or $X$ were reducible, $S$ would be also; but it is not, so $Y$ and $X$ are irreducible. If $n \in N$, we have

$$1 \otimes T(n) = S(n) = Y(n) \otimes X(n) = Y(n) \otimes T(n),$$

so $Y(n) = 1_U$.

*Second Proof of Itô's Theorem 22.3*   We use induction on $|G|$. By Clifford's theorem, $\chi|_H = \sum_{i=1}^{s} \lambda_i$, the $\lambda_i$ $G$-conjugate linear characters of $H$. If the $\lambda_i$ are not all the same, then Clifford's theorem shows that $\chi = \mu^G$, $\mu$ an irreducible complex character of a proper subgroup $K$ of $G$, $H \subseteq K \neq G$. By induction $\mu(1)\big|\,|K:H|$, so $\chi(1) = |G:K|\mu(1)$ divides $|G:H|$. If the $\lambda_i$ are all the same, then by Theorem 25.9, $s = \chi(1)$ is the degree of an irreducible projective representation of $G/H$. By Corollary 25.6, $s\big|\,|G/H|$, and we are done.

*Remarks*   The theory of projective representations originates with Schur [1, 2]. Recent papers on the subject include Iwahori and Matsumoto [1], Yamazaki [1, 2], Reynolds [1], Mangold [1], Alperin and Kuo [1], Pahlings [1, 2], DeMeyer and Janusz [1], and Fein [1].

EXERCISES

**1**   Prove that Maschke's theorem holds for projective representations of finite groups.

**2**   Prove that if $N \lhd G$, and $V$ is an irreducible $\mathbf{C}G$-module such that $V_N$ is the direct summand of $s$ irreducible $\mathbf{C}N$-modules, then $s$ divides $|G:N|$.

# §26

## The Finite Two-Dimensional Linear Groups

As before, $\mathbf{C}$ denotes the complex numbers and $S_n$ the symmetric permutation group on $n$ letters. $A_n$, the *alternating group* of degree $n$, is the subgroup of $S_n$ of index 2 consisting of even permutations. Thus $|A_4| = 12$, $|S_4| = 24$, $|A_5| = 60$; $A_n$ is known to be simple for $n \geq 5$.

**Theorem 26.1**     *Let $G$ be a finite nonabelian subgroup of $GL(2, \mathbf{C})$. Then one of the following holds.*

(1)   *$G$ has a normal abelian subgroup of index 2.*

(2)   *$G/Z(G) \cong A_4$, $S_4$ or $A_5$, and $Z(G)$ consists of scalar matrices.*

*Proof, Step 1*   Let $N = G \cap \{\text{scalar matrices}\}$. Then for any $x \in G - N$, $C_G(x)$ is abelian. $N = Z(G)$.

*Proof of Step 1*   The finite group $\langle x \rangle$ is completely reducible, so there is a $w \in GL(2, \mathbf{C})$ with

$$x^w = w^{-1}xw = \begin{pmatrix} \alpha & 0 \\ 0 & \beta \end{pmatrix}, \qquad \text{some} \quad \alpha \neq \beta \text{ in } \mathbf{C}.$$

$C_G(x)^w \subseteq C_{GL(2,\mathbf{C})}(x^w)$, so it is enough to show the group of all matrices centralizing $x^w$ is abelian. But any matrix centralizing $\begin{pmatrix} \alpha & 0 \\ 0 & \beta \end{pmatrix}$ is diagonal, and the diagonal matrices form an abelian group.

In particular, $x \notin Z(G)$, so $N = Z(G)$.

*Step 2*   If $x, y \in G - N$, then $C_G(x) \cap C_G(y) = N$ or $C_G(x) = C_G(y)$. Each $C_G(x)$ is a maximal abelian subgroup of $G$.

*Proof*   If $C_G(x) \cap C_G(y) \neq N$, choose $z$ in $C_G(x) \cap C_G(y) - N$.

144

Then $x, y \in C_G(z)$ and $C_G(z)$ is abelian, so $C_G(x) = C_G(z) = C_G(y)$. Clearly no $C_G(x)$ can be in a larger abelian subgroup of $G$.

*Step 3* If $x \in G - N$ and $C = C_G(x)$, then $|N_G(C): C| \leq 2$.

*Proof* Choose a matrix $w$ with $w^{-1}xw$ diagonal. Then $C^w = C_{G^w}(x^w)$ consists of diagonal matrices, and $N_{G^w}(C^w) = N_G(C)^w$. Replacing $x$ by $x^w$ and $G$ by $G^w$, we may assume $C$ consists of diagonal matrices.

Let $D$ be the group of diagonal matrices in $GL(2, \mathbf{C})$, and let $x = \begin{pmatrix} \alpha & 0 \\ 0 & \beta \end{pmatrix}$, $\alpha \neq \beta$ in $\mathbf{C}$. We see that for any $\begin{pmatrix} a & b \\ c & d \end{pmatrix} \in GL(2, \mathbf{C})$,

$$\begin{pmatrix} a & b \\ c & d \end{pmatrix}^{-1} \begin{pmatrix} \alpha & 0 \\ 0 & \beta \end{pmatrix} \begin{pmatrix} a & b \\ c & d \end{pmatrix} = \frac{1}{ad - bc} \begin{pmatrix} ad\alpha - bc\beta & bd(\alpha - \beta) \\ ac(\beta - \alpha) & ad\beta - bc\alpha \end{pmatrix}$$

is diagonal only if $bd = 0$, $ac = 0$. Hence $\begin{pmatrix} a & b \\ c & d \end{pmatrix}$ has form $\begin{pmatrix} a & 0 \\ 0 & d \end{pmatrix}$ or $\begin{pmatrix} 0 & b \\ c & 0 \end{pmatrix}$, and

$$N_0 = \{y \in GL(2, \mathbf{C}) | y^{-1}xy \in D\}$$

is a group satisfying $|N_0: D| = 2$.

$N_G(C) \subseteq G \cap N_0$, so we have

$$|N_G(C): C| = |N_G(C): G \cap D| \leq |G \cap N_0: G \cap D|$$
$$= |G \cap N_0: (G \cap N_0) \cap D|$$
$$= |(G \cap N_0)D: D| \leq |N_0: D| = 2.$$

*Step 4* $G$ has nonconjugate maximal abelian subgroups $H_1, \ldots,$ $H_r$ and $K_1, \ldots, K_s$ with $N_G(H_i) = H_i$, $|N_G(K_j): K_j| = 2$, such that $G$ is the disjoint union

$$G = N \cup \bigcup_{i=1}^{r} \{H_i^g - N | g \in G\} \cup \bigcup_{j=1}^{s} \{K_j^g - N | g \in G\}.$$

If $g = |G: N|$, $h_i = |H_i: N|$, $k_j = |K_j: N|$, then

$$1 = \frac{1}{g} + \sum_{i=1}^{r} \left(1 - \frac{1}{h_i}\right) + \sum_{j=1}^{s} \frac{1}{2}\left(1 - \frac{1}{k_j}\right).$$

In this equation, $r = 0$ or $r = 1$.

*Proof* Existence of the $H_i$ and $K_j$ follows from Steps 2 and 3. If $n = |N|$, then our disjoint union shows

$$gn = n + \sum_{i=1}^{r} \frac{g}{h_i}(nh_i - n) + \sum_{j=1}^{s} \frac{g}{2k_j}(nk_j - n).$$

Dividing by $gn$, we get the equation of Step 4. Each term in the equation is positive, and each $1 - 1/h_i$ is at least $\frac{1}{2}$, so $r \le 1$.

*Step 5* If $r = 1$, then $s = 1$. Either $G$ has a normal abelian subgroup of index 2, or $G/N \cong A_4$.

*Proof* If $r = 1$, then

$$\sum_{j=1}^{s} \frac{1}{2}\left(1 - \frac{1}{k_j}\right) < \frac{1}{2}.$$

Each term $\frac{1}{2}(1 - 1/k_j)$ is at least $\frac{1}{4}$, so $s \le 1$. If $s = 0$, the equation reduces to $g = h_1$, meaning $G$ is abelian, a contradiction.

If $s = 1$, the equation reduces to

$$\frac{1}{h_1} + \frac{1}{2k_1} = \frac{1}{2} + \frac{1}{g}.$$

$1/2k_1 \le 1/4$ so $1/h_1 > 1/4$, meaning $h_1$ is 2 or 3. $h_1 = 2$ forces $g = 2k_1$, so $G$ has a normal abelian subgroup $K_1$ of index 2. On the other hand, $h_1 = 3$ forces $1/2k_1 > 1/6$ so $k_1 = 2$, and then $g = 12$. $G/N$ then has four 3-Sylow subgroups (the conjugates of $H_1/N$), $|G/N| = 12$. $G/N$ is faithfully represented by conjugation as a permutation group on its four 3-Sylow subgroups, so $G/N$ is a subgroup of $S_4$ of order 12 and must be $A_4$.

*Step 6* If $r = 0$, then $s = 3$.

*Proof* If $r = 0$, the equation is

$$1 = \frac{1}{g} + \sum_{j=1}^{s} \frac{1}{2}\left(1 - \frac{1}{k_j}\right).$$

Each term $\frac{1}{2}(1 - 1/k_j)$ is at least $\frac{1}{4}$, so $s \le 3$.

$s = 0$ implies $g = 1$, a contradiction to the hypothesis $G$ nonabelian. $s = 1$ means $1/2k_1 + 1/2 = 1/g$ so $g < 2$, a contradiction.

If $s = 2$, the equation reduces to $1/2k_1 + 1/2k_2 = 1/g$. Choosing the notation so $k_1 \leqq k_2$, we have

$$g = \frac{2k_1k_2}{k_1 + k_2} = 2\frac{k_1k_2}{k_1(1 + k_2/k_1)} \leq k_2,$$

forcing $G$ to be abelian, a contradiction.

*Step 7*  If $r = 0$ and $s = 3$, choose the notation so $k_1 \leq k_2 \leq k_3$. Then $k_1 = 2$ and $k_2 \in \{2, 3\}$.

*Proof*  The equation becomes

$$\frac{1}{2k_1} + \frac{1}{2k_2} + \frac{1}{2k_3} = \frac{1}{g} + \frac{1}{2}.$$

Therefore $1/2k_1 > 1/6$, forcing $k_1 = 2$, and

$$\frac{1}{2k_2} + \frac{1}{2k_3} = \frac{1}{g} + \frac{1}{4}.$$

Hence $1/2k_2 > 1/8$, so $k_2 \in \{2, 3\}$.

*Step 8*  If $k_2 = 2$ in Step 7, then $G$ has a normal abelian subgroup of index 2.

*Proof*  We get $1/2k_3 = 1/g$, $g = 2k_3$.

*Step 9*  If $k_2 = 3$, then $k_3$ is 4 or 5. If $k_3$ is 4 then $G/N \cong S_4$, and if $k_3$ is 5 then $G/N \cong A_5$.

*Proof*  The equation is $1/2k_3 = 1/g + 1/12$, so $1/2k_3 > 1/12$. $k_3 \geq k_2$, so $3 \leq k_3 \leq 5$.

$k_3 = 3$ means $g = 12$. Then $|G/N| = 12$ and $G/N$ has a 3-Sylow subgroup with normalizer of order 6, a contradiction to Sylow's theorem.

$k_3 = 4$ means $g = 24$, $|G/N| = 24$. $k_2 = 3$, so $G/N$ has four 3-Sylow subgroups. Consider $G/N$ as a permutation group on them acting by conjugation. Any one of them permutes the other three, so the kernel of this permutation representation has order at most 2. If it had order 2, $G/N$ would have center of order 2 and hence have elements of order 6, a contradiction to the fact that $G/N$ is a union of subgroups of orders 2, 3, and 4. So $G/N \cong S_4$.

$k_3 = 5$ means $g = 60$, $|G/N| = 60$. $G/N$ is a union of three conjugacy classes of subgroups, of orders 2, 3, and 5, all of index 2 in their normalizers.

If $G/N$ is solvable, it has a nontrivial abelian normal subgroup, and hence a normal $p$-subgroup $P/N$. That subgroup cannot have order 2, 3, or 5, so $|P/N| = 4$. Then $5 \nmid |\mathrm{Aut}(P/N)|$ so a 5-element centralizes $P/N$, meaning that $G/N$ has an element of order 10, a contradiction. So $G/N$ is not solvable.

We just saw that $G/N$ has more than one 2-Sylow subgroup $P/N$. Hence by Sylow's theorem, $G/N$ has 3, 5, or 15 2-Sylow subgroups. By Theorem 18.7 it cannot have 15 2-Sylow subgroups. If it has 3 2-Sylow subgroups then $|N_{G/N}(P/N)| = 20$ so 5-elements centralize 2-elements, a contradiction. We conclude that $|N_{G/N}(P/N)| = 12$.

Let $P_1/N$ be a second 2-Sylow subgroup. If there were an $x \in N_{G/N}(P/N) \cap N_{G/N}(P_1/N)$ of order 2, it would be in $P/N \cap P_1/N$ and have centralizer of order at least 8, a contradiction to $|N_{G/N}(K_1/N)| = 4$. Therefore

$$|N_{G/N}(P/N) \cap N_{G/N}(P_1/N)| = 3;$$

since $G/N$ has no normal subgroup of order 3, this means

$$\bigcap_{y \in G/N} N_{G/N}(P/N)^y = 1.$$

Therefore $G/N$ is faithfully represented as a permutation group on its five 2-Sylow subgroups. Thus $G/N$ is a subgroup of $S_5$ of index 2, $G/N \cong A_5$.

*Remarks*  The cases $A_4, S_4, A_5$ in (2) of Theorem 26.1 all actually occur. Since $A_4$, $S_4$, and $A_5$ do not have faithful complex irreducible representations of degree 2, this means that $Z(G) \neq 1$ in each occurrence of (2). By Lemma 25.4, we see that $A_4, S_4, A_5$ all do have faithful complex irreducible projective representations of degree 2. Theorems analogous to 26.1 for $GL(3, \mathbf{C})$ and $GL(4, \mathbf{C})$ were proved by Jordan [1] and Blichfeldt [1]. Finite subgroups of $GL(n, \mathbf{C})$, $5 \leq n \leq 7$, have been recently determined by Brauer and his students (see Brauer [10] and Wales [1, 2, 3]). The theorem of Jordan in §30 shows that such a classification could theoretically be carried out for any $GL(n, \mathbf{C})$. Several of the recently discovered finite simple groups have projective representations of degree $\leq 7$.

# §27

## Special Conjugacy Classes

**Lemma 27.1**    *Let $K$ be a finite Galois extension of $\mathbf{Q}$, let $V = \{(a_1, \ldots, a_s) \,|\, \text{all } a_i \in K\}$, and let $W$ be a $K$-subspace of $V$ which is invariant under all automorphisms $\sigma$ in the Galois group $\mathrm{Gal}(K/\mathbf{Q})$ (where $\sigma: (a_1, \ldots, a_s) \to (\sigma(a_1), \ldots, \sigma(a_s)))$. Then $W$ has a basis in $\{(r_1, \ldots, r_s) \,|\, r_i \in \mathbf{Q}\}$.*

*Proof*  By induction on $s$, true for $s = 1$ since then $W$ is $\{0\}$ or $V$. Assume the result true for $s - 1$, and let $V_0 = \{(a_1, \ldots, a_{s-1}, 0) \,|\, a_i \in K\}$, an $(s - 1)$-dimensional $K$-subspace of $V$. By induction, $W_0 = W \cap V_0$ has a basis

$$b_1 = (r_{11}, \ldots, r_{1,s-1}, 0), \ldots, b_k = (r_{k1}, \ldots, r_{k,s-1}, 0), \qquad \text{all } r_{ij} \in \mathbf{Q},$$

so we are done if $W_0 = W$. If not, choose some

$$\alpha = (c_1, \ldots, c_{s-1}, c_s) \in W - W_0, \qquad c_s \neq 0.$$

Setting $c_i' = c_i c_s^{-1}$, we have

$$\alpha' = (c_1', \ldots, c_{s-1}', 1) \in W.$$

If $\mathrm{Gal}(K/\mathbf{Q}) = \{1, \sigma_2, \ldots, \sigma_n\}$, then set

$$b_{k+1} = \alpha' + \sigma_2(\alpha') + \cdots + \sigma_n(\alpha') = (r_1, \ldots, r_{s-1}, n),$$

where all $r_i$ are rational. $\{b_1, \ldots, b_{k+1}\}$ is the desired basis of $W$.

*Remark*  The following lemma, due to M. Suzuki [3] and G. Higman [1], has been used in several recent characterizations of simple groups.

**Lemma 27.2**    *Let $H$ be a subgroup of $G$, and assume we have elements $h_1, \ldots, h_n$ of $H$ such that:*

(i)   *For all $1 \leq i \leq n$, $C_G(h_i) \subseteq H$.*

(ii)   *If $i \neq j$, then $h_i$ and $h_j$ are not conjugate in $G$.*

(iii)   *If $\mathscr{D}_i$ denotes the conjugacy class of $h_i$ in $H$, and $h \in H$ satisfies $\langle h \rangle = \langle h_i \rangle$, then $h \in \mathscr{D}_j$ for some $j$. $\mathscr{D}_1, \ldots, \mathscr{D}_n$ are called the* special classes *of $H$ in $G$. Then:*

*(1)   $H$ has $n$ linearly independent generalized complex characters $\theta_1, \ldots, \theta_n$ which vanish on $H - \bigcup_{i=1}^{n} \mathscr{D}_i$.*

*(2)   $D = \bigcup_{i=1}^{n} \mathscr{D}_i$ is a T. I. set in $G$ with normalizer $H$.*

*(3)   If $1_G = \chi_1, \ldots, \chi_r$ are the irreducible complex characters of $G$ and $1_H = \phi_1, \ldots, \phi_s$ the irreducible complex characters of $H$, let*

$$\theta_i = \sum_{j=1}^{s} a_{ij}\phi_j, \qquad \theta_i^G = \sum_{j=1}^{r} b_{ij}\chi_j.$$

*Then we have*

$$b_{i1} = a_{i1}, \qquad 1 \leq i \leq n$$

*and*

$$\sum_{k=1}^{r} b_{ik}b_{jk} = \sum_{k=1}^{s} a_{ik}a_{jk}, \qquad 1 \leq i, j \leq n.$$

*(4)   There are uniquely determined numbers $c_{jk}$ with*

$$\phi_i(h_j) = \sum_{k=1}^{n} c_{jk}a_{ki},$$

*and these numbers $c_{jk}$ satisfy*

$$\chi_i(h_j) = \sum_{k=1}^{n} c_{jk}b_{ki}.$$

*(5)   The equations in (3) determine, for each $k$, the integers $\{b_{1k}, \ldots, b_{nk}\}$ up to finitely many possibilities. There are only finitely many possibilities for the $n$ columns of the character table of $G$ corresponding to $h_1, \ldots, h_n$. (Except that we do not know how many irreducible characters of $G$ vanish on all of $h_1, \ldots, h_n$.)*

*Proof of (1)*   Let $\varepsilon$ be a primitive $|H|$th root of 1, $F = \mathbf{Q}(\varepsilon)$, and let $V$ be the $F$-vector space $\{(a_1, \ldots, a_s) \,|\, a_i \in F\}$. Let $W = \{(a_1, \ldots, a_s) \in V \,|\, a_1\phi_1 + \cdots + a_s\phi_s$ vanishes on $H - D\}$, so $W$ is an $n$-dimensional $F$-subspace of $V$. We wish to apply Lemma 27.1. Choose any $h \in H - D$, any $\sigma \in \mathrm{Gal}(F/\mathbf{Q})$, and any $(a_1, \ldots, a_s) \in W$. Then $\sigma: \varepsilon \to \varepsilon^k$, some integer $k$ with $(|H|, k) = 1$. Choose $h' \in \langle h \rangle$ with $(h')^k = h$; by (iii), $h' \in H - D$, and we have

$$0 = \sigma(0) = \sigma[a_1\phi_1(h') + \cdots + a_s\phi_s(h')]$$
$$= \sigma(a_1)\phi_1(h) + \cdots + \sigma(a_s)\phi_s(h),$$

so $(\sigma(a_1), \ldots, \sigma(a_s)) \in W$. By Lemma 27.1, $W$ has a basis

$$\{(r_{11}, \ldots, r_{1s}), \ldots, (r_{n1}, \ldots, r_{ns})\}, \qquad \text{all} \quad r_{ij} \in \mathbf{Q}.$$

The $n$ class functions

$$r_{11}\phi_1 + \cdots + r_{1s}\phi_s, \ldots, r_{n1}\phi_1 + \cdots + r_{ns}\phi_s$$

are linearly independent and vanish on $H - D$, so multiplying by the common denominator of all $r_{ij}$ we have the generalized characters needed for (1).

*Proof of (2)*   Suppose $x \in G$ satisfies $x^{-1}Dx \cap D \neq \phi$, say $x^{-1}dx \in D$. By (ii), $x^{-1}dx = h^{-1}dh$, some $h \in H$, so $hx^{-1} \in C_G(d)$. By (i), $hx^{-1} \in H$ so $x \in H$.

*Proof of (3)*   By Frobenius reciprocity,

$$b_{i1} = (\theta_i^G, 1_G)_G = (\theta_i, 1_H)_H = a_{i1}.$$

By Theorem 12.1, $(\theta_i, \theta_j)_H = (\theta_i^G, \theta_j^G)_G$. We see that

$$(\theta_i, \theta_j)_H = \left(\sum_{k=1}^{s} a_{ik}\phi_k, \sum_{t=1}^{s} a_{jt}\phi_t\right)_H = \sum_{k=1}^{s}\sum_{t=1}^{s} a_{ik}a_{jt}\delta_{kt} = \sum_{k=1}^{s} a_{ik}a_{jk},$$

$$(\theta_i^G, \theta_j^G)_G = \left(\sum_{k=1}^{r} b_{ik}\chi_k, \sum_{t=1}^{r} b_{jt}\chi_t\right)_G = \sum_{k=1}^{r}\sum_{t=1}^{r} b_{ik}b_{jt}\delta_{kt} = \sum_{k=1}^{r} b_{ik}b_{jk}.$$

*Proof of (4)*   Define $\mu_j = \sum_{i=1}^{s} \phi_i(h_j)\phi_i$. If $h \in H - D$, then $h^{-1} \in H - D$, and by the orthogonality relations we have

$$\mu_j(h) = \sum_{i=1}^{s} \phi_i(h_j)\phi_i(h) = \sum_{i=1}^{s} \phi_i(h_j)\phi_i((h^{-1})^{-1}) = 0.$$

Hence $\mu_j$ vanishes on $H - D$, and by (1) there are uniquely determined constants $c_{jk}$ with $\mu_j = \sum_{k=1}^{n} c_{jk}\theta_k$. Therefore

$$\sum_{i=1}^{s} \phi_i(h_j)\phi_i = \mu_j = \sum_{k=1}^{n} c_{jk}\theta_k = \sum_{k=1}^{n} c_{jk} \sum_{i=1}^{s} a_{ki}\phi_i,$$

implying that $\phi_i(h_j) = \sum_{k=1}^{n} c_{jk}a_{ki}$. If also $\phi_i(h_j) = \sum_{k=1}^{n} d_{jk}a_{ki}$, then

$$\mu_j = \sum_{i=1}^{s} \sum_{k=1}^{n} d_{jk}a_{ki}\phi_i = \sum_{k=1}^{n} d_{jk}\theta_k,$$

so $d_{jk} = c_{jk}$.

We now set out to show that $\chi_i(h_j) = \sum_{k=1}^{n} c_{jk}b_{ki}$. By Theorem 12.1 (1) we have, for any $\theta_i$ and $h_k$,

$$\sum_{j=1}^{r} b_{ij}\chi_j(h_k) = \theta_i^G(h_k) = \theta_i(\tilde{h_k}) = \sum_{j=1}^{s} a_{ij}\phi_j(h_k).$$

By the orthogonality relations, we have

$$\sum_{i=1}^{r} \chi_i(h_j)\chi_i(h_k) = |C_G(h_j)| \qquad \text{if} \quad h_k^{-1} \in \mathscr{D}_j$$

$$= \quad 0 \qquad \text{otherwise};$$

$$\sum_{i=1}^{s} \phi_i(h_j)\phi_i(h_k) = |C_H(h_j)| \qquad \text{if} \quad h_k^{-1} \in \mathscr{D}_j$$

$$= \quad 0 \qquad \text{otherwise}.$$

Since $C_G(h_j) = C_H(h_j)$ by (i), we have two sets of equations for the unknowns $\chi_i(h_j)$:

(a)  $\sum_{j=1}^{r} b_{ij}\chi_j(h_k) = \sum_{j=1}^{s} a_{ij}\phi_j(h_k).$

(b)  $\sum_{i=1}^{r} \chi_i(h_j)\chi_i(h_k) = \sum_{i=1}^{s} \phi_i(h_j)\phi_i(h_k).$

We first show that $\chi_i(h_j) = \sum_{k=1}^{n} c_{jk}b_{ki}$ is a solution to (a) and (b). Using (3), we have

$$\sum_{j=1}^{r} b_{ij}\left( \sum_{l=1}^{n} c_{kl}b_{lj} \right) = \sum_{l=1}^{n} c_{kl}\left( \sum_{j=1}^{r} b_{ij}b_{lj} \right)$$

$$= \sum_{l=1}^{n} c_{kl} \sum_{j=1}^{s} a_{ij}a_{lj} = \sum_{j=1}^{s} a_{ij}\phi_j(h_k),$$

$$\sum_{i=1}^{r} (\sum_{l=1}^{n} c_{jl}b_{li})(\sum_{m=1}^{n} c_{km}b_{mi}) = \sum_{i=1}^{s} \sum_{l=1}^{n} c_{jl}a_{li} \sum_{m=1}^{n} c_{km}a_{mi}$$

$$= \sum_{i=1}^{s} \phi_i(h_j)\phi_i(h_k).$$

We next show that $\chi_i(h_j) = \sum_{k=1}^{n} c_{jk}b_{ki}$ is the actual value $\chi_i(h_j)$. If in general

$$\chi_i(h_j) = \sum_{k=1}^{n} c_{jk}b_{ki} + \varepsilon_{ij},$$

we will show all $\varepsilon_{ij} = 0$. We have

$$\sum_{j=1}^{r} b_{ij}(\sum_{l=1}^{n} c_{kl}b_{lj} + \varepsilon_{jk}) = \sum_{j=1}^{s} a_{ij}\phi_j(h_k) = \sum_{j=1}^{r} b_{ij}(\sum_{l=1}^{n} c_{kl}b_{lj}),$$

proving that $\sum_{j=1}^{r} b_{ij}\varepsilon_{jk} = 0$. Also,

$$\sum_{i=1}^{r} (\sum_{l=1}^{n} c_{jl}b_{li} + \varepsilon_{ij})(\sum_{m=1}^{n} c_{km}b_{mi} + \varepsilon_{ik}) = \sum_{i=1}^{s} \phi_i(h_j)\phi_i(h_k)$$

$$= \sum_{i=1}^{r} (\sum_{l=1}^{n} c_{jl}b_{li})(\sum_{m=1}^{n} c_{km}b_{mi}),$$

so

$$\sum_{i=1}^{r} (\sum_{l=1}^{n} c_{jl}b_{li}\varepsilon_{ik}) + \sum_{i=1}^{r} (\sum_{m=1}^{n} c_{km}b_{mi}\varepsilon_{ij}) + \sum_{i=1}^{r} \varepsilon_{ij}\varepsilon_{ik} = 0.$$

Using the fact $\sum_{j=1}^{r} b_{ij}\varepsilon_{jk} = 0$ for any $i$ and $k$, this means

$$\sum_{i=1}^{r} \varepsilon_{ij}\varepsilon_{ik} = 0.$$

Fix $j$, and choose $k$ so $h_k$ is conjugate to $h_j^{-1}$. Then

$$\sum_{l=1}^{n} c_{kl}a_{li} = \phi_i(h_k) = \overline{\phi_i(h_j)} = \sum_{l=1}^{n} \overline{c_{jl}}a_{li},$$

so by uniqueness of the $c_{jk}$'s, $c_{kl} = \overline{c_{jl}}$. Therefore

$$\sum_{l=1}^{n} c_{kl}b_{li} + \varepsilon_{ik} = \chi_i(h_k) = \overline{\chi_i(h_j)} = \overline{\sum_{l=1}^{n} c_{jl}b_{li} + \varepsilon_{ij}}$$

$$= \sum_{l=1}^{n} \overline{c_{jl}}b_{li} + \overline{\varepsilon_{ij}} = \sum_{l=1}^{n} c_{kl}b_{li} + \overline{\varepsilon_{ij}}.$$

We have shown $\varepsilon_{ik} = \overline{\varepsilon_{ij}}$, so

$$0 = \sum_{i=1}^{r} \varepsilon_{ij}\varepsilon_{ik} = \sum_{i=1}^{r} |\varepsilon_{ij}|^2,$$

and all $\varepsilon_{ij} = 0$. This completes the proof of (4).

*Proof of* (5)   The $b_{ik}$ are integers and (3) determines $\sum_{k=1}^{r} b_{ik}^2$. By (4), the $b_{ik}$ determine $\chi_i(h_j)$.

EXERCISE

Let $A$ be a subgroup of the finite group $G$, and assume that for all $x \in A - \{1\}$, $C_G(x) = A$.

(a)   Show that $A$ is a T. I. set in $G$.

(b)   Let $H = N_G(A)$, $n = |H: A|$. Show that $A - \{1\}$ contains $(|A| - 1)/n$ conjugacy classes of $H$, and these classes are special in the sense of Lemma 27.2.

# §28

## A Characterization Via Centralizers of Involutions

**Definition**    If $F$ is a field, $GL(n, F)$ is the group of all nonsingular $n \times n$ matrices over $F$. If $|F| = q < \infty$, we denote $GL(n, F) = GL(n, q)$. We denote

$$SL(n, F) = \{M \in GL(n, F) | \det M = 1\},$$

$$PSL(n, F) = SL(n, F)/Z(SL(n, F)).$$

Our object in this section is to prove

**Theorem 28.1**    *If the finite simple group $G$ has an involution $t$ with $C_G(t) \cong GL(2, 3)$, then $|G| = 5616$ or $|G| = 7920$.*

*Remarks*    $PSL(3, 3)$ is a simple group of order 5616, and the simple Mathieu permutation group on 11 symbols has order 7920; these both have involutions with centralizers isomorphic to $GL(2, 3)$. Theorem 28.1 may be thought of as a characterization of these two groups.

Such characterizations of known simple groups are basic tools in attempts to find or classify new simple groups. Because of the theorem of Feit and Thompson [4], noncyclic simple groups contain involutions, and their centralizers form convenient proper subgroups with which to work. Many characterizations via centralizers of involutions have been achieved recently; they are surveyed in Suzuki [10]. The paper of Brauer and Fowler [1] exhibits methods used in several characterizations. Besides Theorem 28.1, due to Brauer, other characterizations using ordinary character theory extensively are found in Brauer, Suzuki, and Wall [1], Feit [4], Brauer [6], Wong [1, 3], Janko [2, 3, 4], Glauberman [2], and G. Higman [1].

**Lemma 28.2**    $|GL(2, 3)| = 48$. *$GL(2, 3)$ has eight conjugacy classes,*

155

*with representatives*

$$e = \begin{pmatrix} 1 & 0 \\ 0 & 1 \end{pmatrix}, \quad t = \begin{pmatrix} -1 & 0 \\ 0 & -1 \end{pmatrix}, \quad u = \begin{pmatrix} 0 & 1 \\ -1 & 0 \end{pmatrix},$$

$$v_0 = \begin{pmatrix} 1 & 0 \\ 1 & 1 \end{pmatrix}, \quad v = \begin{pmatrix} -1 & 0 \\ -1 & -1 \end{pmatrix}, \quad w_0 = \begin{pmatrix} 1 & 0 \\ 0 & -1 \end{pmatrix},$$

$$w_1 = \begin{pmatrix} 1 & -1 \\ 1 & 1 \end{pmatrix}, \quad w_2 = \begin{pmatrix} -1 & 1 \\ -1 & -1 \end{pmatrix}.$$

*We have*

| $x =$ | $e$ | $t$ | $u$ | $v_0$ | $v$ | $w_0$ | $w_1$ | $w_2$ |
|---|---|---|---|---|---|---|---|---|
| $|\langle x \rangle| =$ | 1 | 2 | 4 | 3 | 6 | 2 | 8 | 8 |
| $|C_{GL(2,3)}(x)| =$ | 48 | 48 | 8 | 6 | 6 | 4 | 8 | 8 |

*If* $1_{GL(2,3)} = \phi_1, \phi_2, \ldots, \phi_8$ *are the irreducible characters of GL(2, 3), then the character table of GL(2, 3) is*

|          | $e$ | $t$ | $u$ | $v_0$ | $v$ | $w_0$ | $w_1$ | $w_2$ |
|----------|-----|-----|-----|-------|-----|-------|-------|-------|
| $\phi_1$ | 1   | 1   | 1   | 1     | 1   | 1     | 1     | 1     |
| $\phi_2$ | 1   | 1   | 1   | 1     | 1   | $-1$  | $-1$  | $-1$  |
| $\phi_3$ | 2   | 2   | 2   | $-1$  | $-1$| 0     | 0     | 0     |
| $\phi_4$ | 3   | 3   | $-1$| 0     | 0   | 1     | $-1$  | $-1$  |
| $\phi_5$ | 3   | 3   | $-1$| 0     | 0   | $-1$  | 1     | 1     |
| $\phi_6$ | 2   | $-2$| 0   | $-1$  | 1   | 0     | $i\sqrt{2}$  | $-i\sqrt{2}$ |
| $\phi_7$ | 2   | $-2$| 0   | $-1$  | 1   | 0     | $-i\sqrt{2}$ | $i\sqrt{2}$  |
| $\phi_8$ | 4   | $-4$| 0   | 1     | $-1$| 0     | 0     | 0     |

*Proof* It is easy to check that the given elements and their centralizers have the given orders. After checking that $w_1$ and $w_2$ are not conjugate, the given elements are all not conjugate and their conjugacy classes exhaust $H = GL(2, 3)$. $H$ has four Sylow 3-subgroups, and $H$ permutes them by conjugation. The subgroup fixing one has order 12, and the subgroup fixing two has order 4; it cannot fix all four as $H$ has no normal subgroup of order 4, so $\langle t \rangle$ is the subgroup fixing all Sylow 3-subgroups. Since $|H/\langle t \rangle| = 24$, this proves $H/\langle t \rangle \cong S_4$.

$S_4$, of course, has a normal subgroup $V$ of order 4, and $S_4/V \cong S_3$; using this fact and the character of $S_4$ given by its double transitivity, we easily get the characters of $S_4$ and hence the characters $\phi_1$, $\phi_2$, $\phi_3$, $\phi_4$, $\phi_5$ of $H$.

The number $\gamma = (1 + i)/\sqrt{2}$ is a primitive 8th root of 1, and the number $-\omega = (1 - i\sqrt{3})/2$ is a primitive 6th root of 1. We see that $w_1$ and $w_1^3$ are conjugate, and so are $w_2 = w_1^5$ and $w_1^7$. $\lambda: w_1 \to \gamma$ and $\lambda_1: w_1 \to -\gamma$ are linear characters of $\langle w_1 \rangle$, and $\mu: v \to -\omega$ is a linear character of $\langle v \rangle$. We compute that the induced characters are

| | $e$ | $t$ | $u$ | $v_0$ | $v$ | $w_0$ | $w_1$ | $w_2$ |
|---|---|---|---|---|---|---|---|---|
| $\lambda^H$ | 6 | $-6$ | 0 | 0 | 0 | 0 | $i\sqrt{2}$ | $-i\sqrt{2}$ |
| $\lambda_1^H$ | 6 | $-6$ | 0 | 0 | 0 | 0 | $-i\sqrt{2}$ | $i\sqrt{2}$ |
| $\mu^H$ | 8 | $-8$ | 0 | $-1$ | 1 | 0 | 0 | 0 |

We then compute that $(\lambda^H, \lambda^H)_H = 2$, $(\lambda_1^H, \lambda_1^H)_H = 2$, $(\mu^H, \mu^H)_H = 3$, $(\lambda^H, \mu^H)_H = 2$, $(\lambda_1^H, \mu^H)_H = 2$. Therefore $\phi_6 = \mu^H - \lambda_1^H$, $\phi_7 = \mu^H - \lambda^H$, $\phi_8 = \mu^H - \phi_6 - \phi_7$ are the other irreducible characters of $H$.

*Proof of Theorem 28.1, Step 1*  Denote by $t$, $u$, $v_0$, $v$, $w_0$, $w_1$, $w_2$ the elements in $C_G(t) = H$ corresponding to the elements $t$, $u$, $v_0$, $v$, $w_0$, $w_1$, $w_2$ of $GL(2, 3)$ in Lemma 28.2. Then the five conjugacy classes of $t$, $u$, $v$, $w_1$, $w_2$ are special classes of $H$, in the sense of Lemma 27.2.

*Proof*  Elements $x$ of the classes have $t \in \langle x \rangle$, so $C_G(x) \subseteq C_G(t) = H$ and (i) of Lemma 27.2 holds. The only two elements in $\{t, u, v, w_1, w_2\}$ of the same order are $w_1$ and $w_2$. If $w_1^g = w_2$ for some $g \in G$, then since $w_2 = w_1^5$ we have $t^g = (w_1^4)^g = (w_1^4)^5 = t$, $g \in C_G(t) = H$, a contradiction; we have proved that (ii) of Lemma 27.2 holds. (iii) of that Lemma is clear, as our classes contain all $x \in H$ with $t \in \langle x \rangle$.

*Step 2*  The generalized characters $\theta_1 = \phi_1 + \phi_3 - \phi_4$, $\theta_2 = \phi_2 + \phi_3 - \phi_5$, $\theta_3 = \phi_3 - \phi_6$, $\theta_4 = \phi_3 - \phi_7$, $\theta_5 = \phi_1 + \phi_2 + \phi_3 - \phi_8$ (notation of Lemma 28.2) form a basis for the class functions of $H$ vanishing on $H - D$.

*Proof*  $\theta_1, \ldots, \theta_5$ are clearly linearly independent, and vanish on $H - D$.

*Step 3*  Note that, if $\{b_{ij}\}$ is a set of solutions to the equations (3) of Lemma 27.2, then so is $\{\varepsilon_j b_{i\sigma(j)}\}$, where all $\varepsilon_j = \pm 1$, $\varepsilon_1 = 1$, and $\sigma$ is a permutation with $\sigma(1) = 1$. We now assert that, except for these modifications, $b_{ij} = a_{ij}$ is the *only* solution to the equations (3).

*Proof*  The $a_{ij}$ are defined by $\theta_i = \sum_{j=1}^{8} a_{ij}\phi_j$, so we have here

$$(a_{ij}) = \begin{pmatrix} 1 & 0 & 1 & -1 & 0 & 0 & 0 & 0 \\ 0 & 1 & 1 & 0 & -1 & 0 & 0 & 0 \\ 0 & 0 & 1 & 0 & 0 & -1 & 0 & 0 \\ 0 & 0 & 1 & 0 & 0 & 0 & -1 & 0 \\ 1 & 1 & 1 & 0 & 0 & 0 & 0 & -1 \end{pmatrix}.$$

We will fill in the matrix $(b_{ij})$, one row at a time. The permutation $\sigma$ of the columns means that we may always put new nonzero entries in the first previously unused column by permuting the characters $\chi_j$. By multiplying columns by $\varepsilon_j = \pm 1$, we may choose the first nonzero entry in each column to be positive. The first of equations (3), Lemma 27.2, tells us that

$$(b_{ij}) = \begin{pmatrix} 1 \\ 0 \\ 0 & & ? \\ 0 \\ 1 \end{pmatrix}.$$

$\sum_{k=1}^{r} b_{1k}b_{1k} = \sum_{k=1}^{8} a_{1k}a_{1k} = 3$, so we must complete the first row as

$$(b_{ij}) = \begin{pmatrix} 1 & 1 & 1 & 0 & 0 & \cdots \\ 0 \\ 0 & & & ? \\ 0 \\ 1 \end{pmatrix}.$$

$$\sum_{k=1}^{r} b_{1k}b_{2k} = \sum_{k=1}^{8} a_{1k}a_{2k} = 1 \quad \text{and} \quad \sum_{k=1}^{r} b_{2k}b_{2k} = \sum_{k=1}^{8} a_{2k}a_{2k} = 3,$$

so the second row must give us

$$(b_{ij}) = \begin{pmatrix} 1 & 1 & 1 & 0 & 0 & 0 & \cdots \\ 0 & 1 & 0 & 1 & 1 & 0 & \cdots \\ 0 & & & ? \\ 0 \\ 1 \end{pmatrix}.$$

$$\sum_{k=1}^{r} b_{1k}b_{3k} = 1, \qquad \sum_{k=1}^{r} b_{2k}b_{3k} = 1, \qquad \sum_{k=1}^{r} b_{3k}b_{3k} = 2.$$

We see there are two ways to fill the third row.

$$(b_{ij}) = \begin{pmatrix} 1 & 1 & 1 & 0 & 0 & 0 & 0 & \cdots \\ 0 & 1 & 0 & 1 & 1 & 0 & 0 & \cdots \\ 0 & 1 & 0 & 0 & 0 & 1 & 0 & \cdots \\ 0 & & & & ? & & & \\ 1 & & & & & & & \end{pmatrix} \quad \text{or} \quad \begin{pmatrix} 1 & 1 & 1 & 0 & 0 & 0 & \cdots \\ 0 & 1 & 0 & 1 & 1 & 0 & \cdots \\ 0 & 0 & 1 & 1 & 0 & 0 & \cdots \\ 0 & & & ? & & & \\ 1 & & & & & & \end{pmatrix}.$$

$$\sum_{k=1}^{r} b_{1k}b_{4k} = 1, \qquad \sum_{k=1}^{r} b_{2k}b_{4k} = 1, \qquad \sum_{k=1}^{r} b_{3k}b_{4k} = 1, \qquad \sum_{k=1}^{r} b_{4k}b_{4k} = 2.$$

There is one way to fill each of the fourth rows:

$$(b_{ij}) = \begin{pmatrix} 1 & 1 & 1 & 0 & 0 & 0 & 0 & \cdots \\ 0 & 1 & 0 & 1 & 1 & 0 & 0 & \cdots \\ 0 & 1 & 0 & 0 & 0 & 1 & 0 & \cdots \\ 0 & 1 & 0 & 0 & 0 & 0 & 1 & \cdots \\ 1 & & & & ? & & & \end{pmatrix} \quad \text{or} \quad \begin{pmatrix} 1 & 1 & 1 & 0 & 0 & 0 & \cdots \\ 0 & 1 & 0 & 1 & 1 & 0 & \cdots \\ 0 & 0 & 1 & 1 & 0 & 0 & \cdots \\ 0 & 0 & 1 & 0 & 1 & 0 & \cdots \\ 1 & & & ? & & & \end{pmatrix}.$$

$$\sum_{k=1}^{r} b_{1k}b_{5k} = \sum_{k=1}^{r} b_{2k}b_{5k} = 2, \qquad \sum_{k=1}^{r} b_{3k}b_{5k} = \sum_{k=1}^{r} b_{4k}b_{5k} = 1,$$

$$\sum_{k=1}^{r} b_{5k}b_{5k} = 4.$$

Clearly there is a unique way to fill the first matrix:

$$(b_{ij}) = \begin{pmatrix} 1 & 1 & 1 & 0 & 0 & 0 & 0 & 0 & 0 & \cdots \\ 0 & 1 & 0 & 1 & 1 & 0 & 0 & 0 & 0 & \cdots \\ 0 & 1 & 0 & 0 & 0 & 1 & 0 & 0 & 0 & \cdots \\ 0 & 1 & 0 & 0 & 0 & 0 & 1 & 0 & 0 & \cdots \\ 1 & 1 & 0 & 1 & 0 & 0 & 0 & 1 & 0 & \cdots \end{pmatrix}.$$

Can we fill the fifth row of the second matrix? Besides $b_{51} = 1$, there must be three more entries $\pm 1$, the rest 0's. $b_{52} = -1$ would force $\sum_k b_{1k}b_{5k} \le 1$, a contradiction. $b_{52} = +1$ would mean, since $\sum_k b_{1k}b_{5k} = 2$, that $b_{53} = 0$. Since $\sum_k b_{2k}b_{5k} = 2$, this would force $b_{54} = 1, b_{55} = 0$ or $b_{54} = 0, b_{55} = 1$. The former would give the contradiction $\sum_k b_{4k}b_{5k} = 0$ and the latter the contradiction $\sum_k b_{3k}b_{5k} = 0$.

So the second matrix can only be finished with $b_{52} = 0$. Since $\sum_k b_{1k}b_{5k} = 2$ this forces $b_{53} = 1$. $\sum_k b_{3k}b_{5k} = 1$ now forces $b_{54} = 0$ and $\sum_k b_{4k}b_{5k} = 1$ forces $b_{55} = 0$. But now $\sum_k b_{2k}b_{5k} = 0$, the final contradiction.

Thus the only solution for $(b_{ij})$ is the last matrix written. It can clearly be made equal to $(a_{ij})$ by permuting columns and multiplying columns by $\pm 1$.

*Step 4* A portion of the character table of $G$ is

|  | 1 | $t$ | $u$ | $v$ | $w_1$ | $w_2$ |
|---|---|---|---|---|---|---|
| $1_G = \chi_1$ | 1 | 1 | 1 | 1 | 1 | 1 |
| $\varepsilon_2\chi_2$ | $x$ | 1 | 1 | 1 | $-1$ | $-1$ |
| $\varepsilon_3\chi_3$ | $y$ | 2 | 2 | $-1$ | 0 | 0 |
| $\varepsilon_4\chi_4$ | $y+1$ | 3 | $-1$ | 0 | $-1$ | $-1$ |
| $\varepsilon_5\chi_5$ | $x+y$ | 3 | $-1$ | 0 | 1 | 1 |
| $\varepsilon_6\chi_6$ | $y$ | $-2$ | 0 | 1 | $i\sqrt{2}$ | $-i\sqrt{2}$ |
| $\varepsilon_7\chi_7$ | $y$ | $-2$ | 0 | 1 | $-i\sqrt{2}$ | $i\sqrt{2}$ |
| $\varepsilon_8\chi_8$ | $x+y+1$ | $-4$ | 0 | $-1$ | 0 | 0 |
| other $\chi_j$'s | $\chi_j(1)$ | 0 | 0 | 0 | 0 | 0 |

Here all $\varepsilon_j = \pm 1$, and $x$ and $y$ are unknown integers.

*Proof* With a proper arrangement of the $\chi_j$'s, we have $b_{ij} = \varepsilon_j a_{ij}$ by Step 3. Hence

$$\varepsilon_i\chi_i(h_j) = \sum_{k=1}^{n} c_{jk}b_{ki}\varepsilon_i = \sum_{k=1}^{n} c_{jk}a_{ki} = \phi_i(h_j)$$

in the notation of Lemma 27.2, so $\varepsilon_i\chi_i$ and $\phi_i$ have the same values on $t, u, v, w_1, w_2$. This verifies all of Step 4 except the values $\varepsilon_i\chi_i(1)$. Denote $\varepsilon_1 = 1$. Now

$$\theta_i^G = \sum_{j=1}^{r} b_{ij}\chi_j = \sum_{j=1}^{8} a_{ij}\varepsilon_j\chi_j \quad \text{and} \quad \theta_i^G(1) = |G:H|\theta_i(1) = 0,$$

so we have

$$0 = \sum_{j=1}^{8} a_{ij}\varepsilon_j\chi_j(1), \quad \text{all} \quad 1 \le i \le 5.$$

Putting in the values $\{a_{ij}\}$, we get five equations:

$$0 = \chi_1(1) + \varepsilon_3\chi_3(1) - \varepsilon_4\chi_4(1),$$
$$0 = \varepsilon_2\chi_2(1) + \varepsilon_3\chi_3(1) - \varepsilon_5\chi_5(1),$$
$$0 = \varepsilon_3\chi_3(1) - \varepsilon_6\chi_6(1),$$
$$0 = \varepsilon_3\chi_3(1) - \varepsilon_7\chi_7(1),$$
$$0 = \chi_1(1) + \varepsilon_2\chi_2(1) + \varepsilon_3\chi_3(1) - \varepsilon_8\chi_8(1).$$

Setting $\chi_1(1) = 1$, $\varepsilon_2\chi_2(1) = x$, $\varepsilon_3\chi_3(1) = y$, these equations give us $\varepsilon_i\chi_i(1)$ for $4 \le i \le 8$.

*Step 5* For any $y \in G$, denote by $\beta(y)$ the number of ordered pairs $(t_1, t_2)$ of conjugates of $t$ in $G$ such that $t_1t_2 = y$. The values of $\beta(y)$ on special classes are as follows:

| | $\beta(t)$ | $\beta(u)$ | $\beta(v)$ | $\beta(w_1)$ | $\beta(w_2)$ |
|---|---|---|---|---|---|
| Case I. $w_0$ conjugate to $t$ in $G$ | 12 | 4 | 6 | 0 | 0 |
| Cast II. $w_0$ not conjugate to $t$ in $G$ | 0 | 0 | 0 | 0 | 0 |

*Proof* Let $h$ denote $t$, $u$, $v$, $w_1$ or $w_2$. Suppose $t^a$ and $t^b$ are conjugates of $t$ satisfying $t^at^b = h$. Then

$$h^{-1} = (t^b)^{-1}(t^a)^{-1} = t^bt^a = (t^a)^{-1}ht^a,$$

so $t^a$ inverts $h$ and hence centralizes $t \in \langle h \rangle$. Thus $t^a$, $t^b \in C_G(t)$, and we can compute $\beta(h)$ by our knowledge of the structure of $C_G(t)$.

In Case II, $t$ is itself the only conjugate of $t$ contained in $C_G(t)$, and $tt = 1 \ne h$, so $\beta(h) = 0$ for all $h$.

In Case I, any conjugate $w_0^x$ of $w_0$ in $H$ satisfies $(tw_0^x)^2 = 1$, and $tw_0^x \ne t$ so $tw_0^x \ne h$. So in Case I, $\beta(h)$ is the number of pairs $(w_0^x, w_0^y)$ of conjugates of $w_0$ in $H$ such that $w_0^xw_0^y = h$. By Lemma 19.2 in the group $H$,

$$\beta(h) = \frac{|H|}{|C_H(w_0)|^2} \sum_{t=1}^{8} \frac{\phi_t(w_0)\phi_t(w_0)\overline{\phi_t(h)}}{\phi_t(1)} = 3 \sum_{t=1}^{8} \frac{\phi_t(w_0)^2\overline{\phi_t(h)}}{\phi_t(1)}.$$

Substituting the values of $\phi_t$ from Lemma 28.2, we easily complete the proof of Step 5.

*Step 6* 
$$\beta(t) = \frac{|G|}{48^2}\left(1 + \frac{1}{x} + \frac{27}{y+1} + \frac{27}{x+y} - \frac{8}{y} - \frac{64}{x+y+1}\right),$$

$$\beta(u) = \frac{|G|}{48^2}\left(1 + \frac{1}{x} + \frac{8}{y} - \frac{9}{y+1} - \frac{9}{x+y}\right),$$

$$\beta(v) = \frac{|G|}{48^2}\left(1 + \frac{1}{x} + \frac{4}{y} - \frac{16}{x+y+1}\right),$$

$$\beta(w_1) = \beta(w_2) = \frac{|G|}{48^2}\left(1 - \frac{1}{x} - \frac{9}{y+1} + \frac{9}{x+y}\right).$$

*Proof* Lemma 19.2 in $G$ tells us that

$$\beta(h) = \frac{|G|}{|C_G(t)|^2} \sum_{i=1}^{r} \frac{\chi_i(t)\chi_i(t)\overline{\chi_i(h)}}{\chi_i(1)}.$$

$|C_G(t)| = |H| = 48$; by Step 4, $\chi_i(t) = 0$ for $i > 8$, and we see

$$\beta(h) = \frac{|G|}{48^2} \sum_{i=1}^{8} \frac{\chi_i(t)^2\overline{\chi_i(h)}}{\chi_i(1)}.$$

Substituting the values from Step 4, we see that all $\varepsilon_i$ drop out and we get the equations of Step 6.

*Step 7*  Case I holds. $|G| = y(y + 1)\, 2(48)^2/(y - 2)^2$, where $(y - 2)|48$ and $(y - 8)|72$.

*Proof*  If Case II holds, then Steps 5 and 6 give

$$0 + 0 = \beta(u) + \beta(w_1) = \frac{|G|}{48^2}\left(2 + \frac{8}{y} - \frac{18}{y + 1}\right).$$

Therefore

$$0 = 2 + \frac{8}{y} - \frac{18}{y + 1} = \frac{2y^2 + 2y + 8y + 8 - 18y}{y(y + 1)} = \frac{2(y - 2)^2}{y(y + 1)},$$

so $y = 2$. By Theorem 26.1, our simple $G$ cannot have an irreducible representation of degree 2, so this is a contradiction; hence Case I holds.

By Steps 5 and 6, we therefore have

$$4 + 0 = \beta(u) + \beta(w_1) = \frac{|G|}{48^2}\left(2 + \frac{8}{y} - \frac{18}{y + 1}\right),$$

so

$$4 = \frac{|G|}{48^2} \cdot \frac{2(y - 2)^2}{y(y + 1)} \qquad \text{or} \qquad |G| = y(y + 1)\frac{2(48)^2}{(y - 2)^2}.$$

$|y|$ and $|y + 1|$ are character degrees, so $y(y + 1)\big|\,|G|$. Therefore $2(48)^2/(y - 2)^2$ is an integer, and $(y - 2)|48$.

$0 = \beta(w_1)$ from Step 5, so by Step 6,

$$0 = 1 - \frac{1}{x} - \frac{9}{y + 1} + \frac{9}{x + y} = \frac{(x - 1)(y + 1)(x + y) - 9x(x - 1)}{x(y + 1)(x + y)},$$

and

$$(x - 1)[(y + 1)(x + y) - 9x] = 0.$$

$G$ is not abelian, so $x \neq 1$ and we have

$$9x = (y + 1)(x + y) = xy + x + y + y^2,$$

$$x(y - 8) = -y - y^2,$$

$$x = -\frac{y(y + 1)}{y - 8} = -y - 9 + \frac{72}{y - 8}.$$

Therefore $(y - 8)|72$.

*Step 8*  $y \equiv 10 \pmod{16}$.

*Proof*  $P = \langle w_0, w_1 \rangle$ is a Sylow 2-subgroup of $G$ with defining relations $w_0^2 = w_1^8 = 1$, $w_1^{w_0} = w_1^3$. $P' = \langle w_1^2 \rangle$, so $|P:P'| = 4$ and $P$ has 4 linear characters. $P$ also has three irreducible characters of degree 2, induced from linear characters of $\langle w_1 \rangle$. Let $\varepsilon = (1 + i)/\sqrt{2}$, so $\varepsilon$ is a primitive 8th root of 1 with $\varepsilon^2 = i$. Let $\lambda_1, \lambda_2, \lambda_3$ be the linear characters of $\langle w_1 \rangle$ defined by $\lambda_1 : w_1 \to \varepsilon$, $\lambda_2 : w_1 \to \varepsilon^5$, $\lambda_3 : w_1 \to \varepsilon^2$. Then the character table of $P$ is

|            | 1 | $t$ | $u$ | $w_1$ | $w_2$ | $w_0$ | $w_0 w_1$ |
|------------|---|-----|-----|-------|-------|-------|-----------|
| $\tau_1$   | 1 | 1   | 1   | 1     | 1     | 1     | 1         |
| $\tau_2$   | 1 | 1   | 1   | 1     | 1     | $-1$  | $-1$      |
| $\tau_3$   | 1 | 1   | 1   | $-1$  | $-1$  | 1     | $-1$      |
| $\tau_4$   | 1 | 1   | 1   | $-1$  | $-1$  | $-1$  | 1         |
| $\tau_5$   | 2 | $-2$| 0   | $i\sqrt{2}$  | $-i\sqrt{2}$ | 0 | 0 |
| $\tau_6$   | 2 | $-2$| 0   | $-i\sqrt{2}$ | $i\sqrt{2}$  | 0 | 0 |
| $\tau_7$   | 2 | 2   | $-2$| 0     | 0     | 0     | 0 .       |

(Here, of course, $t = w_1^4$, $u = w_1^2$, $w_2 = w_1^5$, $\tau_5 = \lambda_1^P$, $\tau_6 = \lambda_2^P$, $\tau_7 = \lambda_3^P$.)

$\varepsilon_6 \chi_6|_P$ is a generalized character of $P$, and by Step 4 it has values $\varepsilon_6 \chi_6(1) = y$, $\varepsilon_6 \chi_6(t) = -2$, $\varepsilon_6 \chi_6(u) = 0$, $\varepsilon_6 \chi_6(w_1) = i\sqrt{2}$, $\varepsilon_6 \chi_6(w_2) = -i\sqrt{2}$, $\varepsilon_6 \chi_6(w_0) = \varepsilon_6 \chi_6(t) = -2$, $\varepsilon_6 \chi_6(w_0 w_1) = \varepsilon_6 \chi_6(u) = 0$.

Setting $\varepsilon_6 \chi_6|_P = \sum_{i=1}^{7} n_i \tau_i$ for integers $n_i$, and evaluating on the elements $1, t, u, w_1, w_2, w_0, w_0 w_1$, we get equations

$$y = n_1 + n_2 + n_3 + n_4 + 2n_5 + 2n_6 + 2n_7$$

$$-2 = n_1 + n_2 + n_3 + n_4 - 2n_5 - 2n_6 + 2n_7$$

$$0 = n_1 + n_2 + n_3 + n_4 - 2n_7$$

$$\sqrt{2}i = n_1 + n_2 - n_3 - n_4 + \sqrt{2}in_5 - \sqrt{2}in_6$$

$$-\sqrt{2}i = n_1 + n_2 - n_3 - n_4 - \sqrt{2}in_5 + \sqrt{2}in_6$$

$$-2 = n_1 - n_2 + n_3 - n_4$$

$$0 = n_1 - n_2 - n_3 + n_4.$$

The last six equations are easily solved to give $n_2 = n_1 + 1$, $n_3 = n_1$, $n_4 = n_1 + 1$, $n_5 = 2n_1 + 2$, $n_6 = 2n_1$, $n_7 = 2n_1 + 1$, and the first equation then becomes $y = 16n_1 + 10$, proving $y \equiv 10 \pmod{16}$.

*Step 9, Conclusion* Since $(y - 2)|48$, we have $|y| \leq 50$. Since $y \equiv 10 \pmod{16}$, $y$ is either 42, 26, 10, $-6$, $-22$, or $-38$. Only the cases $y = 26$ and $y = 10$ satisfy $(y - 8)|72$.

If $y = 26$, $\quad |G| = y(y + 1)\dfrac{2(48)^2}{(y - 2)^2} = 26 \cdot 27 \cdot 8 = 5616$;

if $y = 10$, $\quad |G| = y(y + 1)\dfrac{2(48)^2}{(y - 2)^2} = 10 \cdot 11 \cdot 72 = 7920.$

# §29

## Primitive Complex Linear Groups

**Definitions**    For $\alpha \in \mathbf{C}$, let $\bar{\alpha}$ denote the complex conjugate of $\alpha$. Let $V$ be an $n$-dimensional $\mathbf{C}$-vector space. A function $f: V \times V \to \mathbf{C}$ is a *hermitian form* on $V$ if

$$f(u + v, w) = f(u, w) + f(v, w),$$

$$f(u, v + w) = f(u, v) + f(u, w),$$

$$\alpha f(u, v) = f(\alpha u, v), f(u, v) = \overline{f(v, u)}, \qquad \text{all} \quad \alpha \in \mathbf{C}, \quad u, v, w \in V.$$

Clearly if $f$ is hermitian then $f(v, v)$ is real, all $v \in V$. If $f(v, v) > 0$ for all $0 \neq v \in V$, $f$ is called *positive definite*.

An example of a positive definite hermitian form is

$$f((a_1, \ldots, a_n), (b_1, \ldots, b_n)) = a_1 \bar{b}_1 + \cdots + a_n \bar{b}_n.$$

In the following definition and lemma, $f$ denotes a positive definite hermitian form on $V$.

**Definition**    $U(V) = U_f(V) = \{T \in GL(V) \mid f(Tu, Tv) = f(u, v), \text{all } u, v \in V\}$ is the *unitary group*. A *unitary basis* of $V$ is a basis $v_1, \ldots, v_n$ with $f(v_i, v_j) = \delta_{ij}$. A matrix $M \in GL(n, \mathbf{C})$ is a *unitary matrix* if $M^t \bar{M} = I$. $U(n, \mathbf{C})$ is the group of all $n \times n$ unitary matrices.

**Lemma 29.1**    (*a*)    *Unitary bases exist.*

(*b*)    *If $f'$ is a second positive definite hermitian form on $V$, $U_{f'}(V)$ the unitary group defined from $f'$, then $U_f(V)$ and $U_{f'}(V)$ are conjugate in $GL(n, \mathbf{C})$.*

(*c*)    *Let $\{v_1, \ldots, v_n\}$ be a unitary basis of $V$, $T \in GL(V)$. Then $T \in U(V)$ if and only if $\{T(v_1), \ldots, T(v_n)\}$ is a unitary basis of $V$.*

(*d*)    *Let $\{v_1, \ldots, v_n\}$ be a unitary basis of $V$, $T \in GL(V)$, and let $M = (a_{ij})$ be the matrix of $T$ with respect to $\{v_1, \ldots, v_n\}$, so*

$T(v_j) = \sum_{i=1}^{n} a_{ij} v_i$. Then $T \in U(V)$ if and only if $M$ is a unitary matrix. Thus $U(V) \cong U(n, \mathbf{C})$.

    (e)   *Every permutation matrix is unitary.*

    (f)   *A triangular unitary matrix is diagonal.*

*Proof*   All parts are easy exercises.

**Lemma 29.2**    (1)  *If $A$ is an $n \times n$ complex matrix, then there is $U \in U(n, \mathbf{C})$ such that $U^{-1}AU$ is triangular.*

    (2)  *If $A \in U(n, \mathbf{C})$, then for some $U \in U(n, \mathbf{C})$, $U^{-1}AU$ is diagonal.*

    (3)  *If $A \in U(n, \mathbf{C})$, then all eigenvalues of $A$ have absolute value 1.*

*Proof*   (1)  $A$ is the matrix with respect to some unitary basis of some linear transformation $T$; by Lemma 29.1(c), $U^{-1}AU$ is the matrix of $T$ under change of unitary basis. Choose a characteristic vector $v \neq 0$ of $T$, and choose $v_1 = \beta v$, some $\beta \in \mathbf{C}$ such that $f(v_1, v_1) = 1$. If $W = \{u \in V | f(u, v_1) = 0\}$, then $W$ has a unitary basis $v_2, \ldots, v_n$, and $v_1, \ldots, v_n$ is a unitary basis of $V$. $U^{-1}AU$ has form

$$\begin{pmatrix} \alpha & * & \cdots & * \\ 0 & & & \\ \vdots & & B & \\ 0 & & & \end{pmatrix},$$

and by induction $B$ may be triangularized.

    (2)  Immediate from (1) and Lemma 29.1(f).

    (3)  If $U$ is unitary with $C = U^{-1}AU$ diagonal, then $A$ has the same eigenvalues as $C$; but $C$ is unitary so $I = C^t \bar{C} = C\bar{C}$, and eigenvalues of $C$ have absolute value 1.

**Lemma 29.3**   *Let $S$ be a finite set of commuting unitary matrices. Then there is a unitary matrix $U$ such that for all $M \in S$, $U^{-1}MU$ is diagonal.*

*Proof*   Let $S = \{M_1, \ldots, M_t\}$. By induction on $t$, we may assume $M_1, \ldots, M_{t-1}$ diagonal. Conjugating by a permutation matrix (see Lemma 29.1(e)), we may assume $M_1, \ldots, M_{t-1}$ written so that

$$M_i = \begin{pmatrix} \alpha_{i1}I_1 & & 0 \\ & \ddots & \\ 0 & & \alpha_{ir}I_r \end{pmatrix},$$

$I_j$ the $n_j \times n_j$ identity matrix, where if $r_1 \neq r_2$ then for some $i$, $\alpha_{ir_1} \neq \alpha_{ir_2}$. The fact that $M_t$ commutes with $M_1, \ldots, M_{t-1}$ means

$$
M_t = \begin{pmatrix} A_1 & & 0 \\ & \ddots & \\ 0 & & A_r \end{pmatrix},
$$

the $A_j$ $n_j \times n_j$ unitary matrices. Choosing unitary $U_j$'s with $U_j^{-1} A_j U_j$ diagonal, we see that

$$
U = \begin{pmatrix} U_1 & & 0 \\ & \ddots & \\ 0 & & U_r \end{pmatrix}
$$

is the required matrix.

**Lemma 29.4**    *Let $V$ be an $n$-dimensional $\mathbf{C}$-vector space, $G$ a finite subgroup of $GL(V)$. Then there is a positive definite hermitian form $f$ on $V$ such that $G \subseteq U_f(V)$.*

*Proof*  Let $f^*$ be any positive definite hermitian form on $V$, and define $f$ by

$$
f(u, v) = \sum_{g \in G} f^*(gu, gv), \qquad \text{all} \quad u, v \in V.
$$

It is easy to verify that $f$ is the desired form.

**Definition**    Let $k$ be any field, $V$ an $n$-dimensional $k$-vector space. Any subgroup $G$ of $GL(V)$ is a *linear group* of *degree* $n$. $G$ is *irreducible, completely reducible*, etc., if the representation of $G$ given by the identity map is irreducible, completely reducible, etc.

**Definition**    A $kG$-module $V$ is called *imprimitive* if $V$ is a direct sum $V = V_1 \oplus \cdots \oplus V_s$, $s > 1$, of nontrivial subspaces $V_i$, such that for each $g \in G$ and each $i$, $gV_i$ is some $V_j$. $\{V_1, \ldots, V_s\}$ is called a *system of imprimitivity* of $V$. An irreducible $kG$-module $V$ is *primitive* if it is not imprimitive.

**Lemma 29.5**    *Let $V$ be an irreducible imprimitive $kG$-module with* *system of imprimitivity* $\{V_1, \ldots, V_s\}$. *Then G is transitive on* $\{V_1, \ldots, V_s\}$, *and if* $H = \{g \in G | gV_1 = V_1\}$ *then* $V \cong V_1^G$, $V_1$ *a kH-module.*

*Proof* Irreducibility of $V$ forces $G$ to be transitive on the $V_i$. If $V_i = g_i V_1$ for $g_i \in G$, then $G = g_1 H \cup \cdots \cup g_s H$ (disjoint union). For any $g \in G$, any $v \in V_1$, if $gg_j = g_k h$ then

$$g \cdot g_j v = (gg_j)v = g_k h v = g_k \cdot hv,$$

exactly the action of $g$ on

$$V_1^G = (g_1 \otimes V_1) \oplus \cdots \oplus (g_s \otimes V_1),$$

so $V \cong V_1^G$.

**Definition**    Let $k$ be a field, $V$ an $n$-dimensional $k$-vector space, $G \subseteq GL(V)$. If $V$ is irreducible and primitive as $kG$-module, $G$ is a *primitive linear group*.

**Lemma 29.6**    *Let k be an algebraically closed field, V an n-dimensional k-vector space, $G \subseteq GL(V)$ a finite primitive linear group, A an abelian normal subgroup of G. Then A is cyclic and $A \subseteq Z(G)$.*

*Proof* By Clifford's theorem $V = W_1 \oplus \cdots \oplus W_m$, the $W_i$ conjugate irreducible $kA$-modules. Since $G$ is primitive, all $W_i$ are isomorphic, all faithful $kA$-modules. By Theorem 4.2, $\mathrm{End}_{kA}(W_1) = k$, so $\mathrm{End}_{kA}(W_1)$ consists of scalar transformations. $A$ is abelian so $A \subseteq \mathrm{End}_{kA}(W_1)$, $A$ consists of scalars on $W_1$. $W_1$ is irreducible so dim $W_1 = 1$, $m = n$. Since all $W_i$ are isomorphic, $A$ consists of scalars on $V$ and $A \subseteq Z(G)$. $Z(G)$, a finite multiplicative subgroup of $k$, is cyclic.

**Definition**    Let $S$ be a unitary $n \times n$ matrix. If the distinct eigenvalues of $S$ are $e^{i\theta_1}, \ldots, e^{i\theta_m}$, then the angles $\theta_1, \ldots, \theta_m$ are the *phases* of $S$.

**Lemma 29.7**    *Let T be a unitary matrix, S a unitary diagonal matrix, and assume that the phases of S all lie in an arc of the unit circle of less than $180°$. Let $U = T^{-1}ST$, and denote $S = (\alpha_i \delta_{ij})$, $T = (a_{ij})$, $U = (b_{ij})$. Then:*
   (1)  *all $b_{ii} \neq 0$.*
   (2)  *if $b_{jj}$ is a root of unity, and i, l are integers with $a_{ij} \neq 0$, $a_{lj} \neq 0$, then $\alpha_i = \alpha_l$.*

*Proof*   $T^{-1} = (\bar{a}_{ji})$, so

$$b_{ij} = \sum_{k=1}^{n} \sum_{l=1}^{n} \bar{a}_{ki} \delta_{kl} \alpha_k a_{lj} = \sum_{k=1}^{n} \bar{a}_{ki} a_{kj} \alpha_k;$$

in particular,

$$b_{ii} = \sum_{k=1}^{n} |a_{ki}|^2 \alpha_k.$$

The fact $T$ is unitary implies $1 = \sum_{k=1}^{n} |a_{ki}|^2$. Since all $\alpha_k$ are on the same side of the unit circle, $b_{ii} \neq 0$, (1) holds. The two equations

$$b_{jj} = \sum_{k=1}^{n} |a_{kj}|^2 \alpha_k, \qquad 1 = \sum_{k=1}^{n} |a_{kj}|^2$$

imply that $b_{jj}$ can only have absolute value 1 when the condition of (2) holds.

**Theorem 29.8** (*Blichfeldt* [1])     *Let $V$ be an $n$-dimensional $\mathbf{C}$-vector space, $f$ a positive definite hermitian form on $V$, $\{v_1, \ldots, v_n\}$ a unitary basis of $V$, $G$ a finite primitive subgroup of $U(V) = U_f(V)$. Let $\varphi(g)$ be the matrix of $g \in G$ in the basis $\{v_1, \ldots, v_n\}$, so $\varphi(g) = (a_{ij})$ where $gv_j = \sum_{i=1}^{n} a_{ij} v_i$; $\varphi(G)$ is a subgroup of $U(n, \mathbf{C})$.*

*Assume that for some $s \in G$, $\varphi(s)$ is a diagonal matrix with phases $\theta_1, \ldots, \theta_m$ such that for some $i$, $|\theta_i - \theta_j| \leq 60°$ for all $j$. Then $s \in Z(G)$.*

*Proof, Step 1*   We choose the notation so $|\theta_1 - \theta_j| \leq 60°$ for all $j$. Denote $\alpha_j = e^{i\theta_j}$, where $\alpha_1$ has multiplicity $k$ as an eigenvalue of $\varphi(s)$. We may assume the diagonal entries of $\varphi(s)$ occur in the order $\alpha_1, \ldots, \alpha_1, \alpha_2, \ldots, \alpha_2, \ldots, \alpha_m, \ldots, \alpha_m$.

*Proof of Step 1*   Permutation matrices are unitary, so by conjugating $\varphi(G)$ by a unitary matrix we can arrange the eigenvalues along the diagonal of $\varphi(s)$ in any way we like.

*Step 2*   An $n \times n$ unitary matrix will be called *k-reduced* for our fixed integer $k < n$, if it has block form

$$\begin{pmatrix} C & 0 \\ 0 & D \end{pmatrix},$$

$C$ a $k \times k$ matrix, $D$ an $(n - k) \times (n - k)$ matrix, the 0's zero matrices. For any unitary matrix $(a_{ij})$, define $f((a_{ij})) = |a_{11} + \cdots + a_{kk}|$. Let

$$S = \{g^{-1}sg | g \in G \text{ and } \varphi(g^{-1}sg) \text{ is not } k\text{-reduced}\}.$$

If $S \neq \varnothing$, choose $t_0 \in S$ with $f(\varphi(t_0))$ maximal. Then $t_0^{-1}st_0 \notin S$.

*Proof* Let $\varphi(s) = (\gamma_i \delta_{ij})$, so $\gamma_1 = \cdots = \gamma_k = \alpha_1$, $\gamma_{k+1} = \alpha_2$, etc. Set $\varphi(t_0) = (a_{ij})$, $\varphi(t_0^{-1}st_0) = (b_{ij})$. Denote $\beta_l = \gamma_l e^{-i\theta_l}$, so if $\beta_l = \cos \phi_l + i \sin \phi_l$, then by our hypothesis all $\cos \phi_l \geq \frac{1}{2}$; if $l \leq k$, then $\beta_l = \cos \phi_l = 1$. For any $j$,

$$\varphi(t_0^{-1}st_0) = \varphi(t_0)^{-1}\varphi(s)\varphi(t_0)$$

so we have

$$b_{jj} = \sum_{l=1}^{n} \sum_{m=1}^{n} \bar{a}_{lj}\delta_{lm}\gamma_l a_{mj} = \sum_{l=1}^{n} \bar{a}_{lj}a_{lj}\gamma_l = \sum_{l=1}^{n} |a_{lj}|^2 \gamma_l.$$

The fact $(a_{ij})$ is unitary implies that

$$1 = \sum_{l=1}^{n} |a_{lj}|^2 = \sum_{l=1}^{n} |a_{jl}|^2,$$

for any $j$. Since $(a_{ij})$ is not $k$-reduced, $|a_{r_0r_0}| < 1$ for some $r_0 \in \{1, \ldots, k\}$. We have:

$$f(\varphi(t_0^{-1}st_0)) = |b_{11} + \cdots + b_{kk}| = |\sum_{r=1}^{k} \sum_{j=1}^{n} |a_{jr}|^2 \gamma_j| = |\sum_{r=1}^{k} \sum_{j=1}^{n} |a_{jr}|^2 \beta_j|$$

$$\geq \sum_{r=1}^{k} \sum_{j=1}^{n} |a_{jr}|^2 \cos \phi_j \geq \sum_{r=1}^{k} (\sum_{j=1}^{k} |a_{jr}|^2 + \frac{1}{2} \sum_{j=k+1}^{n} |a_{jr}|^2)$$

$$= \sum_{r=1}^{k} (\sum_{j=1}^{k} |a_{jr}|^2 + \frac{1}{2}(1 - \sum_{j=1}^{k} |a_{jr}|^2)) = \sum_{r=1}^{k} \frac{1}{2}(1 + \sum_{j=1}^{k} |a_{jr}|^2)$$

$$\geq \sum_{r=1}^{k} \frac{1}{2}(1 + |a_{rr}|^2) > \sum_{r=1}^{k} |a_{rr}| \geq |a_{11} + \cdots + a_{kk}| = f(\varphi(t_0)).$$

By maximality of $f(\varphi(t_0))$, then, $t_0^{-1}st_0 \notin S$.

(We used here the fact that $\frac{1}{2}(1 + x^2) \geq x$ for all real numbers $x$, with equality only when $x = 1$.)

*Step 3* $S = \varnothing$, so for all conjugates $t$ of $s$ in $G$, $\varphi(t)$ is $k$-reduced.

*Proof* If not, choose $t_0$ as in Step 2; we know that $\varphi(t_0^{-1}st_0)$ is $k$-reduced. By Lemma 29.2(2), let $U$ be a $k \times k$ unitary matrix such that

$$\begin{pmatrix} U & 0 \\ 0 & I \end{pmatrix}^{-1} \varphi(t_0^{-1}st_0)\begin{pmatrix} U & 0 \\ 0 & I \end{pmatrix} = \begin{pmatrix} D & 0 \\ 0 & E \end{pmatrix},$$

$D$ a diagonal $k \times k$ matrix.

Conjugation of $\varphi(G)$ by $\begin{pmatrix} U & 0 \\ 0 & I \end{pmatrix}$ amounts to change of the first $k$ unitary basis elements. Since $s$ acts as a scalar on these basis elements, this does not change $\varphi(s)$. For any $g \in G$, suppose

$$\varphi(g) = \begin{pmatrix} A & B \\ C & F \end{pmatrix}$$

where $A$ is a $k \times k$ matrix and $F$ is an $(n - k) \times (n - k)$ matrix. Then the matrix of $g$ in the new basis is

$$\begin{pmatrix} U & 0 \\ 0 & I \end{pmatrix}^{-1}\begin{pmatrix} A & B \\ C & F \end{pmatrix}\begin{pmatrix} U & 0 \\ 0 & I \end{pmatrix} = \begin{pmatrix} U^{-1}AU & U^{-1}B \\ CU & F \end{pmatrix}.$$

This is $k$-reduced if and only if $B = C = 0$, which is exactly the statement that $\varphi(g)$ is $k$-reduced. So our change of basis does not change the set $S$.

Making this change of basis, we may therefore assume

$$\varphi(t_0^{-1}st_0) = \begin{pmatrix} D & 0 \\ 0 & E \end{pmatrix},$$

where $D$ is a $k \times k$ diagonal unitary matrix $(\tau_i \delta_{ij})$. In Step 2, we saw that if $\varphi(t_0) = (a_{ij})$, then $\tau_i = \sum_{j=1}^{n} |a_{ji}|^2 \gamma_j$.

All the $\tau_i$ are roots of unity. By Lemma 29.7(1), all $a_{ii} \neq 0$. Fix $i \leq k$. We use Lemma 29.7(2) with $i = j$, $l > k$. When $l > k$, $\gamma_l$ is some $\alpha$ other than $\alpha_1$ so $\gamma_l \neq \gamma_i$. Lemma 29.7(2) says that $a_{li} = 0$. This proves that $\varphi(t_0)$ has form

$$\varphi(t_0) = \begin{pmatrix} A & B \\ 0 & D \end{pmatrix}, \qquad 0 \text{ an } (n - k) \times k \text{ zero matrix.}$$

$\varphi(t_0)$ has inverse ${}^t\overline{\varphi(t_0)}$, so

$$I = \begin{pmatrix} {}^t\overline{A} & 0 \\ {}^t\overline{B} & {}^t\overline{D} \end{pmatrix} \begin{pmatrix} A & B \\ 0 & D \end{pmatrix} = \begin{pmatrix} ({}^t\overline{A})A & ({}^t\overline{A})B \\ ({}^t\overline{B})A & ({}^t\overline{B})B + ({}^t\overline{D})D \end{pmatrix},$$

forcing $B = 0$. Therefore $\varphi(t_0)$ is $k$-reduced, $t_0 \notin S$. This contradicts the choice of $t_0 \in S$, so we must have $S = \varnothing$, Step 3 holds.

*Step 4, Conclusion*   Let $G_0$ be the normal subgroup of $G$ generated by all conjugates of $s$. By Step 3, $G_0$ is reducible. Let $V = V_1 \oplus \cdots \oplus V_t$, the $V_i$ irreducible $\mathbf{C}G_0$-modules. Since $G$ is primitive, Clifford's Theorem 14.1 implies that all the $V_i$ are isomorphic faithful $\mathbf{C}G_0$-modules. But we constructed a $\mathbf{C}G_0$-submodule $W$ of $V$ of dimension $k$ on which $s$ is a scalar. So $s$ is a scalar on one, hence all, of the $V_i$'s; $s \in Z(G_0) \triangleleft G$, so all conjugates $g^{-1}sg$ of $s$ are in $Z(G_0)$. These conjugates generate $G_0$, so $G_0$ is abelian. By Lemma 29.6, $G_0 \subseteq Z(G)$, $s \in Z(G)$.

**Theorem 29.9**     *Let $V$ be an $n$-dimensional $\mathbf{C}$-vector space, $n > 1$, $G$ a finite primitive subgroup of $GL(V)$, $A$ a maximal abelian subgroup of $G$. Then $|A : Z(G)| < 6^{n-1}$.*

*Proof*   By Lemma 29.4 we may assume $G \subseteq U(V)$. Let $\varphi$ be the matrix representation of $G$ afforded by $V$, so $\varphi(g)$ is unitary for all $g \in G$. By Lemma 29.3 we may assume that, for all $a \in A$, $\varphi(a)$ is a diagonal matrix. Since $V$ is absolutely irreducible as $\mathbf{C}G$-module, $Z(G)$ is the group of scalar transformations in $G$.

Assume $|A : Z(G)| \geq 6^{n-1}$, and choose elements $a_1, \ldots, a_{6^{n-1}}$ of $A$ in different cosets of $Z(G)$. Each $\varphi(a_j)$ is a diagonal unitary matrix; let

$$\varphi(a_j) = \begin{pmatrix} e^{i\theta_{j1}} & & 0 \\ & \ddots & \\ 0 & & e^{i\theta_{jn}} \end{pmatrix}.$$

Let $H$ be the group of all scalars $e^{i\theta_{jk}}I$, all $j$, all $k$. $GH$ is still a finite group, $Z(GH) = Z(G)H$, and

$$|AH : Z(G)H| = \frac{|A||H|}{|A \cap H|} \cdot \frac{|Z(G) \cap H|}{|Z(G)||H|} = |A : Z(G)|,$$

so we may replace $G$ by $GH$ and assume all $e^{i\theta_{jk}}I \in G$. Replacing $a_j$ by $a_j(e^{-i\theta_{j1}}) \in a_j Z(G)$, we may assume all $e^{i\theta_{j1}} = 1$, all $\theta_{j1} = 0°$.

At least $6^{n-1}/6$ of the $a_j$'s, say $a_1, \ldots, a_{6^{n-2}}$, have $\theta_{j2}$ in some arc $A_2$ of length $60°$ in the unit circle. At least $6^{n-2}/6$ of these $a_j$'s, say $a_1, \ldots, a_{6^{n-3}}$, have $\theta_{j3}$ in some arc $A_3$ of length $60°$ in the unit circle. Continuing, we find $a_1, \ldots, a_6$ with $|\theta_{jm} - \theta_{lm}| \leq 60°$, all $m \in \{2, \ldots, n-1\}$, $j, l \in \{1, \ldots, 6\}$. Two of $a_1, \ldots, a_6$ have $\theta_{jn}$ in some arc $A_n$ of length $60°$, so we may assume that for all $2 \leq l \leq n$, $|\theta_{1l} - \theta_{2l}| \leqq 60°$. Then

$$\varphi(a_1 a_2^{-1}) = \begin{pmatrix} 1 & & & & 0 \\ & e^{i(\theta_{12}-\theta_{22})} & & & \\ & & \cdot & & \\ & & & \cdot & \\ 0 & & & & e^{i(\theta_{1n}-\theta_{2n})} \end{pmatrix}$$

has all its phases $\phi_j = \theta_{1j} - \theta_{2j}$ satisfying $|0° - \phi_j| \leq 60°$. By Theorem 29.8, $a_1 a_2^{-1} \in Z(G)$, $a_1 Z(G) = a_2 Z(G)$, contradiction, done.

# §30

## Jordan's Theorem à la Blichfeldt

**Definition**   Let $\pi$ be a set of primes. An integer is a *$\pi$-number* if it is divisible only by primes in $\pi$, and a *$\pi'$-number* if it is divisible by no prime in $\pi$. A subgroup $H$ of a finite group $G$ is a *Hall $\pi$-subgroup* of $G$, if $|H|$ is a $\pi$-number and $|G : H|$ a $\pi'$-number.

**Lemma 30.1**   *Let $V$ be an $n$-dimensional $\mathbf{C}$-vector space, $G$ a finite irreducible subgroup of $GL(V)$, $p$ and $q$ primes such that $p\,\|\,|G|$, $q\,\|\,|G|$, $p > n + 1$, $q > n + 1$. Then $G$ contains an element of order $pq$.*

*Proof*   Let $\chi(g) = \operatorname{tr} g$, all $g \in G$, so $\chi$ is a faithful irreducible character of $G$ of degree $n$. Let $g_1 \in G$ have order $p$, $g_2 \in G$ have order $q$. $\chi(g_1)$ is a sum of $n$ $p$th roots of 1, at least one of which is a primitive $p$th root $\varepsilon_p$ of 1 since $g_1 \notin \ker \chi$. $\mathbf{Q}(\varepsilon_p)$ has degree $p - 1 > n$ over $\mathbf{Q}$, so $\chi(g_1)$ requires the $p$th roots of 1. Similarly $\chi(g_2)$ requires the $q$th roots of 1, so by Theorem 23.5, $G$ has an element of order $pq$.

**Theorem 30.2**   *Let $V$ be an $n$-dimensional $\mathbf{C}$-vector space, $G$ a finite subgroup of $GL(V)$, $\pi$ the set of all primes $p$ such that $p\,\|\,|G|$ and $p > n + 1$. Then $G$ contains an abelian Hall $\pi$-subgroup.*

*Proof, Step 1*   Using induction, we assume the theorem true for groups of degree $< n$. The theorem is true if $G$ is reducible.

*Proof of Step 1*   Suppose $V = V_1 \oplus V_2$, $V_1$ and $V_2$ proper $\mathbf{C}G$-submodules of $V$. By Lemma 29.4, we may assume $G \subseteq U(V)$. If $K_i = \{g \in G \mid gv = v \text{ for all } v \in V_i\}$, then $K_1 \lhd G$, $K_2 \lhd G$, $K_1 \cap K_2 = 1$. By induction, $G/K_1$ has an abelian Hall $\pi$-subgroup $H_1/K_1$. By Lemma 29.3, we can find a basis $v_1, \ldots, v_m$ of $V_1$ such that all $x \in H_1$ fix $\mathbf{C}v_1, \ldots,$

$Cv_m$. $H_1/H_1 \cap K_2$ is a linear group on $V_2$, so by induction $H_1/H_1 \cap K_2$ has an abelian Hall $\pi$-subgroup $L_1/H_1 \cap K_2$. By Lemma 29.3, we choose a basis $v_{m+1}, \ldots, v_n$ of $V_2$ such that all $x \in L_1$ fix $Cv_{m+1}, \ldots, Cv_n$. $L_1$ fixes $Cv_1, \ldots, Cv_n$, so is isomorphic to a group of diagonal matrices and is abelian. $|G : L_1| = |G : H_1| |H_1 : L_1|$ is a $\pi'$-number, so $L_1$ contains an abelian Hall $\pi$-subgroup of $G$.

*Step 2*    We may assume $\det g = 1$, all $g \in G$.

*Proof*    Assume the result true in this case for groups of degree $n$, and take a general $G$. Let $N$ be the group of all scalar transformations $\varepsilon I$, $\varepsilon$ an $(n|G|)$th root of 1, and let $G_1$ be the subgroup of $GN$ consisting of elements of determinant 1. For any $g \in G$, $\det g$ is an $|G|$th root of 1 so there is $h \in N$ with $\det h = (\det g)^{-1}$, $\det hg = 1$; therefore $G_1 N = GN$. By the hypothesis in this step, $G_1$ has an abelian Hall $\pi$-subgroup $H$. $HN$ is abelian.

$$|G : G \cap HN| = |GHN : HN| = |GN : HN| = |G_1 N : HN|$$

$$= \frac{|G_1||N|}{|G_1 \cap N|} \cdot \frac{|H \cap N|}{|H||N|} = \frac{|G_1 : H|}{|G_1 \cap N : H \cap N|}$$

is a divisor of $|G_1 : H|$ and so a $\pi'$-number. Therefore the abelian group $G \cap HN$ contains a Hall $\pi$-subgroup of $G$, done.

*Step 3*    If $p, q \in \pi$, then $G$ has an abelian Hall $\{p, q\}$-subgroup.

*Proof*    By Lemma 30.1, $G$ has elements $x$ of order $p$ and $y$ of order $q$, $xy = yx$. Let $x \in P$, $P$ a Sylow $p$-subgroup of $G$. Since $p > n$ and $V_P$ is completely reducible, $V_P$ is a direct sum of one-dimensional $CP$-modules. Hence $P$ is isomorphic to a group of diagonal matrices, $P$ is abelian. Let $H = \langle P, y \rangle$. If $x$ is a scalar $\varepsilon I$, then $\varepsilon^p = 1$, $\varepsilon$ is a primitive $p$th root of 1. But by Step 2, $1 = \det x = \varepsilon^n$, a contradiction. Therefore $x$ is not a scalar, $x \in Z(H)$.

For any $g \in G$, let $\varphi(g)$ be the matrix of $g$ on $V$. In a suitable basis,

$$\varphi(x) = \begin{pmatrix} \alpha_1 I & & 0 \\ & \ddots & \\ 0 & & \alpha_r I \end{pmatrix},$$

the $\alpha$'s all different, $r > 1$. For any $h \in H$, $\varphi(h)\varphi(x) = \varphi(x)\varphi(h)$, so $\varphi(h)$ has form

$$\varphi(h) = \begin{pmatrix} A_1 & & 0 \\ & \ddots & \\ 0 & & A_r \end{pmatrix}.$$

In particular, $H$ is reducible. By Step 1, $H$ has an abelian subgroup $H_0$ of order $|P|/q$. We may assume $y \in H_0$. Let $P_0$ be a Sylow $p$-subgroup of $H_0$.

Let $y$ be in the Sylow $q$-subgroup $Q$ of $G$. Just as $x$ is not scalar, $y$ is not scalar; just as $P$ was abelian, $Q$ is abelian. $y \in Z(\langle P_0, Q \rangle)$, so $\langle P_0, Q \rangle$ is reducible and by Step 1 contains an abelian Hall $\{p, q\}$-subgroup of $G$.

*Step 4, Conclusion*   Let $\pi = \{p, q_1, \ldots, q_m\}$, and let $P$ be a Sylow $p$-subgroup of $G$. By Step 3, $P$ is abelian and $C_G(P)$ contains a Sylow $q_i$-subgroup of $G$, all $q_i$. As in Step 3, $C_G(P)$ is reducible, so by Step 1, $C_G(P)$ contains an abelian Hall $\pi$-subgroup of $G$.

**Theorem 30.3**   *Let $V$ be an $n$-dimensional $\mathbf{C}$-vector space, $n > 1$, $G \subseteq GL(V)$ a finite primitive linear group. Denote by $\pi(x)$ the number of primes $\leq x$, any number $x$. Then*

$$|G : Z(G)| < n!(6^{n-1})^{\pi(n+1)+1}.$$

*Proof*   Let $p$ be any prime $\leq n + 1$, $P$ a Sylow $p$-subgroup of $G$. Let $V = V_1 \oplus \cdots \oplus V_t$, the $V_i$ irreducible $\mathbf{C}P$-modules. By Corollary 15.6, each $V_i$ has a basis in which the matrices of elements of $P$ are monomial. So $P$ is isomorphic to a subgroup of the group $M$ of $n \times n$ monomial matrices over $\mathbf{C}$. $M$ has a normal subgroup $D$ of diagonal matrices, $M/D \cong S_n$. So $P \cap D$ is an abelian normal subgroup of $P$ with $P/P \cap D$ isomorphic to a $p$-subgroup of $S_n$. If $[n!]_p$ is the largest power of $p$ dividing $n!$, then by Theorem 29.8,

$$|P : P \cap Z(G)| = |P : P \cap \cdot D| \, |P \cap D : P \cap Z(G)| < [n!]_p (6^{n-1}).$$

Let

$$\pi = \{\text{primes } p \mid p \mid |G| \text{ and } p > n + 1\}.$$

By Theorem 30.2, $G$ has an abelian Hall $\pi$-subgroup $H$, and by Theorem 29.9, $|HZ(G) : Z(G)| < 6^{n-1}$. By the previous paragraph $|G : HZ(G)| < n!(6^{n-1})^{\pi(n+1)}$, so our theorem holds.

**Theorem 30.4** *Let $V$ be an $n$-dimensional $\mathbf{C}$-vector space, $n > 1$, $G$ a finite subgroup of $GL(V)$. Then $G$ has an abelian normal subgroup $A$,*

$$|G : A| < n!(6^{n-1})^{\pi(n+1)+1}.$$

*Proof* We use induction on $n$. Theorem 30.3 gives the result when $G$ is primitive.

Next suppose $G$ is irreducible but imprimitive. Then for some $d > 1$ we have $d|n$, say $n = dm$, where $V = V_1 \oplus \cdots \oplus V_d$, the $V_i$ spaces of imprimitivity of $G$. Let $G_i = \{g \in G | g V_i = V_i\}$ so $|G : G_i| = d$, and let $N = \bigcap_{i=1}^{d} G_i$ so $N \lhd G$ and $|G : N| \, |d!$. We may assume each $G_i$ primitive on $V_i$; for if not, then by Lemma 29.5 and transitivity of induction we could choose a larger $d$ where all $G_i$ are primitive.

Denote

$$H_i = \{h \in G_i | hg|_{V_i} = gh|_{V_i}, \text{ all } g \in G_i\}.$$

We have, by Theorem 30.3,

$$|G_i : H_i| > m!(6^{m-1})^{\pi(m+1)+1},$$

all $H_i \lhd G_i$, $\bigcap_{i=1}^{d} H_i \subseteq Z(N)$.

We have

$$|N : N \cap H_i| = |NH_i : H_i| \leq |G_i : H_i|,$$

so

$$|N : Z(N)| \leq |N : \bigcap_{i=1}^{d} H_i| = |N : \bigcap_{i=1}^{d} (N \cap H_i)|$$

$$\leq \prod_{i=1}^{d} |N : N \cap H_i| \leq \prod_{i=1}^{d} |G_i : H_i|.$$

Therefore $Z(N)$ is an abelian normal subgroup of $G$ and satisfies

$$|G : Z(N)| = |G : N| \, |N : Z(N)| < (d!)(m!(6^{m-1})^{\pi(m+1)+1})^d$$

$$= d!(m!)^d(6^{n-d})^{\pi(m+1)+1} < n!(6^{n-1})^{\pi(n+1)+1}.$$

Finally, suppose $V$ reducible, say $V = V_1 \oplus V_2$, the $V_i$ $\mathbf{C}G$-modules. Then $G$ has normal subgroups $A_1$ and $A_2$ with $A_i|_{V_i}$ abelian. Denote dim $V_i = n_i$, so $n_1 + n_2 = n$. $A_1 \cap A_2$ is an abelian normal subgroup of

$G$. The theorem holds for $G/A_1$ and $G/A_2$ by induction on $n$, so

$$|G: A_1 \cap A_2| \leq |G: A_1|\, |G: A_2|$$
$$< (6^{n_1-1})^{\pi(n_1+1)+1}(n_1!)(6^{n_2-1})^{\pi(n_2+1)+1}(n_2!)$$
$$< n!(6^{n-1})^{\pi(n+1)+1}.$$

*Remark* Jordan [1] first proved that for some function $f(n)$, finite subgroups of $GL(n, \mathbf{C})$ must have abelian normal subgroups of index $\leq f(n)$. This is known as *Jordan's theorem*. The proof we have given, due to Blichfeldt [1], gives the best value for $f(n)$ known even today. (Blichfeldt stated, but did not prove, that he can replace 6 by 5 in Theorem 30.4.) The simplest known proof of Jordan's theorem is probably the one in §36 of Curtis and Reiner [1]; that proof does not obtain the information about primitive linear groups and their abelian subgroups and Sylow subgroups that we have here.

An easy *lower* bound for $f(n)$ is $(n + 1)!$, since by Theorem 11.3(d) the symmetric group $S_{n+1}$ has a faithful irreducible representation of degree $n$. Brauer [12] has asked if better upper bounds for $f(n)$ can be obtained and has shown that a much better upper bound is likely. A sharp upper bound for finite *solvable* groups will be given in §36.

An analog of Jordan's theorem for fields of characteristic $p$ is given in Brauer and Feit [1].

# §31

## Extra-Special p-Groups

**Definition**    A finite $p$-group $P$ is *extra-special* if $P' = Z(P)$, $|P'| = p$, and $P/P'$ is elementary abelian.

*Remark*  If $p$ is an odd prime, there are two nonisomorphic non-abelian groups of order $p^3$, both extra-special. One has exponent $p$ and one has exponent $p^2$. There are two nonisomorphic nonabelian groups of order $2^3 = 8$, both extra-special of exponent $2^2 = 4$. One is the quaternion group $Q$ of §19, and the other is the *dihedral group* $D$ with generators $c$, $d$ and relations $c^2 = d^4 = 1$, $d^c = d^{-1}$.

**Lemma 31.1**    *Let $G$ be a group satisfying $G' \subseteq Z(G)$. If $x$, $y$, $z \in G$ and $n$ is a positive integer, then we have*

$$[xy, z] = [x, z][y, z], \ [x, yz] = [x, y][x, z],$$
$$[x^n, y] = [x, y]^n = [x, y^n], \qquad (xy)^n = x^n y^n [y, x]^{n(n-1)/2}.$$

*Proof*  For any elements $u$, $v$, $w$ of any group, one can verify that

$$[uv, w] = [u, w][[u, w], v][v, w]$$
$$\text{and} \qquad [u, vw] = [u, w][u, v][[u, v], w].$$

The first two equations follow directly from these, using $G' \subseteq Z(G)$. The third equation follows from the first two by induction on $n$. The last equation was Lemma 19.4.

**Definition**    A finite group $G$ is a *central product* of subgroups $H_1, \ldots, H_n$ if $G = H_1 \cdots H_n$ and, for any $x \in H_i$, $y \in H_j$, $i \neq j$ implies $[x, y] = 1$.

*Remark*  This certainly forces all $H_i \lhd G$, and $H_i \cap H_j \subseteq Z(G)$

whenever $i \neq j$. Notice that any central product $H_1 \cdots H_n$ is a homomorphic image of the direct product $H_1 \times \cdots \times H_n$ via the natural map $(h_1, \ldots, h_n) \rightarrow h_1 \cdots h_n$.

**Lemma 31.2**    Let $P_1, \ldots, P_n$ be extra-special $p$-groups of order $p^3$. Then there is one and up to isomorphism only one central product of $P_1, \ldots, P_n$ with center of order $p$. It is extra-special of order $p^{2n+1}$, denoted by $P_1 \cdots P_n$, and called *the* central product of $P_1, \ldots, P_n$.

*Proof of Existence*    Choose $x_i \in Z(P_i)$ of order $p$, and let $N$ be the subgroup of $Z(P_1 \times \cdots \times P_n)$ generated by $(x_1, x_2, 1, 1, \ldots, 1)$, $(x_1, 1, x_3, 1, \ldots, 1)$, ..., $(x_1, 1, \ldots, 1, x_n)$. Then $|N| = p^{n-1}$, and $(1, \ldots, 1, x_i, 1, \ldots, 1) \notin N$, any $i$. We take $H = (P_1 \times \cdots \times P_n)/N$. If $(y_1, \ldots, y_n)N \in Z(H)$, then for any $z_i \in P_i$,

$$1 = [(y_1, \ldots, y_n)N, (1, \ldots, 1, z_i, 1, \ldots, 1)N]$$

$$= [(1, \ldots, y_i, \ldots, 1)N, (1, \ldots, z_i, \ldots, 1)N]$$

$$= (1, \ldots, [y_i, z_i], \ldots 1)N,$$

so $[y_i, z_i] = 1$, $y_i \in Z(P_i)$. Thus $(y_1, \ldots, y_n) \in Z(P_1) \times \cdots \times Z(P_n)$, proving $|Z(H)| = p$. $H$ is the desired group.

*Proof of Uniqueness*    Consider the case $p$ odd, and suppose $P_1, \ldots, P_m$ have exponent $p$, $P_{m+1}, \ldots, P_n$ have exponent $p^2$. Let $H$ be a central product of $P_1, \ldots, P_n$ with center $\langle c \rangle$ of order $p$.

If $i \leq m$, $P_i$ has generators $x_i$, $y_i$ with relations $x_i^p = y_i^p = 1$, $[x_i, y_i] = c^{n_i}$, some integer $n_i$ not divisible by $p$. If $s_i n_i \equiv 1 \pmod{p}$, then $[x_i^{s_i}, y_i] = c$ by Lemma 31.1, so replacing $x_i$ by $x_i^{s_i}$ we can assume $x_i^p = y_i^p = 1$, $[x_i, y_i] = c$.

If $i > m$, $P_i$ has generators $x_i$, $y_i$ with $x_i^p = 1$, $y_i^p = c^{m_i}$, $[x_i, y_i] = c^{n_i}$, some integers $m_i$, $n_i$ not divisible by $p$. If $t_i m_i \equiv 1 \pmod{p}$, then $(y_i^{t_i})^p = c^{m_i t_i} = c$; and if $s_i n_i t_i \equiv 1 \pmod{p}$, then $[x_i^{s_i}, y_i^{t_i}] = c$. Replacing $x_i$ by $x_i^{s_i}$, $y_i$ by $y_i^{t_i}$, we see that $P_i$ has relations $x_i^p = 1$, $y_i^p = c$, $[x_i, y_i] = c$.

$H$ is now completely determined as the group with generators $x_1, \ldots, x_n, y_1, \ldots, y_n, c$ and the above relations, so $H$ is unique.

The proof is similar (but easier) if $p = 2$.

*Remark*    If $Z$ is a cyclic $p$-group, $E$ an extra-special $p$-group, there is clearly one and only one central product of $Z$ and $E$ which is not a direct product. We denote it by $ZE$.

**Lemma 31.3** *$QQ$ and $DD$ are isomorphic groups of order 32, not isomorphic to $DQ$. If $Z$ is a cyclic 2-group of order $\geq 4$, then $ZQ \cong ZD$.*

*Proof* An exercise.

**Lemma 31.4** *Every extra-special $p$-group $P$ is the central product of nonabelian $p$-groups of order $p^3$, and so has order $p^{2m+1}$, some $m$.*

*Proof* By induction on $|P|$. Choose $x \in P - P'$, and define $\varphi_x \colon z \to [x, z]$, all $z \in P$. By Lemma 31.1, $\varphi_x$ is a homomorphism. Since $x \notin Z(P)$ and $|P'| = p$, $\varphi_x$ is onto $P'$, with kernel $C_P(x)$. Therefore $|P \colon C_P(x)| = p$. Choose $y \in P - C_P(x)$. $x^p \in P'$, $y^p \in P'$ since $P$ is extra-special, so $|\langle xP', yP' \rangle| = p^2$ and $|\langle x, y \rangle| = p^3$, $\langle x, y \rangle \supset P'$. Let $P_0 = C_P(\langle x, y \rangle) = C_P(x) \cap C_P(y)$.

$C_P(x) \neq C_P(y)$ and $|P \colon C_P(x)| = |P \colon C_P(y)| = p$, so $|P \colon P_0| = p^2$. If $|P_0| = p$ then $P_0 = P'$, $P = \langle x, y \rangle$, $|P| = p^3$, done. If $|P_0| > p$, consider any $z \in Z(P_0)$.

$z \in C_P(\langle P_0, x, y \rangle) = C_P(P) = Z(P) = P'$, so $Z(P_0) = P'$, $P_0$ is not abelian. Therefore $P_0' = P'$ and $P_0$ is extra-special. By induction, $P_0$ is the central product of nonabelian $p$-groups of order $p^3$. But $P$ is the central product of $P_0$ and $\langle x, y \rangle$, done.

*Remark* In the special case $|Z| = p$, the following theorem completely describes the irreducible complex characters of extra-special $p$-groups.

**Theorem 31.5** *Let $E$ be an extra-special $p$-group of order $p^{2m+1}$, $Z$ a cyclic $p$-group of order $p^k$, so $|ZE| = p^{2m+k}$. Then $ZE$ has exactly the following irreducible complex characters:*

(i) *$p^{2m+k-1}$ linear characters.*

(ii) *$p^k - p^{k-1}$ faithful irreducible characters $\chi_i$ of degree $p^m$, which vanish outside $Z$ and satisfy $\chi_i|_Z = p^m \lambda_i$, $\lambda_i$ a faithful linear character of $Z$.*

*Proof* Certainly the characters (i) are present, since $P = ZE$ satisfies $|P \colon P'| = p^{2m+k-1}$. Let these linear characters be $\mu_1, \ldots, \mu_t$, $t = p^{2m+k-1}$. Let the other irreducible characters be $\chi_1, \ldots, \chi_s$. By the orthogonality relations, for any $x \in P - Z(P) = P - Z$ we have

$$p^{2m+k-1} \geq |C_P(x)| = \sum_{i=1}^{t} |\mu_i(x)|^2 + \sum_{i=1}^{s} |\chi_i(x)|^2$$

$$= p^{2m+k-1} + \sum_{i=1}^{s} |\chi_i(x)|^2.$$

Therefore all $\chi_i$ vanish outside $Z$.

The $\chi_i$ do not have $P'$ in their kernels, so $\ker \chi_i \cap Z = 1$ for all $i$. By Lemma 19.5, all $\ker \chi_i = 1$, all $\chi_i$ are faithful. If $\chi_i$ is afforded by the representation $T_i$, then by Theorem 4.2, $T_i(z)$ is a scalar, any $z \in Z$; therefore $|\chi_i(z)| = \chi_i(1)$. Hence

$$1 = (\chi_i, \chi_i)_P = \frac{1}{p^{2m+k}} \sum_{z \in Z} |\chi_i(z)|^2$$

$$= \frac{1}{p^{2m+k}} |Z| \chi_i(1)^2 = \frac{1}{p^{2m}} \chi_i(1)^2,$$

implying $\chi_i(1) = p^m$.

The condition $|\chi_i(z)| = \chi_i(1)$ forces $\chi_i|_Z$ to be a sum of $p^m$ copies of the same linear character $\lambda_i$. The number $s$ of $\chi_i$'s is given by

$$p^{2m+k} = |P| = \sum_{i=1}^{t} \mu_i(1)^2 + \sum_{i=1}^{s} \chi_i(1)^2 = p^{2m+k-1} + sp^{2m},$$

so $s = p^k - p^{k-1}$. ($p^k - p^{k-1}$ is, of course, the number of faithful linear characters $\lambda_i$ of $Z$.)

# §32

## Normal p-Subgroups of Primitive Linear Groups

**Definition**     If $H$ and $K$ are subgroups of $G$, denote

$$[H, K] = \langle [h, k] | h \in H, k \in K \rangle.$$

If $H \lhd G$ and $K \lhd G$, clearly $[H, K] = [K, H] \lhd G$ and $[H, K] \subseteq H \cap K$. For any group $G$, denote $G_1 = G$ and inductively $G_{i+1} = [G, G_i]$; thus $G_2 = G'$. We see that all $G_i \lhd G$ and $G = G_1 \supseteq G_2 \supseteq \cdots$.

**Lemma 32.1**     *If $G$ is a finite group, the following are equivalent:*
   (i)   *$G_n = 1$ for some $n$.*
   (ii)  *$G$ is nilpotent (i.e., the direct product of its Sylow subgroups).*

*Proof* $(i) \rightarrow (ii)$  (i) is satisfied by subgroups and homomorphic images of $G$. By induction on $|G|$, we first show that if $H \subset G$, $H \neq G$, then $H \neq N_G(H)$. We may assume $n$ chosen so $G_{n-1} \neq 1$. Then $G_n = [G, G_{n-1}] = 1$ so $G_{n-1} \subseteq Z(G)$, $Z(G) \neq 1$. If $Z(G) \nsubseteq H$, clearly $H \neq N_G(H)$. If $Z(G) \subseteq H$, then by induction $H/Z(G) \neq L/Z(G)$, $L/Z(G)$ the normalizer of $H/Z(G)$ in $G/Z(G)$. Hence $H \neq L \subseteq N_G(H)$.

If $P$ is a Sylow $p$-subgroup of $G$, set $H = N_G(P)$. If $g \in N_G(H)$, then $g^{-1}Pg$ is a Sylow $p$-subgroup of $H$, $g^{-1}Pg = P$, $g \in H$, $H = N_G(H)$. By the previous paragraph, $H = G$. Therefore all Sylow subgroups of $G$ are normal in $G$, implying (ii).

*Proof* $(ii) \rightarrow (i)$  It is enough to prove that $p$-groups $P$ satisfy (i). We do this by induction on $|P|$. By induction, $(P/Z(P))_m = 1$ for some $m$. For any $k$, $(P/Z(P))_k = P_k Z(P)/Z(P)$, so $1 = P_m Z(P)/Z(P)$, $P_m \subseteq Z(P)$, $P_{m+1} = 1$.

**Lemma 32.2** (Three Subgroups Lemma)     *Let $G$ be any group,  $H \lhd G$,*

183

$K \triangleleft G$, $L \triangleleft G$. *Then*

$$[[H, K], L] \subseteq [[K, L], H][[L, H], K].$$

*Proof* For any elements $x$, $y$, $z$ of any group, one can verify the *Jacobi identity*:

$$[[x, y^{-1}], z]^y[[y, z^{-1}], x]^z[[z, x^{-1}], y]^x = 1.$$

For any $h \in H$, $k \in K$, we set $x = h$, $y = k^{-1}$, $z = l$, in the Jacobi identity, and solve for $[[h, k], l]$. We get

$$[[h, k], l] = \{([[l, h^{-1}], k^{-1}]^{-1})^h([[k^{-1}, l^{-1}], h]^{-1})^l\}^k,$$

which shows that $[[h, k], l]$ is an element of the right-hand side of the desired inequality.

Let $X = \langle [[h, k], l] | h \in H, k \in K, l \in L \rangle$. By the previous paragraph, it is enough to show $[[H, K], L] = X$; clearly $[[H, K], L] \supseteq X$. Since $[[h, k], l]^g = [[h^g, k^g], l^g]$, we know $X \triangleleft G$; also, $X \subseteq [H, K] \cap L$.

Suppose $t \in [H, K]$ and $l \in L$; we must show $t^{-1}l^{-1}tl \in X$. Let $C$ denote the centralizer of $L/X$ in $G/X$. If $h \in H$ and $k \in K$, then $[h, k] X \in C$, which shows $[H, K]/X \subseteq C$. Hence $tX \in C$, so $(tX)(lX) = (lX)(tX)$, $t^{-1}l^{-1}tl \in X$, as required.

**Lemma 32.3**     *For any group $G$, $[G_r, G_s] \subseteq G_{r+s}$.*

*Proof* By induction on $r$; if $r = 1$, then $[G_1, G_s] = [G, G_s] = G_{s+1}$ by definition. In general, using Lemma 32.2 we have

$$[G_r, G_s] = [[G, G_{r-1}], G_s] \subseteq [[G_{r-1}, G_s], G][[G_s, G], G_{r-1}]$$

$$\subseteq [G_{r+s-1}, G][G_{s+1}, G_{r-1}] \subseteq G_{r+s}G_{r+s} = G_{r+s}.$$

**Theorem 32.4**     *If $P$ is a $p$-group with only one subgroup of order $p$, then $P$ is cyclic or generalized quaternion.*

*Proof* For $p = 2$ this was Theorem 19.6, so we may assume $p$ odd and must prove $P$ cyclic. We use induction on $|P|$. Let $U_1$ be a normal subgroup of $P$ of index $p$.

If $U_1$ is the only subgroup of index $p$, then $P/P'$ is cyclic; let $P/P' = \langle xP' \rangle$ so $P = P'\langle x \rangle$. Let $H_i = P_i\langle x \rangle$, so $H_2 = P'\langle x \rangle$. Then for any

$y \in P_i$, $x^y = x[x, y] \in \langle x \rangle P_{i+1} = H_{i+1}$ so $H_{i+1} \triangleleft H_i$. $H_i/H_{i+1} \cong P_i/P_i \cap P_{i+1}\langle x \rangle$ is a homomorphic image of $P_i/P_{i+1}$ and hence is abelian. If $H_j = P$, then $P/H_{j+1}$ is abelian, so $H_{j+1} \supseteq P'$ and $H_{j+1} \supseteq P'\langle x \rangle = P$. Thus $H_t = P$ for all $t$, $P = \langle x \rangle$, $P$ is cyclic.

If $P$ has two subgroups of index $p$, say $U_1$ and $U_2$, then both are normal, so $U_1 \cap U_2 \supseteq P'$; and $U_1$ and $U_2$ are cyclic by induction.

$$D = U_1 \cap U_2 \subseteq C_P(\langle U_1, U_2 \rangle) = Z(P),$$

so $P' \subseteq Z(P)$. If $x \in P$, then $x^p \in D$, $|P : D| = p^2$. Define $\varphi : P \to D$ by $\varphi : x \to x^p$. For any $x$, $y \in P$, $[y, x]^p = [y^p, x] = 1$ by Lemma 31.1, so

$$(xy)^p = x^p y^p [y, x]^{p(p-1)/2} = x^p y^p,$$

proving that $\varphi$ is a homomorphism. $|\ker \varphi| \geq p^2$ and elements of $\ker \varphi$ have order $p$, a contradiction.

**Theorem 32.5**    *A $p$-group $P$ of order $p^n$ is cyclic if it contains only one subgroup $U$ of order $p^m$ for some $m$, $1 < m < n$.*

*Proof*   Choose $V \subseteq P$, $|V : U| = p$. $U$ is the only subgroup of $V$ of index $p$, so as in the second paragraph of the previous proof we have $V$ cyclic. In particular, $U$ is cyclic. Every subgroup of $P$ of order $p$ or $p^2$ is in a subgroup of order $p^m$, and so is in $U$. $U$ is cyclic, so $P$ has one subgroup of order $p$, one of order $p^2$. The generalized quaternion group has two subgroups of order 4, so by Theorem 32.4 $P$ is cyclic.

**Theorem 32.6** (Rigby [1])    *Let $G$ be a finite primitive linear group of degree $n$ over an algebraically closed field $k$, $P$ a nonabelian normal $p$-subgroup of $G$. Then:*

(1)  *$P$ is a central product $ZE$, $Z = Z(P)$ cyclic, $Z \subseteq Z(G)$, $E$ an extra-special $p$-group.*
(2)  *If $p$ is odd, $E$ has exponent $p$.*
(3)  *If $p = 2$, $P$ is not dihedral of order 8.*
(4)  *If $k = \mathbb{C}$ and $|E| = p^{2m+1}$, then $p^m | n$.*

*Proof, Step 1*   $P' \subseteq Z(P)$.

*Proof of Step 1*   Let $P$ have class $n$, so $P_n \neq 1 = P_{n+1}$. If $n \geq 3$, then by Lemma 32.3,

$$[P_{n-1}, P_{n-1}] \subseteq P_{2n-2} \subseteq P_{n+1} = 1,$$

so $P_{n-1}$ is abelian. $P_{n-1}$ char $P \lhd G$, so by Lemma 29.6 $P_{n-1} \subseteq Z(G)$. Hence $P_n = [P, P_{n-1}] = 1$, the desired contradiction.

*Step 2*   $|P'| = p$. Denote $P' = \langle c \rangle$.

*Proof*   By Lemma 29.6, $Z(P) \subseteq Z(G)$ is cyclic, so $P' \subseteq Z(P)$ is cyclic. Let $|P'| = p^m$, $P' = \langle c \rangle$, $c = [a, b]$ for $a, b \in P$.
Suppose $m > 1$. By Lemma 31.1, if $x, y \in P$ then

$$[x^{p^{m-1}}, y^{p^{m-1}}] = [x, y]^{p^{2m-2}} \in \langle c^{p^{2m-2}} \rangle = 1,$$

so $A = \langle x^{p^{m-1}} | x \in P \rangle$ is abelian. $A$ char $P \lhd G$, so by Lemma 29.6 $A \subseteq Z(P)$. Hence $c^{p^{m-1}} = [a, b]^{p^{m-1}} = [a, b^{p^{m-1}}] = 1$, a contradiction. Therefore $m = 1$.

*Step 3*   $P/Z(P)$ is elementary abelian.

*Proof*   If $x, y \in P$, then by Lemma 31.1, $[x^p, y] = [x, y]^p \in \langle c^p \rangle = 1$, $x^p \in Z(P)$. And $P' \subseteq Z(P)$, so $P/Z(P)$ is elementary abelian.

*Step 4*   If $p$ is odd, then (1) and (2) hold.

*Proof*   If $x, y \in P$ have order $p$, then by Lemma 31.1 and Step 2 we have

$$(xy)^p = x^p y^p [y, x]^{p(p-1)/2} = x^p y^p = 1.$$

Therefore $P_0 = \{x \in P | x^p = 1\}$ is a characteristic subgroup of $P$. If $P_0$ is abelian, then $P_0$ is cyclic by Lemma 29.6 and $P$ is cyclic by Theorem 32.4, a contradiction. Hence $P_0$ is not abelian, and we can pick $x_1, y_1 \in P_0$ with $\langle x_1, y_1 \rangle$ nonabelian of order $p^3$ and exponent $p$.
     Let

$$H_1 = C_P(\langle x_1, y_1 \rangle) = C_P(x_1) \cap C_P(y_1).$$

We have $|P: C_P(x_1)| = |P: C_P(y_1)| = p$, so $|P: H_1| = p^2$. If $H_1$ is abelian, then $H_1 = Z(P)$ is cyclic, $P$ is the central product of $H_1$ and $\langle x_1, y_1 \rangle$ and we are done. If $H_1$ is not abelian, consider $H_{10} = H_1 \cap P_0$. $H_{10}$ abelian would mean $H_{10} = Z(P_0)$, so by Lemma 29.6, $H_{10}$ would be cyclic; Theorem 32.4 would then force $H_1$ cyclic, a contradiction. So $H_{10}$ is not abelian, and we pick $x_2, y_2 \in H_{10}$ with $\langle x_2, y_2 \rangle$ nonabelian of order $p^3$ and exponent $p$. We now define $H_2 = C_P(x_1, y_1, x_2, y_2)$, $H_{20}, \ldots$, continuing the process until the proof of (1) and (2) is complete.

*Step 5*   If $p = 2$, then (1) holds.

*Proof*   The proof is similar to Step 4. However, we cannot in general use Lemma 31.1 to say that $\{x \in P \mid x^2 = 1\}$ is a subgroup. Instead, we set $P_0 = \{x \in P \mid x^4 = 1\}$. By Lemma 31.1, $(xy)^4 = x^4 y^4 [y, x]^6 = x^4 y^4$, so $P_0$ is a characteristic subgroup of $P$. If $P_0$ is abelian, then $P_0$ is cyclic by Lemma 29.6; hence $P$ has only one subgroup of order 4, and $P$ is cyclic by Theorem 32.5, a contradiction. Therefore $P_0$ is not abelian, and we can pick $x_1, y_1 \in P_0$ with $[x_1, y_1] = c$. Also $x_1^2 \in \langle c \rangle$ and $y_1^2 \in \langle c \rangle$, so $\langle x_1, y_1 \rangle$ is nonabelian of order 8. Now we set $H_1 = C_P(\langle x_1, y_1 \rangle)$, $H_{10} = P_0 \cap H_1$, and continue as in Step 4 to complete the proof of (1).

*Step 6*   (3) holds.

*Proof*   $D$ has exactly two elements of order 4, so $P \cong D$ would mean $P$ had a characteristic noncentral cyclic subgroup $A$ of order 4; this would contradict Lemma 29.6.

*Step 7*   (4) holds.

*Proof*   Let $G$ be a linear group on the **C**-vector space $V$ of dimension $n$. By Clifford's theorem and the primitivity of $G$, $V_P = W_1 \oplus \cdots \oplus W_t$, the $W_i$ isomorphic faithful irreducible **C**$P$-modules. By Theorem 31.5, $\dim_{\mathbf{C}} W_i = p^m$.

EXERCISE

Show that if $G$ is a Frobenius group with complement $H$, then all Sylow subgroups of $H$ are cyclic or generalized quaternion.

# §33

## The Frobenius-Schur Count of Involutions

**Lemma 33.1**    *Let $G$ be a finite group, $\varphi_1: G \to Mat_n(\mathbf{C})$ and $\varphi_2: G \to Mat_n(\mathbf{C})$ equivalent irreducible matrix representations of $G$, and assume all matrices $\varphi_1(g)$, $\varphi_2(g)$ are unitary. Then there is a unitary matrix $U$ such that*

$$U^{-1}\varphi_1(g)U = \varphi_2(g), \qquad \text{all} \quad g \in G.$$

*Proof*   At least there is a complex matrix $M$ satisfying

$$M^{-1}\varphi_1(g)M = \varphi_2(g), \qquad \text{all} \quad g \in G.$$

Let $\bar{\phantom{x}}$ denote complex conjugation. Since $\varphi_1(g)M = M\varphi_2(g)$, we have

$$\varphi_2(g)^{-1}({}^t\overline{M}) = {}^t\overline{\varphi_2(g)}\,{}^t\overline{M} = {}^t\overline{(M\varphi_2(g))}$$
$$= {}^t\overline{(\varphi_1(g)M)} = {}^t\overline{M}\,{}^t\overline{\varphi_1(g)} = {}^t\overline{M}\varphi_1(g)^{-1}.$$

Therefore

$$\varphi_1(g)M({}^t\overline{M})\varphi_1(g^{-1}) = M\varphi_2(g)\varphi_2(g^{-1})({}^t\overline{M}) = M({}^t\overline{M}),$$

implying that $\varphi_1(g)M({}^t\overline{M}) = M({}^t\overline{M})\varphi_1(g)$, all $g \in G$. By Theorem 4.2 this means that $M({}^t\overline{M}) = cI$, some $c \in \mathbf{C}$.

If $M = (m_{ij})$, then the $i$, $i$th entry of $M({}^t\overline{M})$ is

$$\sum_{t=1}^{n} m_{it}\overline{m_{it}} = \sum_{t=1}^{n} |m_{it}|^2 = c,$$

implying that $c > 0$ is real. Let $c = m\overline{m}$, $m \in \mathbf{C}$, and $U = (1/m)M$, so $U({}^t\overline{U}) = I$ and $U$ is unitary.

$$U^{-1}\varphi_1(g)U = \left(\frac{1}{m}M\right)^{-1}\varphi_1(g)\left(\frac{1}{m}M\right) = M^{-1}\varphi_1(g)M = \varphi_2(g),$$

done.

*Remark* Let $\varphi$ be a complex matrix representation of $G$ affording the character $\chi$, and assume $\chi = \bar{\chi}$ (i.e., $\chi(g)$ is real for all $g \in G$). Denote $\mathbf{R}$ = real numbers. If $s_{\mathbf{R}}(\chi)$ denotes the Schur index of $\chi$ over $\mathbf{R}$, we saw in §24 that $\varphi$ is equivalent to a real matrix representation of $G$ if and only if $s_{\mathbf{R}}(\chi) = 1$. If $s_{\mathbf{R}}(\chi) \neq 1$, then $s_{\mathbf{R}}(\chi) = 2$ because $\dim_{\mathbf{R}}\mathbf{C} = 2$ (Theorem 24.12(1)).

**Lemma 33.2** *If $M$ is a unitary matrix and $D$ a unitary diagonal matrix such that $MD = DM$, then there is a diagonal unitary matrix $E$ such that $E^2 = D$ and $ME = EM$.*

*Proof* Let $P$ be a permutation matrix such that

$$P^{-1}DP = \begin{pmatrix} \alpha_1 I & & 0 \\ & \ddots & \\ 0 & & \alpha_r I \end{pmatrix},$$

the $\alpha_i$ all different. By Lemma 29.1(e), $P$ is unitary. Then $P^{-1}MP \cdot P^{-1}DP = P^{-1}DP \cdot P^{-1}MP$, so

$$P^{-1}MP = \begin{pmatrix} A_1 & & 0 \\ & \ddots & \\ 0 & & A_r \end{pmatrix}, \qquad \text{each } A_i \text{ unitary.}$$

Choose $\beta_1, \ldots, \beta_r$ with $\beta_i^2 = \alpha_i$ and set

$$E_0 = \begin{pmatrix} \beta_1 I & & 0 \\ & \ddots & \\ 0 & & \beta_r I \end{pmatrix},$$

so $E_0^2 = P^{-1}DP$ and $P^{-1}MPE_0 = E_0 P^{-1}MP$. Since all $|\beta_i| = 1$, $E_0$ is unitary. Set $E = PE_0 P^{-1}$. Then $E_0 = P^{-1}EP$ so $ME = EM$, $E^2 = (PE_0 P^{-1})^2 = PE_0^2 P^{-1} = PP^{-1}DPP^{-1} = D$.

**Lemma 33.3** *Let $\varphi$ be an irreducible complex matrix representation of $G$ with character $\chi = \bar{\chi}$, and let $U$ be a unitary matrix satisfying $U^{-1}\varphi(g)U = \overline{\varphi(g)}$, all $g \in G$. Then ${}^t U = U$ or ${}^t U = -U$. Furthermore, ${}^t U = U$ if and only if $s_{\mathbf{R}}(\chi) = 1$, ${}^t U = -U$ if and only if $s_{\mathbf{R}}(\chi) = 2$.*

*Proof* By Lemma 33.1, such a matrix $U$ exists if $\varphi$ is unitary. $U^{-1} = {}^t\overline{U}$ and $\overline{U^{-1}} = {}^tU$, $\overline{U} = ({}^tU)^{-1}$. Hence

$$\varphi(g) = \overline{U^{-1}\varphi(g)}\overline{U} = {}^tU\overline{\varphi(g)}({}^tU)^{-1} = {}^tUU^{-1}\varphi(g)U({}^tU)^{-1}.$$

Hence ${}^tUU^{-1}$ commutes with all $\varphi(g)$, and by Theorem 4.2 we have ${}^tUU^{-1} = cI$, $c \in \mathbf{C}$. Therefore ${}^tU = cU$, $U = {}^{tt}U = {}^t(cU) = c{}^tU = c^2U$, $c^2 = 1$, $c = \pm 1$. It remains to show that ${}^tU = U$ if and only if $s_\mathbf{R}(\chi) = 1$.

If $s_\mathbf{R}(\chi) = 1$, let $M$ be a complex matrix such that, for all $g \in G$, $M^{-1}\varphi(g)M$ is real. Then

$$M^{-1}\varphi(g)M = \overline{M}^{-1}\overline{\varphi(g)}\overline{M} = \overline{M}^{-1}U^{-1}\varphi(g)U\overline{M}.$$

Therefore $U\overline{M}M^{-1}$ commutes with all $\varphi(g)$, and by Theorem 4.2 we have $U\overline{M}M^{-1} = mI$, $m \in \mathbf{C}$. Then $U = mM\overline{M}^{-1}$ and

$$ {}^tU = \overline{U}^{-1} = \overline{\overline{M}M^{-1}m^{-1}} = \overline{m}^{-1}M\overline{M}^{-1}.$$

${}^tU = -U$ would mean

$$-mM\overline{M}^{-1} = \overline{m}^{-1}M\overline{M}^{-1}, \qquad \overline{m}^{-1} = -m, \qquad 1 = -m\overline{m} = -|m|^2,$$

a contradiction. Therefore ${}^tU = U$.

Conversely, assume ${}^tU = U$. By Lemma 29.2(2), there is a unitary matrix $C$ with $C^{-1}UC = D$ diagonal. Therefore $CDC^{-1} = U = {}^tU = {}^t(C^{-1})D({}^tC)$, implying that $({}^tC)CD = D({}^tC)C$. By Lemma 33.2, choose a diagonal unitary matrix $E$ satisfying $E^2 = D$ and $({}^tC)CE = E({}^tC)C$. If $V = CEC^{-1}$, then $V^2 = CDC^{-1} = U$. We see that

$$ {}^tV = {}^t(C^{-1})E({}^tC) = {}^t(C^{-1})E({}^tC)CC^{-1}$$
$$= {}^t(C^{-1})({}^tC)CEC^{-1} = CEC^{-1} = V,$$

so

$$V^{-1} = {}^t\overline{V} = \overline{V}.$$

For all $g \in G$, we have

$$\overline{V^{-1}\varphi(g)V} = \overline{V}\overline{\varphi(g)}\overline{V} = VU^{-1}\varphi(g)UV^{-1}$$
$$= VV^{-2}\varphi(g)V^2V^{-1} = V^{-1}\varphi(g)V.$$

If we define $\psi(g) = V^{-1}\varphi(g)V$, then $\psi$ is a representation of $G$ equivalent to $\varphi$ with all $\psi(g)$ real; so $s_\mathbf{R}(\chi) = 1$.

**Lemma 33.4**     *Let G be a finite group, $\chi$ a complex irreducible character of G. Then*

$$\frac{1}{|G|} \sum_{g \in G} \chi(g^2) = 1 \qquad if \quad \chi = \bar{\chi} \quad and \quad s_{\mathbf{R}}(\chi) = 1$$

$$= -1 \qquad if \quad \chi = \bar{\chi} \quad and \quad s_{\mathbf{R}}(\chi) = 2$$

$$= 0 \qquad if \quad \chi \neq \bar{\chi}.$$

*Proof*  Let $\chi$ be afforded by the matrix representation $\varphi$, and denote $\varphi(g) = (a_{ij}(g))$. Therefore

$$\frac{1}{|G|} \sum_{g \in G} \chi(g^2) = \frac{1}{|G|} \sum_{g \in G} \operatorname{tr} \varphi(g)\varphi(g)$$

$$= \frac{1}{|G|} \sum_{g \in G} \operatorname{tr} \varphi(g)\,{}^t\overline{\varphi(g^{-1})}$$

$$= \frac{1}{|G|} \sum_{g \in G} \sum_{i,j} a_{ij}(g)\overline{a_{ij}(g^{-1})}.$$

If $\chi \neq \bar{\chi}$, then since $\bar{\chi}$ is afforded by the map $g \to \overline{\varphi(g)}$, Lemma 5.2(a) shows this last sum is 0.

If $\chi = \bar{\chi}$ and $s_{\mathbf{R}}(\chi) = 1$, then we can assume all $a_{ij}(g)$ real. Using Lemma 5.2(b), we have

$$\frac{1}{|G|} \sum_{g \in G} \sum_{i,j} a_{ij}(g)\overline{a_{ij}(g^{-1})} = \frac{1}{|G|} \sum_{g \in G} \sum_{i,j} a_{ij}(g)a_{ij}(g^{-1})$$

$$= \frac{1}{|G|} \sum_{g \in G} \sum_i a_{ii}(g)a_{ii}(g^{-1})$$

$$= \frac{1}{|G|} (\deg \varphi) \frac{|G|}{\deg \varphi} = 1.$$

If $\chi = \bar{\chi}$ and $s_{\mathbf{R}}(\chi) = 2$, then by Lemmas 33.1 and 33.3 there is a unitary matrix $U$ satisfying ${}^tU = -U$ and $U^{-1}\varphi(g)U = \overline{\varphi(g)}$, all $g \in G$. Denote $U = (u_{ij})$; we have $U^{-1} = {}^t\overline{U} = -\overline{U}$. The equation

$$\overline{\varphi(g^{-1})} = U^{-1}\varphi(g^{-1})U = -\overline{U}\varphi(g^{-1})U$$

implies that

$$\overline{a_{ij}(g^{-1})} = -\sum_{s,t} \overline{u_{is}}a_{st}(g^{-1})u_{tj}.$$

Using Lemma 5.2(b) and the fact that $U$ is unitary, we have

$$\frac{1}{|G|} \sum_{g \in G} \sum_{i,j} a_{ij}(g)\overline{a_{ij}(g^{-1})}$$

$$= \frac{-1}{|G|} \sum_{g \in G} \sum_{i,j} \sum_{s,t} a_{ij}(g)\overline{u_{is}} a_{st}(g^{-1}) u_{tj}$$

$$= \frac{-1}{|G|} \sum_{i,j} \sum_{s,t} \overline{u_{is}} u_{tj} \frac{\delta_{js}\delta_{it}}{\deg \varphi} |G|$$

$$= \frac{-1}{\deg \varphi} \sum_{i,j} \overline{u_{ij}} u_{ij}$$

$$= \frac{-1}{\deg \varphi} \sum_{i} (\sum_{j} \overline{u_{ij}} u_{ij}) = \frac{-1}{\deg \varphi} \sum_{i} 1 = -1.$$

**Theorem 33.5** (Frobenius and Schur[1])    *Let $G$ be a finite group, $1_G = \chi_1, \ldots, \chi_h$ the irreducible complex characters of $G$. Define*

$$v(\chi_i) = 1 \qquad if \quad \chi_i = \overline{\chi_i} \quad and \quad s_{\mathbf{R}}(\chi_i) = 1$$

$$= -1 \qquad if \quad \chi_i = \overline{\chi_i} \quad and \quad s_{\mathbf{R}}(\chi_i) = 2$$

$$= 0 \qquad if \quad \chi_i \neq \overline{\chi_i}.$$

*(1)   For any $g \in G$, let $t(g) = |\{x \in G | x^2 = g\}|$. Then*

$$t(g) = \sum_{i=1}^{h} v(\chi_i)\chi_i(g).$$

*(2)   The number of elements of $G$ of order 2 is*

$$\sum_{i=1}^{h} v(\chi_i)\chi_i(1) - 1 \leq \sum_{i=1}^{h} \chi_i(1) - 1.$$

*Equality holds if and only if $\mathbf{R}$ is a splitting field for $G$.*

*Proof (1)    $t$ is a class function of $G$, so there exist elements $a_i \in \mathbf{C}$* with

$$t(g) = \sum_{i=1}^{h} a_i \chi_i(g), \qquad all \quad g \in G.$$

Let $T(g) = \{x \in G | x^2 = g\}$, so $t(g) = |T(g)|$, and choose $g_1, \ldots, g_m \in G$

such that $G = \bigcup_{i=1}^{m} T(g_i)$ (disjoint union). Using Lemma 33.4, we have

$$a_i = (t, \chi_i)_G = \frac{1}{|G|} \sum_{g \in G} t(g)\overline{\chi_i(g)} = \frac{1}{|G|} \sum_{j=1}^{m} t(g_j)\overline{\chi_i(g_j)}$$

$$= \frac{1}{|G|} \sum_{x \in G} \overline{\chi_i(x^2)} = v(\chi_i).$$

*Proof (2)*    For the first statement, take $g = 1$ in (1). Equality holds iff all $v(\chi_i) = 1$, iff all $\chi_i = \bar{\chi}_i$ and all $s_R(\chi_i) = 1$, iff all $\chi_i$ are realizable over **R**, iff **R** is a splitting field for $G$.

To provide an example of the use of Theorem 33.5, we discuss extra-special 2-groups.

**Corollary 33.6**    *Let $E$ be an extra-special 2-group of order $2^{2m+1}$, $Q$ the quaternion group of order 8, $D$ the dihedral group of order 8. Then:*

(*1*)    *$E$ is isomorphic to $DD \cdots D$ ($m$ copies of $D$) or $QD \cdots D$ ($m - 1$ copies of $D$).*

(*2*)    *$E$ has only one nonlinear irreducible character $\chi$ over **C**. $\chi(1) = 2^m$ and $\chi = \bar{\chi}$.*

(*3*)    *If $E \cong DD \cdots D$, then $E$ contains exactly $2^{2m} + 2^m - 1$ involutions. If $E \cong QD \cdots D$, then $E$ contains exactly $2^{2m} - 2^m - 1$ involutions.*

(*4*)    *If $E \cong DD \cdots D$, then $s_R(\chi) = 1$; if $E \cong QD \cdots D$, then $s_R(\chi) = 2$.*

*Proof (1)*    By Lemma 31.4, $E$ is the central product of $D$'s and $Q$'s. By Lemma 31.3, whenever two $Q$'s occur we can replace them by $DD$.

*Proof (2)*    This follows from Theorem 31.5. If $E' = \langle c \rangle$, $|E'| = 2$, then $\chi(1) = 2^m$, $\chi(c) = -2^m$, $\chi = 0$ on $E - E'$.

*Proof (3)*    This is proved by induction on $m$, true if $m = 1$. Suppose $E = DD \cdots D$, so $E = DE_0$ and $E_0 = D \cdots D$ ($m - 1$ copies of $D$). By induction, $E_0$ contains exactly $2^{2m-2} + 2^{m-1}$ solutions of the equation $x^2 = 1$. Let

$$S = \{(x, y) \in D \times E_0 | (xy)^2 = 1 \text{ in } E\}.$$

For any $(x, y) \in D \times E_0$, we have $(x, y) \in S$ iff either $x^2 = y^2 = 1$ or $x^2 \neq 1 \neq y^2$. Hence

$$|S| = 6(2^{2m-2} + 2^{m-1}) + 2(2^{2m-2} - 2^{m-1}) = 2^{2m+1} + 2^{m+1}.$$

Each element of $E$ has exactly two expressions $xy$, $x \in D$, $y \in E_0$. Hence the number of elements of $E$ of order at most 2 is exactly $\frac{1}{2}|S| = 2^{2m} + 2^m$. This proves (3) when $E \cong DD \cdots D$; the proof is similar when $E \cong QD \cdots D$.

*Proof* (4) $E/E'$ is elementary abelian of order $2^{2m}$, so $E$ has $2^{2m}$ linear characters, all real. By Theorem 33.5, the number of involutions in $E$ is $2^{2m} + v(\chi) 2^m - 1$. The result follows from (3) and Theorem 33.5.

# §34

## Primitive Solvable Linear Groups

**Lemma 34.1**  *If $H$ and $K$ are normal nilpotent subgroups of the finite group $G$, then $HK$ is also a normal nilpotent subgroup of $G$.*

*Proof*  By induction on $|G|$. We know at least $HK \lhd G$. If $HK \neq G$ then nilpotence of $HK$ follows by induction, so we may assume $G = HK$. $K$ is nilpotent so $1 \neq Z(K) \lhd G$. Let $N = [H, Z(K)] \lhd G$. If $N = 1$, then $Z(K) \subseteq Z(G)$ so $Z(G) \neq 1$. If $N \neq 1$, then by Lemma 19.5 we have $N \cap Z(H) \neq 1$, $N \cap Z(H) \subseteq Z(G)$, and again $Z(G) \neq 1$. Hence by induction,

$$G/Z(G) = HZ(G)/Z(G) \cdot KZ(G)/Z(G)$$

is nilpotent. Using Lemma 32.1, $G$ is nilpotent.

**Definition**  By Lemma 34.1, the product of all normal nilpotent subgroups of $G$ is a normal nilpotent subgroup of $G$. It is called the *Fitting subgroup* of $G$ and denoted Fit($G$). Fit($G$) is the unique maximal normal nilpotent subgroup of $G$, so Fit($G$) char $G$.

**Lemma 34.2**  *If $G$ is solvable then there is a chain $G = G_0 \supset G_1 \supset \cdots \supset G_n = 1$, all $G_i \lhd G$, all $G_i/G_{i+1}$ abelian.*

*Proof*  Using Lemma 8.3, we can define $G_0 = G$, $G_1 = G'$, ..., $G_{i+1} = (G_i)'$, .... Then all $G_{i+1}$ char $G_i \lhd G$.

**Theorem 34.3**  *If $G$ is a finite solvable group, then $C_G(\text{Fit}(G)) \subseteq \text{Fit}(G)$.*

*Proof*  Set $F = \text{Fit}(G)$, $C = C_G(F)$. Assume the theorem is false, so $C \cap F \neq C$. $C \lhd G$ and $C \cap F \lhd G$, so as in Lemma 34.2 we can

195

find a chain

$$G = G_0 \supset G_1 \supset \cdots \supset G_n = 1,$$

all $G_i \lhd G$, $G_i \neq G_{i+1}$, $C = G_r$, $C \cap F = G_s$, $r < s$, all $G_i/G_{i+1}$ abelian. Then $r \leq s - 1$, so $G_{s-1} \subseteq C$. We have

$$[G_{s-1}, G'_{s-1}] = [G_{s-1}, [G_{s-1}, G_{s-1}]] \subseteq [G_{s-1}, G_s] = 1$$

so $G'_{s-1} \subseteq G_s \subseteq Z(G_{s-1})$. By Lemma 15.4, $G_{s-1}$ is nilpotent. So $G_{s-1} \subseteq F$, $G_{s-1} \subseteq C \cap F = G_s$, a contradiction.

**Definition**     Let $V$ be a vector space over the field $k$, $(\ ,\ ): V \times V \to k$ a bilinear form. We say that $(\ ,\ )$ is *alternating* if $(v, v) = 0$ for all $v \in V$, and *nonsingular* if $(u, v) = 0$ for all $v \in V$ implies $u = 0$.

**Lemma 34.4**     *Let $V$ be a finite-dimensional $k$-vector space, $(\ ,\ )$ a nonsingular alternating form on $V$. Then:*

   *(1)   We can find a $k$-basis $\{u_1, v_1, \ldots, u_m, v_m\}$ of $V$ such that $(u_i, v_i) = 1$, $(v_i, u_i) = -1$, and $(x, y) = 0$ for all other basis elements $x, y$. In particular, $\dim V = 2m$ is even.*

   *(2)   Let*

$$G = \{f \in GL(V) | \ (f(u), f(v)) = (u, v), \text{ all } u, v \in V\}.$$

*Let $[\ ,\ ]$ be a second nonsingular alternating form on $V$, and let*

$$H = \{f \in GL(V) | \ [f(u), f(v)] = [u, v], \text{ all } u, v \in V\}.$$

*Then the groups $G$ and $H$ are conjugate in $GL(V)$.*

   *Proof*   (1) is proved by induction on $\dim V$. Choose $0 \neq u_1 \in V$, and by nonsingularity choose $v'_1$ with $(u_1, v'_1) \neq 0$. Let $v_1$ be a scalar multiple of $v'_1$ with $(u_1, v_1) = 1$. We have

$$0 = (u_1 + v_1, u_1 + v_1) = (u_1, u_1) + (u_1, v_1) + (v_1, u_1) + (v_1, v_1)$$

$$= (u_1, v_1) + (v_1, u_1),$$

so $(v_1, u_1) = -1$. Let $U = \{v \in V | (u_1, v) = (v_1, v) = 0\}$, so $\dim V - \dim U = 2$ and the restriction of $(\ ,\ )$ to $U$ is alternating and nonsingular on $U$. By induction, we choose $u_2, v_2, \ldots, u_m, v_m$ in $U$ to satisfy (1).

   Now, by (1) let $\{x_1, y_1, \ldots, x_m, y_m\}$ be a basis of $V$ satisfying

$[x_i, y_i] = 1$, $[y_i, x_i] = -1$, $[x, y] = 0$ for all other basis elements $x, y$. Define $T \in GL(V)$ by $T(u_i) = x_i$, $T(v_i) = y_i$. Using bilinearity, we have $[T(u), T(v)] = (u, v)$ for all $u, v \in V$. Hence

$$f \in G \quad \text{iff} \quad (u, v) = (f(u), f(v)) \quad \text{for all } u, v \in V,$$
$$\text{iff} \quad [T(u), T(v)] = [Tf(u), Tf(v)] \quad \text{for all } u, v \in V,$$
$$\text{iff} \quad [x, y] = [TfT^{-1}(x), TfT^{-1}(y)] \quad \text{for all } x, y \in V,$$
$$\text{iff} \quad TfT^{-1} \in H.$$

**Definition**     Because of Lemma 34.4(2), we can choose a fixed non-singular alternating form $(\ ,\ )$ on $V$ and define

$$G = Sp(V) = \{T \in GL(V) | (T(u), T(v)) = (u, v) \text{ for all } u, v \in V\},$$

the *symplectic group on $V$*. As a group of matrices, $G$ is also denoted by $Sp(2m, k)$, or by $Sp(2m, q)$ if $|k| = q < \infty$.

**Definition**     An automorphism $\alpha$ of a group $G$ is *inner* if it has form $\alpha: x \to gxg^{-1}$, some $g \in G$. The inner automorphisms of $G$ form a group isomorphic to $G/Z(G)$.

**Lemma 34.5**     *Let $P = ZE$ be the central product of an extra-special $p$-group $E$ and a cyclic $p$-group $Z$. If $p$ is odd, assume $E$ has exponent $p$. Let $\alpha$ be any automorphism of $P$ which is trivial on $Z$ and the factor group $P/Z$. Then $\alpha$ is inner.*

*Proof*   If $|E| = p^{2m+1}$, then $P$ has $|P/Z| = p^{2m}$ inner automorphisms. Let $E = \langle a_1, b_1 \rangle \cdots \langle a_m, b_m \rangle$, each $\langle a_i, b_i \rangle$ nonabelian of order $p^3$.

If $a_i$ has order $p$, then $a_i^\alpha$ has order $p$. Let $a_i^\alpha = a_i x$, some $x \in Z$. $1 = (a_i x)^p = a_i^p x^p = x^p$, so there are only $p$ possibilities for $x$.

If $a_i$ has order 4, then $a_i^\alpha$ has order 4. Let $a_i^\alpha = a_i x$, some $x \in Z$. $1 = (a_i x)^4 = a_i^4 x^4 = x^4$, so $x^4 = 1$. But if $x^2 \neq 1$ then $a_i^2 = x^2 \in Z$ has order 2, so $(a_i x)^2 = a_i^2 x^2 = x^2 x^2 = 1$, a contradiction. Hence $x^2 = 1$, and there are only $p = 2$ possibilities for $x$.

We have shown there are only $p$ possibilities for each of $a_1^\alpha, b_1^\alpha, \ldots, a_m^\alpha, b_m^\alpha$, so the group of automorphisms of $P$ trivial on $Z$ and $P/Z$ has order $p^{2m}$. Hence they are all inner.

**Theorem 34.6**     *Let $k$ be an algebraically closed field, $V$ an $n$-dimensional $k$-vector space, $G$ a finite primitive solvable subgroup of $GL(V)$. Let $Fit(G) =$*

$P_1 \times \cdots \times P_l$, the $P_i$ normal $p_i$-subgroups of $G$, where $P_1, \ldots, P_k$ are nonabelian and $P_{k+1}, \ldots, P_l$ are abelian. Then:

(1)  if $i > k$, $P_i$ is cyclic and $P_i \subseteq Z(G)$.

(2)  if $i \le k$, then $P_i$ is the central product $Z_i E_i$, where $Z_i = Z(P_i)$ is cyclic and $Z_i \subseteq Z(G)$; $E_i$ is an extra-special $p_i$-group of order $p_i^{2m_i+1}$, where $E_i$ has exponent $p_i$ if $p_i$ is odd.

(3)  $G/\mathrm{Fit}(G)$ is isomorphic to a solvable subgroup of $\mathrm{Sp}(2m_1, p_1) \times \cdots \times \mathrm{Sp}(2m_k, p_k)$.

(4)  At least if $k = C$, $\prod_{i=1}^{k} p_i^{m_i}$ divides $n$.

*Remark*  Solvable linear groups are extensively studied in D. Suprunenko's *Soluble and Nilpotent Linear Groups* [1]. In this section we are presenting an alternative approach to main theorems of his. Other recent papers on solvable linear groups include Huppert [3], Dixon [3, 4, 5], Padzerski [1], and Suprunenko [2, 3].

*Proof of Theorem 34.6*  (1) follows from Lemma 29.6, and (2) from Theorem 32.6.

For (3), let $E_i' = \langle c_i \rangle$, of order $p_i$. $V_i = P_i/Z_i$ is a $2m_i$-dimensional $GF(p_i)$-vector space. Consider $GF(p_i)$ as the set of integers modulo $p_i$, and define

$$[ , ]_i : V_i \times V_i \to GF(p_i)$$

by $[uZ_i, vZ_i]_i = n(u, v)$, where $[u, v] = c_i^{n(u,v)}$. By Lemma 31.1, $[ , ]_i$ is a well-defined bilinear form; it is certainly nonsingular and alternating.

If $g \in G$, then conjugation by $g$ performs an automorphism on $\mathrm{Fit}(G)$ which is trivial on $Z(G)$ and so trivial on all $c_i$. This means that

$$[u^g, v^g] = (c_i^g)^{n(u,v)}$$
$$= c_i^{n(u,v)} = [u, v], \qquad \text{all} \quad u, v \in P_i,$$

so $[(uZ_i)^g, (vZ_i)^g]_i = [uZ_i, vZ_i]_i$ and $g$ induces an element of $\mathrm{Sp}(2m_i, p_i)$ on $V_i$. We therefore have a homomorphism from $G$ into $\mathrm{Sp}(2m_1, p_1) \times \cdots \times \mathrm{Sp}(2m_k, p_k)$, and we wish to show $\mathrm{Fit}(G) = \{g \in G | g$ is trivial on all $V_i\}$.

If conjugation by $g$ fixes all $V_i$, then conjugation by $g$ induces an automorphism on each $P_i$ trivial on $Z_i$ and $P_i/Z_i$. By Lemma 34.5, there is $x_i \in P_i$ such that $x_i g$ centralizes $P_i$. Then $x_1 \cdots x_k g$ centralizes $P_1, \ldots, P_k$, $x_1 \cdots x_k g \in C_G(\mathrm{Fit}(G))$. By Theorem 34.3, $x_1 \cdots x_k g \in \mathrm{Fit}(G)$, $g \in \mathrm{Fit}(G)$.

(4)  follows from Theorem 32.6(4).

# §35

## Simplicity of PSL(n, F) and PSp(2m, F)

**Definitions**     Let $F$ be any field, $n \geq 2$ an integer, $V$ an $n$-dimensional $F$-vector space. The *special linear group* $SL(V)$ is $\{g \in GL(V) | \det g = 1\}$, and the *projective special linear group* $PSL(V)$ is the factor group $SL(V)/H$, $H$ the central subgroup of $SL(V)$ of scalar transformations. (It is easy to see $H = Z(SL(V))$.) We defined $SL(n, F)$ and $PSL(n, F)$ in §28; we have $SL(n, F) \cong SL(V)$, $PSL(n, F) \cong PSL(V)$.

Now let ( , ) be a fixed nonsingular alternating form on $V$, so $n = 2m$ is even. The *projective symplectic group on* $V$ is $PSp(V) = Sp(V)/H_1$, $H_1$ the central subgroup of $Sp(V)$ of scalar transformations. A *symplectic basis* of $V$ is a basis $\{u_1, u_2, \ldots, u_m, v_m, \ldots, v_2, v_1\}$ such that $(u_i, v_j) = \delta_{ij}$, $(u_i, u_j) = 0$, $(v_i, v_j) = 0$; note the order of the basis elements. Lemma 34.4 shows symplectic bases exist; $g \in GL(V)$ is in $Sp(V)$ if and only if $g$ sends symplectic bases to symplectic bases. The matrices of the $g \in Sp(V)$, with respect to a symplectic basis, form the group $Sp(2m, F) \cong Sp(V)$. If $H_1'$ is the central subgroup of $Sp(2m, F)$ of scalar matrices, we define

$$PSp(2m, F) = Sp(2m, F)/H_1' \cong PSp(V).$$

If $|F| = q < \infty$, then we denote $SL(n, F) = SL(n, q)$, $PSL(n, F) = PSL(n, q)$, $Sp(2m, F) = Sp(2m, q)$, $PSp(2m, F) = PSp(2m, q)$.

In this section we show that, with very few exceptions, $PSL(n, F)$ and $PSp(2m, F)$ are simple. We need some of this information later. The principal references for structure and simplicity of these groups are Dickson [1] and Dieudonné [1, 2].

**Lemma 35.1**     $Sp(2, F) = SL(2, F)$.

*Proof*  Let $\{u_1, v_1\}$ be a symplectic basis of the $F$-vector space $V$, so $(u_1, v_1) = 1$. If $g \in GL(V)$, write

$$g(u_1) = a_{11}u_1 + a_{21}v_1, \qquad g(v_1) = a_{12}u_1 + a_{22}v_1.$$

199

Then $(g(u_1), \; g(v_1)) = a_{11}a_{22} - a_{21}a_{12} = \det g$, so $g \in \mathrm{Sp}(V)$ iff $(g(u_1), g(v_1)) = 1$ iff $\det g = 1$ iff $g \in SL(V)$.

**Lemma 35.2**    (1)
$$|GL(n, q)| = q^{n(n-1)/2}(q^n - 1)(q^{n-1} - 1) \cdots (q - 1).$$

(2)  $|SL(n, q)| = \dfrac{1}{q-1} |GL(n, q)|.$

(3)  $|PSL(n, q)| = \dfrac{1}{(n, q-1)} |SL(n, q)|.$

(4)  $|\mathrm{Sp}(2m, q)| = q^{m^2}(q^{2m} - 1)(q^{2m-2} - 1) \cdots (q^2 - 1).$

(5)  $|P\mathrm{Sp}(2m, q)| = |\mathrm{Sp}(2m, q)|$  *if  q is even*

$\qquad\qquad\qquad = \tfrac{1}{2}|\mathrm{Sp}(2m, q)|$  *if  q is odd*.

*Proofs*  (1)  The first row of a nonsingular $n \times n$ matrix over $GF(q)$ can be anything except $(0, \ldots, 0)$, so there are $q^n - 1$ possibilities. The second row is not a linear combination of the first, so there are $q^n - q$ possibilities for it. Continuing, we see
$$|GL(n, q)| = (q^n - 1)(q^n - q) \cdots (q^n - q^{n-1})$$
$$= q^{n(n-1)/2}(q^n - 1) \cdots (q - 1).$$

(2)  $\det\colon GL(n, q) \to F^{\#}$ is a homomorphism with kernel $SL(n, q)$, and $|F^{\#}| = q - 1$.

(3)  $(n, q - 1)$ is the order of the group of scalar $n \times n$ matrices over $GF(q)$ of determinant 1.

(4)  If $V$ is a $2m$-dimensional $GF(q)$-vector space with nonsingular alternating form, $|\mathrm{Sp}(2m, q)|$ is the number of symplectic bases of $V$. There are $q^{2m} - 1$ possibilities for $u_1$. Since $|\{v \in V | (u_1, \; v) = 0\}| = q^{2m-1}$, there are
$$\frac{q^{2m} - q^{2m-1}}{q - 1} = q^{2m-1}$$

vectors $v_1$ with $(u_1, v_1) = 1$. Continuing, there are $q^{2m-2} - 1$ possibilities for $u_2$, $q^{2m-3}$ possibilities for $v_2$, etc. Hence $|\mathrm{Sp}(2m, q)|$ must be
$$(q^{2m} - 1)q^{2m-1}(q^{2m-2} - 1)q^{2m-3} \cdots (q^2 - 1)q$$
$$= q^{m^2}(q^{2m} - 1)(q^{2m-2} - 1) \cdots (q^2 - 1).$$

(5) If $V$ is as in (4) and $g$ is a scalar transformation in $\text{Sp}(V)$, let $g(v) = \lambda v$, all $v \in V$, where $\lambda \in F$. Then

$$1 = (u_1, v_1) = (g(u_1), g(v_1)) = (\lambda u_1, \lambda v_1) = \lambda^2,$$

so $\lambda = \pm 1$. Therefore $\{\pm 1\}$ is the group of scalars in $\text{Sp}(V)$.

**Lemma 35.3**     $\text{Sp}(4, 2) = PSp(4, 2)$ *is isomorphic to* $S_6$, *the symmetric group on 6 symbols.*

*Proof*  Let $V$ be a 4-dimensional vector space over $GF(2)$, so $V^{\#} = V - \{0\}$ has order 15. Define

$$\mathscr{S} = \{\{x_1, x_2, x_3, x_4, x_5\} \subseteq V^{\#} \,|\, \text{if } i \neq j \text{ then } (x_i, x_j) = 1\}.$$

Let $\{u_1, u_2, v_2, v_1\}$ be a symplectic basis of $V$, so that

$$(u_1, u_2) = (v_1, v_2) = (u_1, v_2) = (u_2, v_1) = 0,$$
$$(u_1, v_1) = (u_2, v_2) = 1.$$

Then

$$S_1 = \{u_1, v_1, u_1 + v_1 + u_2, u_1 + v_1 + v_2, u_1 + v_1 + u_2 + v_2\} \in \mathscr{S},$$

so $\mathscr{S}$ is not empty. In fact, given any ordered pair $u, v$ of vectors of $V$ with $(u, v) = 1$, they extend to a symplectic basis of $V$, so $\{u, v\}$ is contained in a member of $\mathscr{S}$. There are $(2^4 - 1)2^3 = 120$ such ordered pairs, and each member of $\mathscr{S}$ contains 20, so $|\mathscr{S}| = 6$.

$G = \text{Sp}(4, 2)$ must permute the set $\mathscr{S}$. Let $H = \{g \in G | g(S) = S,$ all $S \in \mathscr{S}\}$, so $H \lhd G$.

$$S_2 = \{u_2, v_2, u_2 + v_2 + u_1, u_2 + v_2 + v_1, u_2 + v_2 + u_1 + v_1\} \in \mathscr{S}.$$

Any $h \in H$ fixes $S_1$ and $S_2$, and so fixes $S_1 \cap S_2 = \{u_1 + v_1 + u_2 + v_2\}$. One-element subsets of $V^{\#}$ extend to symplectic bases, so $G$ is transitive on $V^{\#}$. For any $v \in V^{\#}$, choose $g \in G$ with $v = g(u_1 + u_2 + v_1 + v_2)$. Then

$$H(v) = H(g(u_1 + v_1 + u_2 + v_2)) = g(H(u_1 + v_1 + u_2 + v_2))$$
$$= g(u_1 + v_1 + u_2 + v_2) = v,$$

so $H$ is trivial on $V$, $H = 1$, $G$ is isomorphic to a subgroup of $S_6$. $|G| = |S_6|$, so $G \cong S_6$.

**Definition**      Let $F$ be a field, $V$ an $n$-dimensional $F$-vector space, $n \geq 2$. A *transvection* is a $g \in GL(V)$ such that $\dim_F \ker(g - 1) = n - 1$, $\text{im}(g - 1) \subseteq \ker(g - 1)$.

*Notation.* In this section we frequently have to display matrices. When we do, missing entries are always zeros; thus,

$$\begin{pmatrix} 1 & & \\ 1 & & \\ & & 1 \end{pmatrix} \qquad \text{denotes} \qquad \begin{pmatrix} 0 & 1 & 0 \\ 1 & 0 & 0 \\ 0 & 1 & 0 \end{pmatrix}.$$

**Lemma 35.4**      *Let $\dim_F V = n \geq 2$. All transvections $g$ of $GL(V)$ have determinant 1, and so are in $SL(V)$. Any two transvections are conjugate in $GL(V)$.*

*Proof*   Let $\text{im}(g - 1) = Fu_1'$, and let $\{u_1', u_2, \ldots, u_{n-1}\}$ be a basis of $\ker(g - 1)$, $\{u_1', u_2, \ldots, u_n\}$ a basis of $V$. $g$ fixes all basis elements except $u_n$, and $g(u_n) = u_n + \lambda u_1'$, some $0 \neq \lambda \in F$. Setting $u_1 = \lambda u_1'$, $\{u_1, \ldots, u_n\}$ is a basis of $V$ in which $g$ has matrix

$$\begin{pmatrix} 1 & & & & 1 \\ & 1 & & & \\ & & \ddots & & \\ & & & \ddots & \\ & & & & 1 \end{pmatrix}.$$

In particular, $\det g = 1$. Any two transvections have the same matrix in different bases, and so are conjugate in $GL(V)$.

**Lemma 35.5**      *If $\dim_F V \geq 3$, then any two transvections $x$, $y$ of $V$ are conjugate in $SL(V)$. If $\dim_F V = 2$ and $\{v_1, v_2\}$ is a basis of $V$, then each conjugacy class of transvections in $SL(V)$ contains an element with matrix*

$$\begin{pmatrix} 1 & \lambda \\ & 1 \end{pmatrix}$$

*in the basis $\{v_1, v_2\}$, some $0 \neq \lambda \in F$.*

*Proof.* We saw in Lemma 35.4 that, in some basis, $x$ has matrix

$$\begin{pmatrix} 1 & & & & 1 \\ & 1 & & & \\ & & 1 & & \\ & & & \ddots & \\ & & & & 1 \end{pmatrix}.$$

When $n \geq 3$, this matrix is centralized by matrices of any determinant, so $GL(V) = C_{GL(V)}(x)SL(V)$. If $y = g^{-1}xg$ for $g \in GL(V)$, write $g = cn$ where $n \in SL(V)$ and $cx = xc$. Then $y = g^{-1}xg = n^{-1}c^{-1}xcn = n^{-1}xn$, so $y$ and $x$ are conjugate in $SL(V)$.

If $\dim_F V = 2$ and $g \in GL(V)$, let $\varphi(g)$ be the matrix of $g$ in $\{v_1, v_2\}$. If $x$ is any transvection, choose $g$ so $\varphi(g^{-1}xg) = \begin{pmatrix} 1 & 1 \\ 0 & 1 \end{pmatrix}$. If $\lambda = \det g$, choose $h \in GL(V)$ with $\varphi(h) = \begin{pmatrix} \lambda^{-1} & 0 \\ 0 & 1 \end{pmatrix}$. Then $gh \in SL(V)$, and

$$\varphi(h^{-1}g^{-1}xgh) = \begin{pmatrix} \lambda & 0 \\ 0 & 1 \end{pmatrix}\begin{pmatrix} 1 & 1 \\ 0 & 1 \end{pmatrix}\begin{pmatrix} \lambda^{-1} & 0 \\ 0 & 1 \end{pmatrix} = \begin{pmatrix} 1 & \lambda \\ 0 & 1 \end{pmatrix}.$$

**Definition**  A *hyperplane* in an $n$-dimensional vector space is an $(n - 1)$-dimensional subspace.

**Lemma 35.6**  *If $\dim_F V \geq 2$, then $SL(V)$ is generated by transvections.*

*Proof, Step 1.*  If $u, v$ are linearly independent, then there is a transvection $g$, $g(u) = v$.

*Proof of Step 1*  Let $w = v - u$, and choose a hyperplane $H$ of $V$ with $w \in H$, $u \notin H$. Define $g \in GL(V)$ by $g|H = 1$, $g(u) = u + w$.

*Step 2*  If $H_1, H_2$ are distinct hyperplanes of $V$ and $u \in V - (H_1 \cup H_2)$, then there is a transvection $g$ satisfying $g(H_1) = H_2$, $g(u) = u$.

*Proof*  Let $H$ be the hyperplane $(H_1 \cap H_2) + Fu$. Choose $x \in H_1$, $y \in H_2$ so $x + y = -u$, $x + u = -y$. $u \notin H_2$, so $x \notin H_2$ and $H_1 = (H_1 \cap H_2) + Fx$. $u \notin H_1$, so $y \notin H_1$ and $H_2 = (H_1 \cap H_2) + Fy$,

$V = (H_1 \cap H_2) + Fx + Fy$. $x \in H$ would mean $y \in H$, $H = V$, a contradiction; hence $x \notin H$. Define $g$ by $g|_H = 1$, $g(x) = x + u$; then $g(u) = u$ and $g(H_1) = H_2$.

    *Step 3, Conclusion*  Let $H$ be a hyperplane in $V$, $u \in V - H$ so $V = H + Fu$. Let $g \in SL(V)$ be fixed.

    If $u, g(u)$ are linearly independent, choose by Step 1 a transvection $t_1$ with $t_1(g(u)) = u$. If $u, g(u)$ are linearly dependent, let $t_0$ be a transvection with $t_0(g(u))$ and $u$ linearly independent, and by Step 1 let $t_1'$ be a transvection with $t_1'(t_0(g(u))) = u$; set $t_1 = t_1' \circ t_0$.

    By Step 2, choose a transvection $t_2$ with $t_2(u) = u$ and $t_2(t_1(g(H))) = H$. Then $g_1 = t_2 \circ t_1 \circ g$ fixes $u$ and $H$, so $g_1 \in SL(H)$. (If $t_1(g(H)) = H$, take $g_1 = t_1 \circ g$).

    If $n = 2$, this means that $g_1 = 1$, so $g = t_1^{-1} t_2^{-1}$ is a product of transvections. If $n > 2$, then $g_1$ is a product of transvections by induction on $n$, and so is $g$.

**Lemma 35.7**      *Let* $\dim_F V = n$. *If* $n \geq 3$, *then* $SL(V)' = SL(V)$. *If* $n = 2$ *and* $|F| \geq 4$, *then* $SL(V)' = SL(V)$.

    *Proof*  Assume first $n \geq 3$. Fix a hyperplane $H$ of $V$ and $u \in V - H$, so $V = H + Fu$. Define

$$T = \{\text{transvections } t \,|\, \ker(t - 1) = H\} \cup \{1\}.$$

Any $t \in T$ satisfies $t(u) = u + v$, some $v \in H$, and is determined by $v$; set $t = t_v$. We see that $t_{v+w} = t_v + t_w$ and $t_v^{-1} = t_{-v}$, so $T$ is an abelian group, $T \cong H$. Choosing $v, w \in H - \{0\}$ with $-v + w \neq 0$, we see that $g = t_v^{-1} t_w = t_{-v+w}$ is a transvection. By Lemma 35.5, $t_v$ and $t_w$ are conjugate in $SL(V)$, so $g$ is a commutator, $g \in SL(V)'$. By Lemmas 35.5 and 35.6, $SL(V)' = SL(V)$.

    If $n = 2$ and $\mu, \phi \in F^\#$, then the following element of $SL(2, F)$ is a commutator:

$$\begin{pmatrix} 1 & \phi \\ & 1 \end{pmatrix} \begin{pmatrix} \mu & \\ & \mu^{-1} \end{pmatrix} \begin{pmatrix} 1 & -\phi \\ & 1 \end{pmatrix} \begin{pmatrix} \mu^{-1} & \\ & \mu \end{pmatrix} = \begin{pmatrix} 1 & \phi(1 - \mu^2) \\ & 1 \end{pmatrix}.$$

Since $|F| > 3$, we can choose $\mu \in F^\#$ with $1 - \mu^2 \neq 0$. Hence all elements of $SL(2, F)$ of form $\begin{pmatrix} 1 & \lambda \\ & 1 \end{pmatrix}$ are commutators. By Lemmas 35.5 and 35.6, $SL(2, F)' = SL(2, F)$.

**Theorem 35.8**     Let $\dim_F V = n \geq 2$. If $n = 2$, *assume* $|F| \geq 4$. *Then* *PSL(V) is simple.*

*Proof, Step 1*     Denote by **P** the set of all one-dimensional subspaces of $V$. Fix $P_0 = Fv_0 \in \mathbf{P}$, and let $G_0$ be the subgroup of $G$ fixing $P_0$. Then $G = SL(V)$ is doubly transitive on **P**.

*Proof*  We must show $G_0$ transitive on $\mathbf{P} - \{P_0\}$. Suppose $P = Fu$, $Q = Fv \in \mathbf{P} - \{P_0\}$. There is a $g \in GL(V)$ such that $g(P_0) = P_0$, $g(u) = v$. Hence for some $\lambda \in F^\#$, there is a $g_\lambda \in SL(V)$, where $g_\lambda(v_0) = g(v_0)$, $g_\lambda(u) = \lambda v$; then $g_\lambda(P_0) = P_0$, $g_\lambda(P) = Q$.

*Step 2*  $G_0$ has an abelian normal subgroup $A$ containing a transvection from each conjugacy class of transvections.

*Proof*  Let $\{v_0, v_2, \ldots, v_n\}$ be a basis of $V$, $\varphi(g)$ the matrix of $g \in G$ in this basis. If $g \in G_0$, then $\varphi(g)$ has form

$$\varphi(g) = \begin{pmatrix} a & a_2 & \cdots & a_n \\ 0 & & & \\ \vdots & & X(g) & \\ 0 & & & \end{pmatrix}, \qquad X(g) \in GL(n-1, F), \quad \det X(g) = a^{-1}.$$

The composite map $g \to \varphi(g) \to X(g)$ is a homomorphism, say with kernel $A$. Elements $h$ of $A$ have $\varphi(h)$ of form

$$\varphi(h) = \begin{pmatrix} 1 & a_2 & \cdots & a_n \\ & 1 & & \\ & & \ddots & \\ & & & 1 \end{pmatrix},$$

so $A$ is abelian. $A$ contains all transvections $t_\lambda$ with

$$\varphi(t_\lambda) = \begin{pmatrix} 1 & & \lambda \\ & \ddots & \\ & & 1 \end{pmatrix},$$

so by Lemma 35.5, Step 2 holds.

*Step 3, Conclusion*   Let $N$ be any normal subgroup of $G = SL(V)$ not consisting of scalar transformations; it is enough to show $N = G$. $N$ is not trivial on **P**, so choose $P$, $R \in \mathbf{P}$ and $n \in N$ with $n(P) = R \neq P$. We claim $N$ is transitive on **P**; for any $Q \in \mathbf{P}$, choose $g \in G$ by Step 1 with $g(P) = P$ and $g(Q) = R$. Then $g^{-1}ng \in N$ and $g^{-1}ng(P) = g^{-1}n(P) = g^{-1}(R) = Q$, so $N$ is transitive. Hence $G_0 N = G$. Let $A \lhd G_0$ be as in Step 2; $G_0 \subseteq N_G(AN)$, so $AN \lhd G$. By Step 2 and Lemma 35.6, $AN = G$. Hence $G/N \cong A/A \cap N$ is abelian, and $N \supseteq G'$. By Lemma 35.7, $G = G'$, so $N = G$.

**Lemma 35.9**   *Let $V$ be a $2m$-dimensional $F$-vector space, $(\,,)$ a non-singular alternating form on $V$. A transvection $\theta$ of $V$ is called a* symplectic *transvection if $\theta \in \mathrm{Sp}(V)$. If $\theta$ is any transvection, $H = \ker(\theta - 1)$, $\mathrm{im}(\theta - 1) = Fv_0$, then $\theta \in \mathrm{Sp}(V)$ iff $(h, v_0) = 0$, all $h \in H$.*

*Proof*   Let $v_0 = \theta(v) - v$. If $\theta \in \mathrm{Sp}(V)$, then

$$(h, v_0) = (h, \theta(v) - v) = (h, \theta(v)) - (h, v)$$
$$= (\theta(h), \theta(v)) - (h, v) = 0.$$

Conversely, assume $(h, v_0) = 0$, all $h \in H$. For any $x, y \in V$, let

$$x = h_1 + \lambda_1 v, \quad y = h_2 + \lambda_2 v, \quad h_i \in H, \quad \lambda_i \in F.$$

Then

$$(\theta(x), \theta(y)) = (h_1 + \lambda_1 \theta(v), h_2 + \lambda_2 \theta(v))$$
$$= (h_1, h_2) - \lambda_1(h_2, \theta(v)) + \lambda_2(h_1, \theta(v))$$
$$= (h_1, h_2) - \lambda_1(h_2, v) + \lambda_2(h_1, v)$$
$$= (h_1 + \lambda_1 v, h_2 + \lambda_2 v) = (x, y),$$

so $\theta \in \mathrm{Sp}(V)$.

**Lemma 35.10**   *Let $V$ be a $2m$-dimensional $F$-vector space with non-singular alternating form $(\,,)$. Let $B = \{u_1, \ldots, u_m, v_m, \ldots, v_1\}$ be a symplectic basis of $V$, and for any $g \in \mathrm{Sp}(V)$ let $\varphi(g)$ be the matrix of $g$ in this basis. Then any symplectic transvection $\theta$ of $V$ is conjugate in $\mathrm{Sp}(V)$*

*to one (call it x) with matrix*

$$\varphi(x) = \begin{pmatrix} 1 & & & & \\ & 1 & & & \\ & & \ddots & & \\ & & & \ddots & \\ \lambda & & & & 1 \end{pmatrix}, \qquad \text{some } \lambda \in F^{\#}.$$

*If also*

$$\varphi(y) = \begin{pmatrix} 1 & & & & \\ & 1 & & & \\ & & \ddots & & \\ & & & \ddots & \\ \lambda\rho^2 & & & & 1 \end{pmatrix}, \qquad \text{some } \rho \in F^{\#},$$

*then x and y are conjugate in* $\mathrm{Sp}(V)$.


*Proof*  Choose $b_1 \in V$ with $(\theta - 1)V = Fb_1$, and choose $a_1 \in V$ with $(a_1, b_1) = 1$. By Lemma 35.9, $a_1 \notin \ker(\theta - 1)$, so $\theta(a_1) = a_1 + \lambda b_1$, some $\lambda \in F^{\#}$. Extending $\{a_1, b_1\}$ to a symplectic basis $B' = \{a_1, \ldots, a_m, b_m, \ldots, b_1\}$ of $V$, the matrix of $\theta$ in the basis $B'$ is

$$\begin{pmatrix} 1 & & & & \\ & 1 & & & \\ & & \ddots & & \\ & & & \ddots & \\ \lambda & & & & 1 \end{pmatrix}.$$

Changing to the basis $B$ is a change of symplectic basis, so $\theta$ is conjugate to $x$ in $\mathrm{Sp}(V)$.

If we write $\theta$ in the symplectic basis

$$B'' = \{\rho a_1, a_2, \ldots, a_m, b_m, \ldots, b_2, \rho^{-1} b_1\},$$

we find

$$\theta(\rho a_1) = \rho\theta(a_1) = \rho(a_1 + \lambda b_1) = \rho a_1 + (\lambda\rho^2)(\rho^{-1}b_1),$$

so $\theta$ has matrix

in this basis. Thus $\theta$ is also conjugate to $y$.

*Notation*  If $W$ is a $t$-dimensional subspace of the $2m$-dimensional symplectic space $V$, denote by $W^\perp$ the $(2m - t)$-dimensional subspace $\{v \in V \,|\, (v, w) = 0,\ \text{all } w \in W\}$.

**Lemma 35.11**  *Let $V$ be a $2m$-dimensional F-vector space with non-singular alternating form $(\,,\,)$. Then $\mathrm{Sp}(V)$ is generated by symplectic transvections.*

*Proof, Step 1*  If $(u, v) \neq 0$, then there is a symplectic transvection $\theta$ such that $\theta(u) = v$, and $\theta = 1$ on $F(v - u)^\perp$.

*Proof of Step 1*  $0 \neq (u, v) = (u, v - u)$, so we can choose $\lambda \in F^\#$, $w \in V$ with $v - u = \lambda w$, $(u, w) = 1$. We extend $\{u, w\}$ to a symplectic basis $\{u, u_2, \dots, v_2, w\}$, and define $\theta$ by $\theta(u) = u + \lambda w$, $\theta = 1$ on other basis elements. $\theta$ has matrix

$$\begin{pmatrix} 1 & & & \\ & \ddots & & \\ & & \ddots & \\ \lambda & & & 1 \end{pmatrix},$$

so $\theta$ is a symplectic transvection with the required properties.

*Step 2*  If $u, v \in V^\#$, then there is a $\varphi \in \mathrm{Sp}(V)$ such that $\varphi$ is a product of at most two symplectic transvections and $\varphi(u) = v$.

*Proof*  If $(u, v) \neq 0$, use Step 1. If $(u, v) = 0$, then choose $w$ with $(u, w) \neq 0$, $(v, w) \neq 0$. Choose by Step 1 symplectic transvections $\theta_1, \theta_2$ with $\theta_1(u) = w$, $\theta_2(w) = v$, and set $\varphi = \theta_2 \circ \theta_1$.

*Step 3, Conclusion*  We now prove Lemma 35.11, by induction on

$m$. If $\{u_1, \ldots, u_m, v_m, \ldots, v_1\}$ and $\{u'_1, \ldots, u'_m, v'_m, \ldots, v'_1\}$ are two symplectic bases of $V$, we shall find a product $\varphi$ of symplectic transvections with $\varphi(u_i) = u'_i$, $\varphi(v_i) = v'_i$, all $i$. By Step 2, we may assume $u_1 = u'_1$.

We first find $\varphi_0$ with $\varphi_0(u_1) = u_1$, $\varphi_0(v_1) = v'_1$. If $(v_1, v'_1) \neq 0$, let $\varphi_0$ by Step 1 be a symplectic transvection with $\varphi_0(v_1) = v'_1$, $\varphi_0 = 1$ on $F(v_1 - v'_1)^\perp$. $(u_1, v_1 - v'_1) = (u_1, v_1) - (u_1, v'_1) = 1 - 1 = 0$, so $\varphi_0$ is as desired.

If $(v_1, v'_1) = 0$, let $w = u_1 + v_1$. Then $(v_1, w) = (v_1, u_1) = -1 \neq 0$, $(v'_1, w) = (v'_1, u_1) = -1 \neq 0$, $(u_1, w) = (u_1, v_1) = 1$. Extend $\{u_1, w\}$ to a symplectic basis $\{u_1, \ldots, w\}$ of $V$. By the previous paragraph, there are transvections $\theta_1, \theta_2$ with $\theta_1(u_1) = u_1, \theta_1(v_1) = w, \theta_2(u_1) = u_1, \theta_2(w) = v'_1$. Take $\varphi_0 = \theta_2 \circ \theta_1$.

Now $\varphi_0$ satisfies $\varphi_0(u_1) = u'_1$, $\varphi_0(v_1) = v'_1$. If $m = 1$, take $\varphi = \varphi_0$ and we are done. If $m > 1$, then $\{\varphi_0(u_2), \ldots, \varphi_0(u_m), \varphi_0(v_m), \ldots, \varphi_0(v_2)\}$ and $\{u'_2, \ldots, u'_m, v'_m, \ldots, v'_2\}$ are symplectic bases of $(Fu'_1 + Fv'_1)^\perp$, so by induction there is a product $\psi$ of symplectic transvections, $\psi(\varphi_0(u_i)) = u'_i$, $\psi(\varphi_0(v_i)) = v'_i$ for all $i > 1$. Define $\psi$ on all of $V$ by $\psi(\varphi_0(u_1)) = \varphi_0(u_1)$, $\psi(\varphi_0(v_1)) = \varphi_0(v_1)$, and set $\varphi = \psi \circ \varphi_0$.

**Lemma 35.12** *Let $V$ be a $2m$-dimensional $F$-vector space with non-singular alternating form $(\ ,\ )$, $G = Sp(V)$. Then:*

  (1) *if $|F| \geq 4$, $G = G'$.*
  (2) *if $|F| = 3$ and $m \geq 2$, $G = G'$.*
  (3) *if $|F| = 2$ and $m \geq 3$, $G = G'$.*

*Proof* Let $\{u_1, \ldots, u_m, v_m, \ldots, v_1\}$ be a symplectic basis of $V$, $\varphi(g)$ the matrix of $g \in G$, so $\varphi : G \cong Sp(2m, F)$. The matrix of the form $(\ ,\ )$ is

$$J = \begin{pmatrix} 0 & E \\ -E & 0 \end{pmatrix}, \qquad E = \begin{pmatrix} & & 1 \\ & \cdot & \\ 1 & & \end{pmatrix},$$

and we know from linear algebra that a matrix $X$ is in $Sp(2m, F)$ if and only if ${}^t X J X = J$.

*Proof of (1)* Choose $x, y \in G$ with

$$\varphi(x) = \begin{pmatrix} 1 & & \\ & \cdot & \\ \lambda & & 1 \end{pmatrix}, \qquad \varphi(y) = \begin{pmatrix} 1 & & \\ & \cdot & \\ \lambda\rho^2 & & 1 \end{pmatrix}, \qquad \lambda, \rho \in F^\#.$$

By Lemma 35.10, $x$ and $y$ are conjugate in $G$, so $x^{-1}y \in G'$.

$$\varphi(x^{-1}y) = \begin{pmatrix} 1 & & & \\ & \cdot & & \\ & & \cdot & \\ \lambda(\rho^2 - 1) & & & 1 \end{pmatrix}.$$

Since $|F| \geqq 4$, we can choose $\rho$ with $\rho^2 - 1 \neq 0$; hence all elements of $F$ have form $\lambda(\rho^2 - 1)$. By Lemma 35.10, $G'$ contains all symplectic transvections; by Lemma 35.11, $G = G'$.

*Proof of* (2)   Let $A$ and $B$ be the matrices

$$A = \begin{pmatrix} 1 & & & & \\ & 1 & & & \\ 1 & & \cdot & & \\ & & & \cdot & \\ & & 1 & & 1 \end{pmatrix}, \quad B = \begin{pmatrix} 1 & & & & \\ & 1 & & & \\ 1 & & \cdot & & \\ & & & \cdot & \\ 1 & 1 & & & 1 \end{pmatrix}.$$

One easily verifies that $^tAJA = J$, $^tBJB = J$, so $A, B \in \mathrm{Sp}(2m, F)$. If $M$ is the matrix

$$M = \begin{pmatrix} 1 & & & & & & \\ 1 & 1 & & & & & \\ & & 1 & & & & \\ & & & \cdot & & & \\ & & & & 1 & & \\ & & & & & 1 & \\ & & & & & -1 & 1 \end{pmatrix},$$

then $^tMJM = J$ and $M^{-1}BM = A$, so $A$ and $B$ are conjugate in $\mathrm{Sp}(2m, F)$. Hence

$$A^{-1}B = \begin{pmatrix} 1 & & \\ & \cdot & \\ 1 & & 1 \end{pmatrix} \in \mathrm{Sp}(2m, F)'. \quad \begin{pmatrix} 1 & & \\ & \cdot & \\ -1 & & 1 \end{pmatrix} = (A^{-1}B)^2 \in \mathrm{Sp}(2m, F)'.$$

By Lemma 35.10, all symplectic transvections are in $G'$, so by Lemma 35.11, $G' = G$.

*Proof of* (3)   Assume first $m = 3$. By Lemma 35.11, $G$ is generated by elements of order 2. If $x_1, \ldots, x_t$ generate $G$ then $x_1 G', \ldots, x_t G'$ generate $G/G'$, so $G/G'$ is a 2-group. Hence $Sp(6, F)'$ contains all elements of order 3 in $Sp(6, F)$. In the equation

$$\begin{pmatrix} 1 & & & & & \\ & 1 & & & 1 & \\ & & 1 & & & \\ 1 & 1 & 1 & 1 & & \\ 1 & 1 & & & & \\ & & & & & 1 \end{pmatrix} \begin{pmatrix} 1 & & & & & \\ 1 & 1 & & 1 & & \\ & & 1 & & & \\ 1 & 1 & 1 & 1 & & \\ 1 & 1 & 1 & & 1 & \\ 1 & 1 & 1 & & & 1 \end{pmatrix} \begin{pmatrix} 1 & & & & & \\ & 1 & & & & \\ & & 1 & 1 & & \\ & & & 1 & & \\ & & & & 1 & \\ & & & & & 1 \end{pmatrix} \begin{pmatrix} 1 & & & & & \\ & 1 & & & & \\ 1 & & 1 & & 1 & \\ 1 & & & 1 & 1 & 1 \\ & & & & 1 & \\ & & & & & 1 \end{pmatrix} \cdot$$

$$\begin{pmatrix} 1 & & & & & \\ & 1 & & 1 & & \\ & & 1 & & & \\ & & & 1 & & \\ & & & & 1 & \\ & & & & & 1 \end{pmatrix} \begin{pmatrix} 1 & & & & & \\ & 1 & & & 1 & \\ & & 1 & & & 1 \\ & & & 1 & & \\ 1 & 1 & & & 1 & \\ 1 & 1 & & & & 1 \end{pmatrix} = \begin{pmatrix} 1 & & & & & \\ & 1 & & & & \\ & & 1 & & & \\ & & & 1 & & \\ & & & & 1 & \\ 1 & & & & & 1 \end{pmatrix},$$

all six matrices $X$ in the product satisfy ${}^tXJX = J$ and have order 3. By Lemma 35.10, all symplectic transvections are in $G'$, so by Lemma 35.11, $G = G'$.

If $m > 3$ and $V$ has the given symplectic basis, we use the six matrices to define $T_1, \ldots, T_6 \in GL(V)$ as follows. The $i$th matrix defines $T_i$ on the subspace spanned by $u_1, u_2, u_3, v_3, v_2, v_1$, and we define $T_i(u_j) = u_j$, $T_i(v_j) = v_j$ for all $j > 3$. Again all $T_i$ have order 3 in $Sp(V)$, and $T_1 \circ T_2 \circ \cdots \circ T_6$ is a symplectic transvection. By Lemmas 35.10 and 35.11, $G = G'$.

**Theorem 35.13**   $PSp(2m, F)$ *is a simple group, except in the cases* $m = 1, |F| \leq 3$ *and* $m = 2, |F| = 2$.

*Proof, Step 1, Notation*   Let $V$ be a $2m$-dimensional $F$-vector space with nonsingular alternating form $(\,,\,)$, and assume the excluded cases do not occur. By Lemma 35.1 and Theorem 35.8, we may assume $m \geq 2$. Let $G = Sp(V)$, and let **P** be the set of all one-dimensional subspaces of $V$. If $P \in \mathbf{P}$, denote $G_P = \{g \in G | g(P) = P\}$. If $P = Fu$, $Q = Fv \in \mathbf{P}$, then $P \perp Q$ means $(u, v) = 0$, $P \not\perp Q$ means $(u, v) \neq 0$. Let $N$ be a normal subgroup of $G$ not consisting of scalar transformations; to prove the theorem, it is enough to show $N = G$.

*Step 2*   For each $P \in \mathbf{P}$, $G_P N \neq G_P$.

*Proof*   If $G_Q N = G_Q$, then $N \subseteq G_Q$, so $N$ is contained in all conjugates of $G_Q$ in $G$. Any one-vector set $\{v\}$ can be extended to a symplectic basis, so $G$ is transitive on $\mathbf{P}$. Thus all $G_P$ are conjugates of $G_Q$, $N \subseteq G_P$ for all $P \in \mathbf{P}$, $N$ consists of scalars, contradiction.

*Step 3*   If $Q$ is in the $G_P N$-orbit of $P$, then $G_P N = G_Q N$.

*Proof*   We can find $n \in N$, $n(P) = Q$. For any $g \in G_Q$, $n^{-1}gn(P) = P$ so $n^{-1}gn \in G_P$, $g \in G_P N$. Hence $G_Q \subseteq G_P N$; similarly $G_P \subseteq G_Q N$, so $G_P N = G_Q N$.

*Step 4*   For some $P_0 \in \mathbf{P}$, $G_{P_0} N = G$.

*Proof*   For each $P \in \mathbf{P}$, let

$$S_P = \{Q \in \mathbf{P} \,|\, P \perp Q, P \neq Q\}, \qquad T_P = \{Q \in \mathbf{P} \,|\, P \not\perp Q\};$$

then

$$\mathbf{P} = \{P\} \cup S_P \cup T_P \qquad \text{(disjoint union)}.$$

If $Q, Q' \in S_P$, we can find symplectic bases $\{u_1, u_2, \ldots\}$ and $\{u_1, u_2', \ldots\}$ of $V$ with $u_1 \in P$, $u_2 \in Q$, $u_2' \in Q'$, so $G_P$ is transitive on $S_P$. If $Q, Q' \in T_P$, we can find symplectic bases $\{u_1, \ldots, v_1\}$ and $\{u_1, \ldots, v_1'\}$ of $V$ with $u_1 \in P$, $v_1 \in Q$, $v_1' \in Q'$, so $G_P$ is transitive on $T_P$.

Thus $G_P$ has three orbits $\{P\}$, $S_P$, $T_P$ on $\mathbf{P}$. By Step 2, $G_P N \neq G_P$, so $\{P\}$ is not an orbit of $G_P N$. If for some $P \in \mathbf{P}$, $G_P N$ is transitive on $\mathbf{P}$, then for any $g \in G$ we can find $h \in G_P N$, $g(P) = h(P)$; $h^{-1}g \in G_P$, so $g \in G_P N$, $G_P N = G$, done.

Hence we may assume each $G_P N$ not transitive on $\mathbf{P}$. Each $G_P N$ must have exactly two orbits on $\mathbf{P}$, say $S_P'$ and $T_P'$, where $S_P \subseteq S_P'$, $T_P \subseteq T_P'$, and $P \in S_P'$ or $P \in T_P'$.

If $P \in S_P'$, then for some $P \neq R \perp P$ and $x \in G_P N$ we have $x(P) = R$. For any $Q \in \mathbf{P}$, choose $g \in G$, $g(Q) = P$, and denote $R' = g^{-1}(R)$. Then

$$0 = (P, R) = (g(Q), g(R')) = (Q, R'),$$

so $Q \perp R'$.

$$g^{-1}xg \in g^{-1}G_P Ng = g^{-1}G_P gg^{-1}Ng = G_Q N,$$

and

$$g^{-1}xg(Q) = g^{-1}x(P) = g^{-1}(R) = R',$$

so $Q \in S'_Q$. Therefore we have two cases:

Case 1: If $P \in S'_P$ for all $P \in \mathbf{P}$, and $P' \in S_P$, then by Step 3, $G_P N = G_{P'} N$. Thus $G_P N$ has orbit

$$\{Q \in \mathbf{P} | P \perp Q\} = \{Q \in \mathbf{P} | P' \perp Q\};$$

these two sets are *not* the same, contradiction.

Case 2: If $P \in T'_P$ for all $P \in \mathbf{P}$, and $P' \in T_P$, then by Step 3 $G_P N = G_{P'} N$. Thus $G_P N$ has orbit

$$\{Q \in \mathbf{P} | P \neq Q, P \perp Q\} = \{Q \in \mathbf{P} | P' \neq Q, P' \perp Q\};$$

these two sets are *not* the same, contradiction.

*Step 5, Conclusion* Let $P_0 = Fu_1$, and let $\{u_1, \ldots, u_m, v_m, \ldots, v_1\}$ be a symplectic basis of $V$, $\varphi(g)$ the matrix of $g \in G$ in this basis. Elements of $G_0 = G_{P_0}$ fix $H = P_0^{\perp}$; $H$ has basis $\{u_1, \ldots, u_m, v_m, \ldots, v_2\}$, so for any $g \in G_0$, $\varphi(g)$ has form

$$\begin{pmatrix} a_1 & a_2 & \cdots & a_{n-1} & a_n \\ 0 & & & & b_2 \\ \vdots & & X(g) & & \vdots \\ 0 & & & & b_{n-1} \\ 0 & 0 & \cdots & 0 & b_n \end{pmatrix}.$$

$g \to \varphi(g) \to X(g)$ is a homomorphism, say with kernel $A \lhd G_0$. Being isomorphic to a triangular group of matrices, $A$ is solvable. $A \lhd G_0$ implies $AN \lhd G_0 N$; by Step 4, $G_0 N = G$, so $AN \lhd G$. By Lemmas 35.10 and 35.11, $AN = G$. $G/N = AN/N \cong A/A \cap N$ is solvable. By Lemma 35.12, $G = G'$, so $(G/N)' = G'N/N = GN/N = G/N$, implying $G/N = 1$, $G = N$, done.

*Remark* $PSL(2, 2) = PSp(2, 2)$ and $PSL(2, 3) = PSp(2, 3)$ are solvable groups of orders 6 and 12, respectively. By Lemma 35.3, $PSp(4, 2) =$

$Sp(4, 2) \cong S_6$. Thus the exceptional cases in Theorems 35.8 and 35.13 really yield nonsimple groups.

EXERCISES

**1**   Show that $SL(2, 3)$ is isomorphic to the group $G$ of Exercise 1, §15.

**2**   Using Lemma 28.2 and Exercise 1, §15, show that in Theorem 26.1 the exceptional cases $A_4$, $S_4$, $A_5$ all occur.

**3**   Show that if $V$ is a two-dimensional vector space over $GF(q)$ and **P** the set of one-dimensional subspaces of $V$, then $PSL(2, q)$ is a doubly transitive permutation group on **P** in which the subgroup fixing one point is a Frobenius group on the remaining points.

# §36

## Jordan's Theorem for Solvable Groups

**Definition**    Let $\Omega$ be a set, $|\Omega| = n < \infty$, $G$ a transitive permutation group on $\Omega$. If

$$\Omega = \Omega_1 \cup \cdots \cup \Omega_m, \qquad 1 < d = |\Omega_i| < n, \quad n = md,$$

where $G$ permutes the $\Omega_i$, then $G$ is *imprimitive* with the $\Omega_i$ as *sets of imprimitivity*. If $\Omega$ is not such a union, $G$ is *primitive*.

**Lemma 36.1**    *Nontrivial normal subgroups of primitive permutation groups are transitive.*

*Proof*  Let $G$ be primitive on $\Omega$, $1 \neq N \vartriangleleft G$, and let $\Omega_1$ be an orbit of $N$ of length $> 1$. If $g \in G$, then $N(g(\Omega_1)) = g(N(\Omega_1)) = g(\Omega_1)$, so $N$ fixes the set $g(\Omega_1)$. If $\alpha, \beta \in \Omega_1$ and $n(\alpha) = \beta$, $n \in N$, then $gng^{-1}(g(\alpha)) = g(\beta)$, so $g(\Omega_1)$ is an orbit of $N$.

If $N$ is not transitive, we have proved the orbits of $N$ are sets of imprimitivity of $G$, a contradiction; so $N$ is transitive.

**Theorem 36.2** (Dixon [3])    *Let $S_n$ be the symmetric group on $n$ symbols, $G$ a solvable subgroup of $S_n$, $a = 2 \cdot 3^{1/3}$. Then $|G| \leq a^{n-1}$.*

*If $n = 4^k$ ($k = 0, 1, \ldots$), this bound is best possible.*

*Proof*  The proof of the inequality is by induction on $n$, true for $n = 1$.

*Case 1, G intransitive*  Let $\Omega_1, \ldots, \Omega_m$ be the orbits of $G$, $m > 1$, $|\Omega_i| = n_i$, $n_1 + \cdots + n_m = n$. Then $G$ performs a solvable group of permutations $G_i$ on each $i$, and the map

$$x \to (x|_{\Omega_1}, \ldots, x|_{\Omega_m})$$

215

is an isomorphism from $G$ into $G_1 \times \cdots \times G_m$. By induction, $|G_i| \le a^{n_i - 1}$, so

$$|G| \le \prod_{i=1}^{m} a^{n_i - 1} = a^{n-m} < a^{n-1}.$$

*Case 2, G transitive but imprimitive*   Let $\Omega_1, \ldots, \Omega_m$ be the sets of imprimitivity of $G$, $|\Omega_i| = d$, $n = md$, $H = \{g \in G | g(\Omega_i) = \Omega_i, \text{ all } i\}$. Then $H \lhd G$, and $G/H$ is a solvable permutation group on $\{\Omega_1, \ldots, \Omega_m\}$. By induction, $|G/H| \le a^{m-1}$; as in Case 1, $|H| \le (a^{d-1})^m$. Hence

$$|G| \le a^{m-1}(a^{d-1})^m = a^{n-1}.$$

*Case 3, G primitive*   Let $A$ be a minimal normal subgroup of $G$. $G$ is solvable, so $A$ is an elementary abelian $p$-group for some prime $p$; let $|A| = p^t$. Let $H$ be the subgroup of $G$ fixing some symbol; by Lemma 36.1, $A$ is transitive, so $G = HA$. $C_G(A) \lhd G$, and $C_G(A) \cap H \lhd AH = G$. Any normal subgroup of $G$ contained in $H$ is contained in all conjugates of $H$, so fixes all symbols and is trivial; hence $C_G(A) \cap H = 1$. Hence

$$|C_G(A)| \le |G : H| = |AH : H| \le |A|,$$

forcing

$$A = C_G(A), \qquad A \cap H = 1, \qquad p^t = |A| = |G : H| = n.$$

Hence $G/A$ is isomorphic to a subgroup of $GL(t, p)$, the automorphism group of $A$. Therefore

$$|G| \le |GL(t, p)| \, |A| = p^t(p^t - 1)(p^t - p) \cdots (p^t - p^{t-1}) < p^{t^2 + t}.$$

If $t > 3$ or if $t = 3$ and $p > 2$, one can verify that

$$p^{t^2 + t} \le a^{n-1} = a^{p^t - 1},$$

so we are done in these cases. In the remaining cases, one can check numerically that

$$2^3(2^3 - 1)(2^3 - 2)(2^3 - 2^2) < a^{2^3 - 1},$$

$$p^2(p^2 - 1)(p^2 - p) \le a^{p^2 - 1},$$

$$p(p - 1) \le a^{p-1}.$$

It remains to show that if $n = 4^k$, then there is a solvable subgroup of $S_n$ of order

$$a^{n-1} = (2 \cdot 3^{1/3})^{4^k - 1} = 2^{4^k - 1} \, 3^{(4^k - 1)/3} = 2^{4^k - 1} 3^{1 + 4 + \cdots + 4^{k-1}}.$$

This is done by induction on $k$; it is true for $k = 1$ because $S_4$ is solvable and $|S_4| = 2^3 \cdot 3 = a^{4-1}$.

Let

$$\Omega = \{(c, d)| 1 \le c \le 4^{k-1}, \quad 1 \le d \le 4\},$$

so $|\Omega| = 4^k$. By induction, there is a solvable permutation group $H$ of $\{1, 2, \ldots, 4^{k-1}\}$ with $|H| = a^{4^{k-1} - 1}$. Then $H \times H \times H \times H$ acts on $\Omega$ by

$$(h_1, h_2, h_3, h_4)(c, d) = (h_d(c), d).$$

(We have taken the direct product of 4 isomorphic copies of $H$.)

For any $\tau \in S_4$, define $\tau$ on $\Omega$ by

$$\tau((c, d)) = (c, \tau(d)).$$

Then we see that

$$\begin{aligned}
[\tau^{-1}(h_1, h_2, h_3, h_4)\tau](c, d) &= \tau^{-1}(h_1, h_2, h_3, h_4)(c, \tau(d)) \\
&= \tau^{-1}(h_{\tau(d)}(c), \tau(d)) = (h_{\tau(d)}(c), d) \\
&= (h_{\tau(1)}, h_{\tau(2)}, h_{\tau(3)}, h_{\tau(4)})(c, d),
\end{aligned}$$

so $S_4$ normalizes $H \times H \times H \times H$, and $G = S_4(H \times H \times H \times H)$ is a permutation group on $\Omega$. $G$ is solvable, of order

$$|S_4||H|^4 = 24(a^{4^{k-1}})^4 = a^3 a^{4^k - 4} = a^{4^k - 1} = a^{n-1}.$$

(The group we constructed is called the *wreath product* of $H$ by $S_4$.)

**Theorem 36.3**(P. Hall)     *Let $G$ be a finite solvable group, $\pi$ any subset of the set of primes dividing $|G|$. Then $G$ has a $\pi$-Hall subgroup.*

*Proof*   The proof is by induction on $|G|$. Let $|G| = mn$, $m$ a $\pi$-number, $n$ a $\pi'$-number, so $(m, n) = 1$; we must show $G$ has a subgroup of order $m$.

If $G$ has a proper normal subgroup $H$ with $|H| = m_1 n_1$, $m_1 | m$, $n_1 | n$, $n_1 < n$, then by induction $G/H$ has a subgroup $K/H$ of order $m/m_1$; $|K| = mn_1$, so by induction $K$ has a subgroup of order $m$, done.

Minimal normal subgroups of $G$ have prime power order; if no subgroup $H$ of the previous paragraph exists, then all minimal normal subgroups of $G$ have order $n = p^a$, $p$ a prime, and are Sylow $p$-subgroups of $G$. Hence we may assume $G$ has a unique minimal normal subgroup $K$ of order $p^a$. Let $L/K$ be a minimal normal subgroup of $G/K$, $|L/K| = q^b$, $q \neq p$, $q$ a prime. Let $Q$ be a Sylow $q$-subgroup of $L$, so $|Q| = q^b$. Set $M = N_G(Q)$, $T = M \cap K \lhd M$. $[T, Q] \subseteq Q \cap K = 1$, so $T \subseteq Z(L)$. $Z(L)$ char $L \lhd G$, so $Z(L) \lhd G$. Hence $Z(L) \supseteq K$ or $Z(L) = 1$. $Z(L) \supseteq K$ would mean $L = K \times Q$, $Q$ char $L \lhd G$, $Q \lhd G$, contradicting uniqueness of $K$. Therefore $Z(L) = 1$, $T = 1$, $N_L(Q) \cap K = 1$, $N_L(Q) = Q$, and $L$ has $|L: Q| = |K| = p^a$ Sylow $q$-subgroups. Any $G$-conjugate of $Q$ is a Sylow $q$-subgroup of $L$, so $|G: M| = p^a$ and $M$ is the desired subgroup of order $m$.

**Theorem 36.4** (Dornhoff[2,3]).     *Let $V$ be an $n$-dimensional **C**-vector space, $G$ a finite solvable subgroup of $GL(V)$. Then there is an abelian normal subgroup $A$ of $G$ such that*

$$|G: A| \leq 2^{4n/3 - 1} 3^{10n/9 - 1/3}.$$

*When $n = 3 \cdot 4^k$, $k = 0, 1, \ldots$, this bound is best possible.*

*Proof, Step 1*     The inequality holds when $G$ is an irreducible, primitive linear group.

*Proof of Step 1*     In this case, Theorem 34.6 tells us that $\mathrm{Fit}(G) = P_1 \cdots P_k P_{k+1} \cdots P_l$, $P_i$ a Sylow $p_i$-subgroup of $\mathrm{Fit}(G)$, $P_1, \ldots, P_k$ non-abelian, $P_{k+1}, \ldots, P_l$ cyclic and in $Z(G)$. For $1 \leq i \leq k$, $|P_i| = p_i^{2m_i + z_i}$, where $P_i$ is a central product of an extra-special $p_i$-group of order $p_i^{2m_i + 1}$ and a cyclic central group of order $p_i^{z_i}$. $G/\mathrm{Fit}(G)$ is isomorphic to a solvable subgroup of $\mathrm{Sp}(2m_1, p_1) \times \cdots \times \mathrm{Sp}(2m_k, p_k)$. We know $\prod_{i=1}^k p_i^{m_i}$ divides $n$, so

$$|G: Z(G)| = |G: \mathrm{Fit}(G)| \prod_{i=1}^k p_i^{2m_i} \text{ divides } |G: \mathrm{Fit}(G)| n^2.$$

Therefore it is enough to show

$$|G: \mathrm{Fit}(G)| \leq n^{-2} 2^{4n/3 - 1} 3^{10n/9 - 1/3}.$$

$|\mathrm{Sp}(2m_i, p_i)|$ is given in Lemma 35.2; always

$$|\mathrm{Sp}(2m_i, p_i)| < p_i^{m_i(2m_i + 1)},$$

so if $m_0 = \max \{m_1, \ldots, m_k\}$ we have $m_0 \leq \log_2 n$ and

$$|G: \text{Fit}(G)| < \prod_{i=1}^{k} p_i^{m_i(2m_i+1)} \leq \prod_{i=1}^{k} p_i^{m_i(2m_0+1)}$$

$$\leq n^{2m_0+1} \leq n^{2\log_2 n+1} = 2^{\log_2 n(2\log_2 n+1)}.$$

Meanwhile,

$$n^{-2} 2^{4n/3-1} 3^{10n/9-1/3} = 2^{(4/3+(10/9)\log_2 3)n-1-(1/3)\log_2 3 - 2\log_2 n}.$$

If $n \geq 13$ we have

$$\log_2 n(2\log_2 n + 1) < \left(\frac{4}{3} + \frac{10}{9}\log_2 3\right)n - 1 - \frac{1}{3}\log_2 3 - 2\log_2 n,$$

so Step 1 holds for $n \geq 13$. For $n \leq 12$ the approximate values are:

| $n$ | $\Pi_i \text{Sp}(2m_i, p_i)$ | $|\Pi_i \text{Sp}(2m_i, p_i)|$ | $n^{-2}2^{4n/3-1}3^{10n/9-1/3}$ |
|---|---|---|---|
| 2 | Sp(2, 2) | 6 | 6.3 |
| 3 | Sp(2, 3) | 24 | 24 |
| 4 | Sp(4, 2) | 720 | 114.6 |
| 5 | Sp(2, 5) | 120 | 630 |
| 6 | Sp(2, 2) × Sp(2, 3) | 6 · 24 | 3740 |
| 7 | Sp(2, 7) | 7 · 48 | $2.3 \times 10^4$ |
| 8 | Sp(6, 2) | 1451520 | 153360 |
| 9 | Sp(4, 3) | $5 \times 10^4$ | $10^6$ |
| 10 | Sp(2, 2) × Sp(2, 5) | 6 · 120 | $7 \times 10^6$ |
| 11 | Sp(2, 11) | 1320 | $5 \times 10^7$ |
| 12 | Sp(4, 2) × Sp(2, 3) | 720 · 24 | $3.6 \times 10^8$ |

This proves the desired inequality, except when $n = 4$ or $n = 8$. These cases will be proved if we show that Sp(4, 2) has no solvable subgroup of index $\leq 6$ and Sp(6, 2) has no solvable subgroup of index $\leq 9$.

By Lemma 35.3, Sp(4, 2) $\cong S_6$. If $H$ is a solvable subgroup of $S_6$ of index $\leq 6$, then $S_6$ permutes the cosets of $H$ so we must have $|S_6 : H| = 6$, $|H| = 2^3 \cdot 3 \cdot 5$. By Theorem 36.3, $H$ then has a subgroup $H_0$ of order 15. Groups of order 15 are cyclic, so $S_6$ has an element of order 15. No element of $S_6$ can be a 15-cycle or a product of disjoint 3- and 5-cycles, so this is a contradiction.

By Theorem 35.13, Sp(6, 2) is simple. If $H$ is a subgroup of Sp(6, 2) of index $\leq 9$, then by representing Sp(6, 2) on the cosets of $H$, Sp(6, 2) is isomorphic to a subgroup of the symmetric group $S_9$. But $|\text{Sp}(6, 2)| \nmid |S_9|$, contradiction; so Sp(6, 2) has no such subgroup, solvable or not. This completes Step 1.

*Step 2*   The inequality holds when $G$ is irreducible but imprimitive as linear group.

*Proof*   For some $m > 1$ we have $m|n$, $n = dm$, $V = V_1 \oplus \cdots \oplus V_m$, dim $V_i = d$, the $V_i$ spaces of imprimitivity of $G$. Let $G_i = \{g \in G | g(V_i) = V_i\}$, $N = \bigcap_{i=1}^m G_i$, so $N \lhd G$. $G/N$ is a permutation group on $\{V_1, \ldots, V_m\}$, so by Theorem 36.2,

$$|G:N| \leqq 2^{m-1} 3^{(m-1)/3}.$$

We now investigate the abelian group $Z(N) \lhd G$. We may assume each $G_i$ primitive on $V_i$; for if not, we could by Lemma 29.5 choose a larger $m$ where the corresponding $G_i|_{V_i}$ were primitive. Denote

$$H_i = \{h \in G_i | hg|_{V_i} = gh|_{V_i}, \text{all } g \in G_i\}.$$

By Step 1,

$$|G_i : H_i| \leq 2^{4d/3 - 1} 3^{10d/9 - 1/3}.$$

Of course $H_i \lhd G_i$, and $\bigcap_{i=1}^m H_i \subseteq Z(N)$. We have

$$|N : Z(N)| \leq |N : \bigcap_{i=1}^m H_i| = |N : \bigcap_{i=1}^m (N \cap H_i)|$$

$$\leq \prod_{i=1}^m |N : N \cap H_i| = \prod_{i=1}^m |NH_i : H_i| \leq \prod_{i=1}^m |G_i : H_i|.$$

Therefore

$$|G : Z(N)| \leq 2^{m-1} 3^{(m-1)/3} (2^{4d/3-1} 3^{10d/9-1/3})^m$$

$$= 2^{4n/3 - m + m - 1} 3^{10n/9 - m/3 + m/3 - 1/3}$$

$$= 2^{4n/3 - 1} 3^{10n/9 - 1/3}.$$

*Step 3*   The inequality holds.

*Proof*   By the previous cases we may assume $V$ a reducible $G$-module, say $V = V_1 \oplus \cdots \oplus V_k$, dim $V_i = n_i$, $n_1 + \cdots + n_k = n$, the $V_i$ $G$-irreducible. By the previous cases there exist normal subgroups $A_i$ of $G$ with $A_i|_{V_i}$ abelian and

$$|G : A_i| \leq 2^{4n_i/3 - 1} 3^{10n_i/9 - 1/3}.$$

Hence $A = \bigcap_{i=1}^{k} A_i \lhd G$ is abelian, and

$$|G: A| \leq \prod_{i=1}^{k} |G: A_i| \leq 2^{4n/3 - k} 3^{10n/9 - k/3}$$

$$< 2^{4n/3 - 1} 3^{10n/9 - 1/3}.$$

*Step 4*   The bound is best possible when $n = 3$.

*Proof*   Let $P$ be an extra-special 3-group of order 27 and exponent 3. Then $P$ is generated by elements $a$, $b$, $c$ with defining relations $a^3 = b^3 = c^3 = 1$, $[a, b] = c$, $c \in Z(P)$. Let $\mathrm{Aut}_0(P)$ be the group of all automorphisms of $P$ which are trivial on $\langle c \rangle$, and let $\alpha \in \mathrm{Aut}_0(P)$. $\alpha(a) = a^\alpha$ can be any element of $P - Z(P)$, so there are 24 possibilities for $a^\alpha$. $b^\alpha$ is then any element of $P - C_P(a^\alpha)$ with $[a^\alpha, b^\alpha] = c$, so there are $(27 - 9)/2 = 9$ possibilities for $b^\alpha$. Hence $|\mathrm{Aut}_0(P)| = 216$.

By Theorem 31.5, there is a faithful irreducible representation $\varphi: P \to GL(3, \mathbf{C})$, whose character $\chi$ vanishes outside $\langle c \rangle$. We may assume $\varphi(P) \subset SL(3, \mathbf{C})$ (use the fact

$$\left[\begin{pmatrix} 0 & 1 & 0 \\ 0 & 0 & 1 \\ 1 & 0 & 0 \end{pmatrix}, \begin{pmatrix} 1 & 0 & 0 \\ 0 & \omega & 0 \\ 0 & 0 & \omega^2 \end{pmatrix}\right] = \begin{pmatrix} \omega & 0 & 0 \\ 0 & \omega & 0 \\ 0 & 0 & \omega \end{pmatrix}, \qquad \text{where } \omega^3 = 1).$$

Let $\alpha \in \mathrm{Aut}_0(P)$. $\varphi^\alpha$, defined by $\varphi^\alpha(x) = \varphi(x^\alpha)$ for all $x \in P$, is a representation of $P$ with character $\chi^\alpha$, $\chi^\alpha(x) = \chi(x^\alpha)$. $\chi$ vanishes outside $\langle c \rangle$ and $\alpha = 1$ on $\langle c \rangle$, so $\chi = \chi^\alpha$. Therefore the representations $\varphi$ and $\varphi^\alpha$ are equivalent, and there is a matrix $M_\alpha \in GL(3, \mathbf{C})$ satisfying

$$\varphi^\alpha(x) = M_\alpha^{-1} \varphi(x) M_\alpha, \qquad \text{all} \quad x \in P.$$

Multiplying $M_\alpha$ by a scalar matrix if necessary, we may assume $\det M_\alpha = 1$, $M_\alpha \in SL(3, \mathbf{C})$.
   Let

$$G = N_{SL(3,\mathbf{C})}(\varphi(P)), \qquad C = C_{SL(3,\mathbf{C})}(\varphi(P)).$$

We have shown that $\mathrm{Aut}_0(P)$ is isomorphic to a subgroup of $G/C$, so $|G: C| \geq 216$. $\varphi(P)$ is absolutely irreducible, so $C$ is a group of $3 \times 3$ scalar matrices of determinant 1, $|C| \leq 3$, forcing $C = Z(\varphi(P))$, $|C| = 3$. $Z(G) \subseteq C$ so $Z(G) = C$; by Lemma 29.6, $Z(G)$ is the unique maximal abelian normal subgroup of $G$. When $n = 3$, $2^{4n/3-1}3^{10n/9-1/3} = 216$, so the inequality $|G:Z(G)| \geqq 216$ forces $|G:Z(G)| = 216$ and completes the proof of Step 4.

*Step 5*  The bound is best possible when $n = 3 \cdot 4^k$, $k = 1, 2, \ldots$.

*Proof*  Let $G$ be the group of matrices of Step 4, $|G| = 2^3 \cdot 3^4$, and let $C$ be a cyclic group of scalar matrices in $GL(3, \mathbf{C})$, $|C| = 5$; set $K = GC = G \times C \subseteq GL(3, \mathbf{C})$; then $Z(K) = Z(G) \times C$, so $|K:Z(K)| = |G:Z(G)| = 216$. Let $N$ be the group of all $n \times n$ diagonal block matrices

$$\begin{pmatrix} x_1 & & 0 \\ & \ddots & \\ 0 & & x_{4^k} \end{pmatrix}, \qquad \text{each} \quad x_i \in K;$$

then $N \subseteq GL(n, \mathbf{C})$ and

$$|N:Z(N)| = (216)^{4^k} = 2^{3 \cdot 4^k}3^{3 \cdot 4^k} = 2^n 3^n.$$

By Theorem 36.2, there is a solvable subgroup of the symmetric group $S_{4^k}$ with order

$$2^{4^k-1}3^{(4^k-1)/3} = 2^{n/3-1}3^{n/9-1/3}.$$

$S_{4^k}$ is isomorphic to the group of $4^k \times 4^k$ permutation matrices. Hence we may construct a solvable group $H$ of permutation matrices in $GL(n, \mathbf{C})$, made entirely of $3 \times 3$ blocks

$$\begin{pmatrix} 0 & 0 & 0 \\ 0 & 0 & 0 \\ 0 & 0 & 0 \end{pmatrix} \quad \text{and} \quad \begin{pmatrix} 1 & 0 & 0 \\ 0 & 1 & 0 \\ 0 & 0 & 1 \end{pmatrix},$$

$|H| = 2^{n/3-1}3^{n/9-1/3}$. $H$ normalizes $N$ and $H \cap N = 1$. Hence $\bar{G} = HN$ is a solvable subgroup of $GL(n, \mathbf{C})$ with

$$|\bar{G}:Z(N)| = |H||N:Z(N)| = 2^{4n/3-1}3^{10n/9-1/3},$$

the exact bound we desire.

To complete the proof, we shall show that every normal abelian subgroup $A$ of $\bar{G}$ is contained in $Z(N)$. At least $A \subseteq \text{Fit}(\bar{G})$. We first show $\text{Fit}(\bar{G}) \subseteq N$; consider any $x \in \bar{G} - N$. Let $B_0$ be the Sylow 5-subgroup of $\bar{G}$; $B_0 \subseteq Z(N)$, and $B_0$ is elementary abelian of order $5^{4^k}$. Let $B = \langle B_0, x \rangle$.

$x$ performs a nontrivial permutation of the diagonal blocks in $N$, and hence does not centralize $B_0$; therefore $B$ is not nilpotent, $B \not\subseteq \mathrm{Fit}(\bar{G})$. But $B_0 \subseteq \mathrm{Fit}(\bar{G})$, so $x \notin \mathrm{Fit}(\bar{G})$, proving $\mathrm{Fit}(\bar{G}) \subseteq N$.

We now know $A \subseteq N$. Hence

$$A \subseteq \mathrm{Fit}(N) = \mathrm{Fit}(K) \times \cdots \times \mathrm{Fit}(K).$$

But $\mathrm{Fit}(K) = \mathrm{Fit}(G) \times C$, $\mathrm{Fit}(G)$ extra-special of order 27. We are trying to show $A \subseteq Z(N)$. If on the contrary $x \in A - Z(N)$, then we write $x = x_1 y_1 \cdots x_{4^k} y_{4^k}$, $x_i$ in the $i$th factor $\mathrm{Fit}(G)$, $y_i$ in the $i$th factor $C$; $x \notin Z(N)$ means that, for some $j$, $x_j \notin Z(\mathrm{Fit}(G))$. Elements of $G$ perform on $\mathrm{Fit}(G)$ all automorphisms of $\mathrm{Fit}(G)$ fixing $Z(\mathrm{Fit}(G))$ elementwise. Hence we can find a $w$ in the $j$th factor $G$, $\langle x_j, x_j^w \rangle = \mathrm{Fit}(G)$. $A \lhd \bar{G}$, so we have $x, x^w \in A$, but

$$[x, x^w] = [x_1 y_1 \cdots x_j y_j \cdots x_4 \, y_{4^k}, x_1 y_1 \cdots x_j^w y_j \cdots x_{4^k} y_{4^k}]$$
$$= [x_j, x_j^w] \neq 1,$$

using Lemma 31.1 in $\mathrm{Fit}(N)$. $[x, x^w] \neq 1$ contradicts the fact that $A$ is abelian.

We conclude that $A \subseteq Z(N)$, as desired.

# §37

## Itô's Theorem on Characters of Solvable Groups

**Lemma 37.1**    *Let $P$ be a normal abelian Sylow $p$-subgroup of $G$, $Z = P \cap Z(G)$. Then $P = Z \times Q$, some $Q \lhd G$.*

*Proof*    Let $|G:P| = m$, $T = \{g_1, \ldots, g_m\}$ a cross section of $P$ in $G$, and define $\varphi = \varphi_T : P \to P$ by

$$\varphi(x) = \prod_{i=1}^{m} g_i^{-1} x g_i.$$

Since $P$ is normal and abelian, this product is independent of the order of the factors. If $S = \{h_1, \ldots, h_m\}$ is a second cross section with subscripts chosen so that $P h_i = h_i P = g_i P = P g_i$, let $h_i = p_i g_i$, $p_i \in P$. Then

$$\varphi_S(x) = \prod_{i=1}^{m} h_i^{-1} x h_i = \prod_{i=1}^{m} g_i^{-1} p_i^{-1} x p_i g_i = \varphi_T(x)$$

since $P$ is abelian; hence $\varphi$ is independent of the choice of $T$. From its definition, $\varphi(xy) = \varphi(x)\varphi(y)$, so $\varphi$ is a homomorphism.

Denote $Q = \ker \varphi$. If $x, y \in P$ are conjugate in $G$, $y = g^{-1}xg$, then $gg_i$ runs through coset representatives of $P$ as $g_i$ does, so $\varphi(x) = \varphi(y)$; this implies $Q \lhd G$. Also, $g_i g$ runs through coset representatives, so $g^{-1}\varphi(x)g = \varphi(x)$, $\varphi(P) \subseteq Z$.

If $x \in Z \cap Q$, then $1 = \varphi(x) = x^m$; $(m, p) = 1$ implies $x = 1$, so $Z \cap Q = 1$. Hence $Z \times Q \subseteq P$. But then

$$|Z||Q| \leq |P| = |\ker \varphi||\text{im } \varphi| \leq |Q||Z|,$$

forcing $Z \times Q = P$.

**Theorem 37.2** (Itô [3]).    *Let $G$ be a finite solvable group, $P$ a Sylow* *p-subgroup* of *G*, and assume that *G* has a faithful complex character $\chi$ with $\chi(1) < p - 1$. Then *P* is abelian and $P \lhd G$.

*Proof*  We use induction on $|G|$, so let *G* be a minimal counterexample; we seek a contradiction.

*Step 1*  *P* is abelian.

*Proof*  $\chi|_P$ must be a sum of linear characters, and is faithful, so $P' = 1$.

*Step 2*  We may assume $\chi$ is irreducible.

*Proof*  We may at least choose $\chi$ faithful of minimum degree. Let $\chi_1$ be a nonlinear irreducible constituent of $\chi$, $K = \ker \chi_1$. If $K \neq 1$, then by induction the Sylow *p*-subgroup $PK/K$ of $G/K$ is normal, so $PK \lhd G$. $PK = G$ would mean $G/K \cong P/P \cap K$ abelian, a contradiction to the fact $\chi_1$ is nonlinear; hence $PK \neq G$. Then *P* char $PK \lhd G$ so $P \lhd G$, a contradiction to the fact that *G* is a counterexample. Therefore $K = 1$, and by minimality of $\chi(1)$, $\chi = \chi_1$.

*Step 3*  For some prime *q*, $|G| = p^a q^b$.

*Proof*  If not, let primes $q_1, \ldots, q_n$ divide $|G|$, $n > 1$. By Theorem 36.3 we can choose for each *i* a Hall $\{p, q_i\}$-subgroup $H_i$ of *G*; by Sylow's theorem, we can assume $H_i \subseteq N_G(P)$ for all *i*. Hence $P \lhd G$, done.

*Step 4*  $|G: G'|$ is a power of *p*.

*Proof*  If not, then there is a proper normal subgroup *H* of *G*, $|G: H|$ a power of *q*. By induction $P \lhd H$, so *P* char $H \lhd G$, $P \lhd G$.

*Step 5*  *G* has a normal Sylow *q*-subgroup *Q*. $|P| = p$.

*Proof*  By Step 4 we choose $H \lhd G$, $|G: H| = p$. If $P_0$ is a Sylow *p*-subgroup of *H*, then $P_0 \lhd H$ by induction so $P_0$ char *H*, $P_0 \lhd G$. *P* is abelian, so $P \subseteq C_G(P_0)$, $C_G(P_0) \lhd G$. $C_G(P_0) \neq G$ would give by induction $P \lhd C_G(P_0)$, *P* char $C_G(P_0)$, $P \lhd G$; hence we may assume $C_G(P_0) = G$.

By Theorem 18.7, *H* has a normal *p*-complement *Q*; *Q* char *H* so $Q \lhd G$, and *Q* is a Sylow *q*-subgroup of *G*.

$P_0 \subseteq Z(G)$, so $\chi|_{P_0} = \chi(1)\mu_0$, $\mu_0$ a faithful linear character of $P_0$; in particular, $P_0$ is cyclic. $\mu_0$ extends to a linear character $\mu$ of $P \cong G/Q$, so $P_0$ is the kernel of the irreducible character $\bar{\mu}\chi$ of $G$. $P_0 \neq 1$ would by induction give $P/P_0 \lhd G/P_0$, $P \lhd G$; hence we have $P_0 = 1$, $|P| = p$.

*Step 6*   $\chi|_Q$ is irreducible.

*Proof*   By Clifford's theorem, $\chi|_Q = \sum_i \theta^{x_i}$, $\theta$ some irreducible character of $Q$, where the $p$-group $G/Q$ permutes the conjugate characters $\theta^{x_i}$. $\chi|_Q$ reducible would mean $p|\chi(1)$, a contradiction; hence $\chi|_Q$ is irreducible.

*Step 7*   $Q$ is extra-special. $N_G(P) = PQ'$.

*Proof*   Let $Q_0$ be any proper subgroup of $Q$ normal in $G$. Then $PQ_0$ is a proper subgroup of $G$, and by induction $P \lhd PQ_0$. Hence $PQ_0 = P \times Q_0$, $P \subseteq C_G(Q_0) \neq G$ means $P \lhd C_G(Q_0)$, $P$ char $C_G(Q_0)$, $P \lhd G$; hence $C_G(Q_0) = G$, $Q_0 \subseteq Z(G)$.

In particular, let $R/Q'$ be a proper subgroup of $Q/Q'$ such that $R/Q' \lhd G/Q'$. Then $R \lhd G$, so by the preceding paragraph $R \subseteq Z(G)$, $R/Q' \subseteq Z(G/Q')$. $Z(G/Q') = G/Q'$ would imply $G' = Q'$, contradicting Step 4; so let $Z(G/Q') = T/Q'$, $R \subseteq T \subseteq Q$. By Lemma 37.1, $Q/Q' = T/Q' \times S/Q'$, some $S \lhd G$. By the previous paragraph, $T \subseteq Z(G)$, $S \subseteq Z(G)$, $Q \subseteq Z(G)$, a contradiction.

Therefore $Q/Q'$ is a minimal normal subgroup of $G/Q'$, implying that $Z(Q) = Q'$, $Q/Q'$ is elementary abelian. By the first paragraph, $Q' = Z(G)$ so $N_G(P) \supseteq PQ'$. If a proper subgroup $U \supseteq Q'$ of $Q$ normalizes $P$ then $U \lhd PU$ so $PU = P \times U$, $U \lhd G$. By the first paragraph of this step then $U \subseteq Z(G)$, $U = Z(Q) = Q'$. So $N_G(P) = PQ'$.

$Q' \subseteq Z(G)$, so $\chi|_{Q'} = \chi(1)\lambda$ for a linear character $\lambda$ of $Q'$, $\chi|_{Q'}$ is faithful, $Q'$ is cyclic. Let $Q' = \langle [a, b] \rangle, a, b \in Q$. Then $[a, b]^q = [a^q, b] = 1$ since $Q/Q'$ is elementary abelian; so $|Q'| = q$, $Q$ is extra-special.

*Step 8, Conclusion*   By Step 7 and Lemma 31.4, $|Q: Q'| = q^{2n}$ for some integer $n$. By Step 6 and Theorem 31.5, $\chi(1) = q^n$. By Step 7, $G$ has $|G: N_G(P)| = |PQ: PQ'| = q^{2n}$ Sylow $p$-subgroups. By Sylow's theorem, $q^{2n} \equiv 1 \pmod p$. Therefore $p$ divides $q^{2n} - 1 = (q^n - 1)(q^n + 1)$, proving $p \leq q^n + 1 = \chi(1) + 1$, the final contradiction.

*Remarks*   Theorem 37.2 is the best possible result of its type. The quaternion group $Q$ of order 8 has an automorphism of order 3, so there exists a solvable group $G$ of order 24 with $Q \lhd G$, $Q = G'$. (Actually $G \cong SL(2, 3)$; see Exercise 1, §35.) $G$ has an irreducible complex character $\chi$, $\chi(1) = 2$, but no normal Sylow 3-subgroup; with $p = 3$ we have $\chi(1) = p - 1$.

Theorem 37.2 does not hold for general finite groups. We shall see in the next section that $SL(2, p)$ has an irreducible character of degree $(p - 1)/2$; but of course $SL(2, p)$ and $PSL(2, p)$ do not have normal Sylow $p$-subgroups. Using modular representation theory, one can show that if any finite group $G$ has a faithful complex character $\chi$, $\chi(1) < (p - 1)/2$, then $G$ has an abelian normal Sylow $p$-subgroup. We do this in §69 and §70 of Part B.

# §38

## Characters of $SL(2, p^n)$

Our reference in this section is Schur [2].

**Theorem 38.1**    Let $F$ be the finite field of $q = p^n$ elements, $p$ an odd prime, and let $v$ be a generator of the cyclic group $F^\# = F - \{0\}$. Denote

$$1 = \begin{pmatrix} 1 & 0 \\ 0 & 1 \end{pmatrix}, \quad z = \begin{pmatrix} -1 & 0 \\ 0 & -1 \end{pmatrix},$$

$$c = \begin{pmatrix} 1 & 0 \\ 1 & 1 \end{pmatrix}, \quad d = \begin{pmatrix} 1 & 0 \\ v & 1 \end{pmatrix}, \quad a = \begin{pmatrix} v & 0 \\ 0 & v^{-1} \end{pmatrix}$$

in $G = SL(2, F)$. $G$ contains an element $b$ of order $q + 1$.

For any $x \in G$, let $(x)$ denote the conjugacy class of $G$ containing $x$. Then $G$ has exactly $q + 4$ conjugacy classes $(1)$, $(z)$, $(c)$, $(d)$, $(zc)$, $(zd)$, $(a)$, $(a^2), \ldots, (a^{(q-3)/2})$, $(b)$, $(b^2), \ldots, (b^{(q-1)/2})$, satisfying

| $x$ | 1 | $z$ | $c$ | $d$ | $zc$ | $zd$ | $a^l$ | $b^m$ |
|---|---|---|---|---|---|---|---|---|
| $|(x)|$ | 1 | 1 | $\frac{1}{2}(q^2-1)$ | $\frac{1}{2}(q^2-1)$ | $\frac{1}{2}(q^2-1)$ | $\frac{1}{2}(q^2-1)$ | $q(q+1)$ | $q(q-1)$ |

for $1 \le l \le (q-3)/2$, $1 \le m \le (q-1)/2$.

Denote $\varepsilon = (-1)^{(q-1)/2}$. Let $\rho \in \mathbf{C}$ be a primitive $(q-1)$th root of $1$, $\sigma \in \mathbf{C}$ a primitive $(q+1)$th root of $1$. Then the complex character table of $G$ is

| | 1 | $z$ | $c$ | $d$ | $a^l$ | $b^m$ |
|---|---|---|---|---|---|---|
| $1_G$ | 1 | 1 | 1 | 1 | 1 | 1 |
| $\phi$ | $q$ | $q$ | 0 | 0 | 1 | $-1$ |
| $\chi_i$ | $q+1$ | $(-1)^i(q+1)$ | 1 | 1 | $\rho^{il}+\rho^{-il}$ | 0 |
| $\theta_j$ | $q-1$ | $(-1)^j(q-1)$ | $-1$ | $-1$ | 0 | $-(\sigma^{jm}+\sigma^{-jm})$ |
| $\xi_1$ | $\frac{1}{2}(q+1)$ | $\frac{1}{2}\varepsilon(q+1)$ | $\frac{1}{2}(1+\sqrt{\varepsilon q})$ | $\frac{1}{2}(1-\sqrt{\varepsilon q})$ | $(-1)^l$ | 0 |
| $\xi_2$ | $\frac{1}{2}(q+1)$ | $\frac{1}{2}\varepsilon(q+1)$ | $\frac{1}{2}(1-\sqrt{\varepsilon q})$ | $\frac{1}{2}(1+\sqrt{\varepsilon q})$ | $(-1)^l$ | 0 |
| $\eta_1$ | $\frac{1}{2}(q-1)$ | $-\frac{1}{2}\varepsilon(q-1)$ | $\frac{1}{2}(-1+\sqrt{\varepsilon q})$ | $\frac{1}{2}(-1-\sqrt{\varepsilon q})$ | 0 | $(-1)^{m+1}$ |
| $\eta_2$ | $\frac{1}{2}(q-1)$ | $-\frac{1}{2}\varepsilon(q-1)$ | $\frac{1}{2}(-1-\sqrt{\varepsilon q})$ | $\frac{1}{2}(-1+\sqrt{\varepsilon q})$ | 0 | $(-1)^{m+1}$ |

*for* $1 \leq i \leq (q - 3)/2,\ 1 \leq j \leq (q - 1)/2,\ 1 \leq l \leq (q - 3)/2,\ 1 \leq m \leq (q - 1)/2$.

(*The columns for the classes* $(zc)$ *and* $(zd)$ *are missing in this table. These values are obtained from the relations*

$$\chi(zc) = \frac{\chi(z)}{\chi(1)}\chi(c), \qquad \chi(zd) = \frac{\chi(z)}{\chi(1)}\chi(d),$$

*for all irreducible characters* $\chi$ *of* $G$.)

*Proof, Step 1*    $G$ contains an element $b$ of order $q + 1$. Suppose $x \in \langle b \rangle$ has order $> 2$; then $C_G(x) = \langle b \rangle$, and if $x$ is conjugate in $G$ to a power $x^r$, $x^r \in \{x, x^{-1}\}$.

*Proof of Step 1*    Let $K = GF(q^2) \supset F$, a two-dimensional $F$-vector space. Let $K^\# = \langle \tau \rangle$, where $\tau^{q+1} = v$. The map $b: \gamma \to \tau^{q-1}\gamma$ is an $F$-linear map of the $F$-vector space $K$, so $b \in GL(K)$; in fact, $b$ is the $(q - 1)$th power of the map $e: \gamma \to \tau\gamma$, an element of $GL(K)$ of order $q^2 - 1$. $SL(K) \lhd GL(K)$ and $|GL(K): SL(K)| = |F^\#| = q - 1$, so $b \in SL(K)$ has order $q + 1$.

Suppose now $x \in \langle b \rangle$ has order $> 2$, say $x = b^i$ so $x: \gamma \to \tau_0\gamma$ where $\tau_0 = \tau^{i(q-1)}$. Since $x$ has order $> 2$, $i$ is not a multiple of $(q + 1)/2$ and $i(q - 1)$ is not a multiple of $q + 1$; therefore $\tau_0 \notin F$, and $K$ has the $F$-basis $\{1, \tau_0\}$.

If $g \in GL(K)$ satisfies $gxg^{-1} = x^r$, then $gx = x^r g$. Let $\varphi: K \to K$ be the map

$$\varphi: \alpha + \tau_0\beta \to \alpha + \tau_0^r\beta, \qquad \alpha, \beta \in F,$$

so $\varphi(\alpha + x\beta) = \alpha + x^r\beta$. If $\gamma = \alpha + \tau_0\beta$, $\mu \in K$, then

$$g(\gamma\mu) = g((\alpha + x\beta)\mu) = (\alpha + x^r\beta)g(\mu) = \gamma^\varphi g(\mu).$$

Hence $g(\gamma\gamma') = (\gamma\gamma')^\varphi g(1)$ and $g(\gamma\gamma') = \gamma^\varphi g(\gamma') = \gamma^\varphi(\gamma')^\varphi g(1)$, proving that $(\gamma\gamma')^\varphi = \gamma^\varphi(\gamma')^\varphi$. Therefore $\varphi$ is an automorphism of $K$ trivial on $F$, $\varphi \in Gal(K/F)$. This Galois group has order 2, so $\varphi(\tau_0) \in \{\tau_0, \tau_0^q\}$, the only two images of $\tau_0$ under $Gal(K/F)$. $\tau_0^{q+1} = 1$, so $\tau_0^q = \tau_0^{-1}$. Thus $x^r$ is either the map $\gamma \to \tau_0\gamma$ or the map $\gamma \to \tau_0^{-1}\gamma$, showing that $x^r \in \{x, x^{-1}\}$.

$g \in C_{SL(K)}(x)$ implies $gxg^{-1} = x$, and as in the previous paragraph we have $\varphi(\tau_0) = \tau_0$, $\varphi = 1$, $g(\gamma) = \gamma g(1)$. Thus $g$ is a multiplication by the scalar $g(1)$, $g \in \langle e \rangle$, $g \in \langle e \rangle \cap SL(K) = \langle b \rangle$.

*Remark*    The argument in this step can be used more generally to show that $GL(m, q) = H$ has an element $y$ of order $q^m - 1$ with $C_H(y) = \langle y \rangle$, $|N_H(\langle y \rangle): \langle y \rangle| = m$.

**Step 2**   $a = \begin{pmatrix} v & 0 \\ 0 & v^{-1} \end{pmatrix}$ has order $q - 1$. If $x \in \langle a \rangle$ has order $> 2$, then $C_G(x) = \langle a \rangle$, and if $x$ is conjugate in $G$ to a power $x^r$ then $x^r \in \{x, x^{-1}\}$.

*Proof* $\langle v \rangle = F^\#$ so $|\langle v \rangle| = q - 1$, $|\langle a \rangle| = q - 1$. If $x \in \langle a \rangle$ has order $> 2$, then $x$ is a diagonal nonscalar matrix, so $C_G(x)$ consists of diagonal matrices of determinant 1, $C_G(x) = \langle a \rangle$. $H = \{g \in G | g^{-1}xg \text{ is diagonal}\}$ is the set of matrices

$$\left\{ \begin{pmatrix} \alpha & 0 \\ 0 & \alpha^{-1} \end{pmatrix}, \quad \begin{pmatrix} 0 & \beta \\ -\beta^{-1} & 0 \end{pmatrix} \right\},$$

so $H = N_G(\langle x \rangle) = \left\langle C_G(x), \begin{pmatrix} 0 & -1 \\ 1 & 0 \end{pmatrix} \right\rangle$ has $|H: C_G(x)| = 2$.

$$\begin{pmatrix} 0 & -1 \\ 1 & 0 \end{pmatrix}^{-1} \begin{pmatrix} v & 0 \\ 0 & v^{-1} \end{pmatrix} \begin{pmatrix} 0 & -1 \\ 1 & 0 \end{pmatrix} = \begin{pmatrix} v^{-1} & 0 \\ 0 & v \end{pmatrix} = a^{-1},$$

so $x$ is conjugate to $x^{-1}$ in $G$, forcing $x, x^{-1}$ to be the only powers of $x$ conjugate to $x$ in $G$.

**Step 3**   The conjugacy classes of $G$ are as described in the theorem.

*Proof* It is easy to check that for $c$ and $d$ to be conjugate in $G$ would force $v$ to be a square in $F^\#$, a contradiction. $|\langle c \rangle| = |\langle d \rangle| = p$ and $|\langle zc \rangle| = |\langle zd \rangle| = 2p$, so the conjugacy classes of $c, d, zc, zd$ are all different. One can check that

$$C_G(c) = C_G(d) = C_G(zc) = C_G(zd) = \left\{ \begin{pmatrix} \alpha & 0 \\ \beta & \alpha \end{pmatrix} \middle| \alpha, \beta \in F, \alpha^2 = 1 \right\},$$

a group of order $2q$. Hence the conjugacy classes $(c)$, $(d)$, $(zc)$, $(zd)$ have order

$$\frac{|G|}{2q} = \frac{q(q + 1)(q - 1)}{2q} = \frac{1}{2}(q^2 - 1).$$

Of course $|(1)| = |(z)| = 1$. In Steps 1–3 we have described the conjugacy classes of

$$1 + 1 + 4 \cdot \frac{1}{2}(q^2 - 1) + \frac{q - 3}{2} q(q + 1) + \frac{q - 1}{2} q(q - 1) = q^3 - q$$

elements of $G$, but $|G| = q^3 - q$ so we are done.

*Step 4* The functions $\psi$, $\chi_1, \ldots, \chi_{(q-3)/2}$, as listed, are irreducible characters of $G$.

*Proof* Let $S$ be the Sylow $p$-subgroup

$$\left\{ \begin{pmatrix} 1 & 0 \\ \beta & 1 \end{pmatrix} \middle| \beta \in F \right\}$$

of $G$, $H$ the subgroup

$$\left\{ \begin{pmatrix} \alpha & 0 \\ \beta & \alpha^{-1} \end{pmatrix} \middle| \beta \in F, \alpha \in F^\# \right\};$$

then $S \lhd H$ and $H = S\langle a \rangle$, $S \cap \langle a \rangle = 1$, $|H| = |S||\langle a \rangle| = q(q - 1)$. If $F_s^\#$ is the set of nonzero squares in $F$, then $|F_s^\#| = (q - 1)/2$. We see that

$$(c) \cap H = \left\{ \begin{pmatrix} 1 & 0 \\ \alpha & 1 \end{pmatrix} \middle| \alpha \in F_s^\# \right\},$$

$$(d) \cap H = \left\{ \begin{pmatrix} 1 & 0 \\ v\alpha & 1 \end{pmatrix} \middle| \alpha \in F_s^\# \right\},$$

$$(zc) \cap H = \left\{ \begin{pmatrix} -1 & 0 \\ -\alpha & -1 \end{pmatrix} \middle| \alpha \in F_s^\# \right\},$$

$$(zd) \cap H = \left\{ \begin{pmatrix} -1 & 0 \\ -v\alpha & -1 \end{pmatrix} \middle| \alpha \in F_s^\# \right\},$$

all sets of order $\frac{1}{2}(q - 1)$. If $1 \leq l \leq (q - 3)/2$, then

$$(a^l) \cap H = \left\{ \begin{pmatrix} v^l & 0 \\ \alpha & v^{-l} \end{pmatrix}, \begin{pmatrix} v^{-l} & 0 \\ \alpha & v^l \end{pmatrix} \middle| \alpha \in F \right\},$$

a set of order $2q$. If $1 \le m \le (q-1)/2$, then $\langle b^m \rangle$ is irreducible, so $(b^m) \cap H = \emptyset$. Of course $1, z \in H$.

Using this information we can compute $\lambda^G$, $\lambda$ any character of $H$. For $0 \le i \le q-1$, let $\lambda_i$ be the linear character of $H$ satisfying

$$\lambda_i : \begin{pmatrix} v^t & 0 \\ \alpha & v^{-t} \end{pmatrix} \to \rho^{it},$$

any $\alpha \in F$; thus $S \subseteq \ker \lambda_i$ and $\lambda_{q-1} = \lambda_0 = 1_H$. The values of the $\lambda_i^G$ are:

| $x$ | 1 | $z$ | $c$ | $zc$ | $d$ | $zd$ | $a^t$ | $b^m$ |
|---|---|---|---|---|---|---|---|---|
| $\lambda_i^G(x)$ | $q+1$ | $(-1)^i(q+1)$ | 1 | $(-1)^i$ | 1 | $(-1)^i$ | $\rho^{it}+\rho^{-it}$ | 0 |

If $1 \le i \le (q-3)/2$, then $\chi_i = \lambda_i^G$ satisfies $(\chi_i, \chi_i)_G = 1$, so $\chi_i$ is an irreducible character of $G$. $(\lambda_0^G, \lambda_0^G)_G = 2$, and

$$(\lambda_0^G, 1_G)_G = (1_H^G, 1_G)_G = (1_H, 1_H)_H = 1,$$

so $\lambda_0^G = 1_G + \psi$, $\psi$ an irreducible character of $G$.

*Step 5*  $G$ has irreducible characters $\xi_1$ and $\xi_2$ satisfying $\xi_1(z) = \varepsilon\xi_1(1)$, $\xi_2(z) = \varepsilon\xi_2(1)$, and

| $x$ | 1 | $z$ | $c$ | $zc$ | $d$ | $zd$ | $a^t$ | $b^m$ |
|---|---|---|---|---|---|---|---|---|
| $\xi_1(x)\pm\xi_2(x)$ | $q+1$ | $\varepsilon(q+1)$ | 1 | $\varepsilon$ | 1 | $\varepsilon$ | $2(-1)^t$ | 0 |

*Proof*  In the proof of Step 4, $\pi = \lambda_{(q-1)/2}$ satisfies $(\pi, \pi)_G = 2$; we set $\pi = \xi_1 \pm \xi_2$, $\xi_1$ and $\xi_2$ irreducible. $\pi(z) = \varepsilon\pi(1)$ forces $\xi_i(z) = \varepsilon\xi_i(1)$.

*Step 6*  The functions $\theta_1, \theta_2, \ldots, \theta_{(q-1)/2}$, as listed, are irreducible characters of $G$.

*Proof*  Let $\varphi_j$ be the linear character of $\langle b \rangle$ satisfying

$$\varphi_j : b^t \to \sigma^{jt}.$$

Then $\varphi_j^G(1) = q(q-1)$, $\varphi_j^G(z) = (-1)^j q(q-1)$, $\varphi_j^G(b^m) = \sigma^{jm} + \sigma^{-jm}$, and $\varphi_j^G$ is 0 on other conjugacy classes of $G$. For $1 \le j \le (q+1)/2$, let $\theta_j$ be the generalized character $\theta_j = \psi\lambda_j^G - \lambda_j^G - \varphi_j^G$. Then the values of $\theta_j$ are

| $x$ | 1 | $z$ | $c$ | $zc$ | $d$ | $zd$ | $a^t$ | $b^m$ |
|---|---|---|---|---|---|---|---|---|
| $\theta_j(x)$ | $q-1$ | $(-1)^j(q-1)$ | $-1$ | $(-1)^{j+1}$ | $-1$ | $(-1)^{j+1}$ | 0 | $-(\sigma^{jm}+\sigma^{-jm})$ |

If $1 \leq j \leq (q - 1)/2$, then $(\theta_j, \theta_j)_G = 1$. $\theta_j(1) > 0$, so $\theta_j$ is an irreducible character of $G$.

*Step 7*   The irreducible characters of $G$ are $1_G, \psi, \chi_1, \ldots, \chi_{(q-3)/2}$, $\theta_1, \ldots, \theta_{(q-1)/2}, \xi_1, \xi_2$, plus two additional characters $\eta_1, \eta_2$ satisfying $\eta_1(z) = -\varepsilon\eta_1(1), \eta_2(z) = -\varepsilon\eta_2(1)$, and

| $x$ | $1$ | $z$ | $c$ | $zc$ | $d$ | $zd$ | $a^l$ | $b^m$ |
|---|---|---|---|---|---|---|---|---|
| $\eta_1(x)+\eta_2(x)$ | $q-1$ | $-\varepsilon(q-1)$ | $-1$ | $\varepsilon$ | $-1$ | $\varepsilon$ | $0$ | $2(-1)^{m+1}$ |

We have

$$\xi_1(1) = \xi_2(1) = \tfrac{1}{2}(q + 1) \qquad \text{and} \qquad \eta_1(1) = \eta_2(1) = \tfrac{1}{2}(q - 1).$$

*Proof*   In Step 6, set $\kappa = \theta_{(q+1)/2}$. $(\kappa, \kappa)_G = 2$, so $\kappa = \eta_1 \pm \eta_2$, some irreducible characters $\eta_1$ and $\eta_2$ of $G$. We can compute that all inner products among $1_G, \psi, \chi_i, \theta_j, \pi, \kappa$ are zero. Hence all are distinct, except possibly the $\xi_i$ and $\eta_j$. But $\kappa(z) = -\varepsilon\kappa(1)$, forcing $\eta_1(z) = -\varepsilon\eta_1(1), \eta_2(z) = -\varepsilon\eta_2(1)$. By Step 5, the $\xi_i$ and $\eta_j$ are different.

We must show that $\kappa$ is $\eta_1 + \eta_2$, not $\eta_1 - \eta_2$. We do this while we simultaneously determine the degrees of the $\xi_i$ and $\eta_j$.

If $\varepsilon = +1$, then $\eta_1(z) = -\eta_1(1), \eta_2(z) = -\eta_2(1)$, so $\ker \eta_1$, $\ker \eta_2$ do not contain $\langle z \rangle$, while $\ker \xi_1$, $\ker \xi_2$ do contain $\langle z \rangle$. The irreducible characters of $G/\langle z \rangle = PSL(2, F)$ are $1_G, \psi, \chi_2, \chi_4, \ldots, \chi_{(q-5)/2}, \theta_2, \theta_4, \ldots,$ $\theta_{(q-1)/2}, \xi_1, \xi_2$. The sum of the squares of their degrees is $|G/\langle z \rangle| = \tfrac{1}{2}q(q^2 - 1)$, so we have

$$1 + q^2 + \frac{q - 5}{4}(q + 1)^2 + \frac{q - 1}{4}(q - 1)^2 + \xi_1(1)^2 + \xi_2(1)^2$$

$$= \frac{q(q^2 - 1)}{2},$$

or $\xi_1(1)^2 + \xi_2(1)^2 = \tfrac{1}{2}(q + 1)^2$. $\xi_1(1)$ and $\xi_2(1)$ are positive integers and $\xi_1(1) \pm \xi_2(1) = q + 1$ by Step 5, so these equations force $\xi_1(1) = \xi_2(1) = \tfrac{1}{2}(q + 1)$.

The fact that $|G|$ is the sum of the squares of the degrees of all irreducible characters of $G$ now yields the equation $\eta_1^2(1) + \eta_2^2(1) = \tfrac{1}{2}(q - 1)^2$. So far we know $q - 1 = \kappa(1) = \eta_1(1) \pm \eta_2(1)$, $\eta_1(1)$ and $\eta_2(1)$ positive integers. These equations force $\kappa(1)$ to be $\eta_1(1) + \eta_2(1)$, $\kappa = \eta_1 + \eta_2$, $\eta_1(1) = \eta_2(1) = \tfrac{1}{2}(q - 1)$.

If $\varepsilon = -1$, the proof is similar. (In this case $\eta_1$ and $\eta_2$ are characters of $G/\langle z \rangle$ and $\xi_1, \xi_2$ are not.)

*Step 8*   The theorem holds if $\varepsilon = +1$.

*Proof*   $\varepsilon = +1$ means $q \equiv 1$ (mod 4). Hence $4 \| F^\# |$, so the element $-1$ of order 2 in $F^\#$ is a square in $F^\#$. Therefore

$$c^{-1} = \begin{pmatrix} 1 & 0 \\ -1 & 1 \end{pmatrix} \quad \text{and} \quad c = \begin{pmatrix} 1 & 0 \\ 1 & 1 \end{pmatrix}$$

are conjugate in $G$, $c^{-1} \in (c)$. Also, $-v$ is not a square in $F^\#$ so $d^{-1} \in (d)$. These results imply $\{zc\}^{-1} \in (zc)$ and $\{zd\}^{-1} \in (zd)$, so every element of $G$ is conjugate to its inverse. By Theorem 23.3, every complex character of $G$ is real.

For each $x \in G - (a^{(q-1)/4})$, $xz \notin (x)$, so the conjugacy class of $x\langle z \rangle$ in $G/\langle z \rangle$ has order $(x)$ and

$$|C_{G/\langle z \rangle}(x\langle z \rangle)| = \frac{1}{2}|C_G(x)|.$$

If $x \in (a^{(q-1)/4})$, then $xz \in (x)$ and $|C_{G/\langle z \rangle}(x\langle z \rangle)| = |C_G(x)|$. $\xi_1$ and $\xi_2$ are the only two irreducible characters of $G/\langle z \rangle$ not fully known; the orthogonality relations 5.3(3) in the group $G/\langle z \rangle$ give the values

$$|\xi_1(x)|^2 + |\xi_2(x)|^2 = \xi_1(x)^2 + \xi_2(x)^2.$$

Since we also know $\xi_1(x) + \xi_2(x)$ by Step 5, and $(\xi_i, 1_G)_G = 0$, we complete the table of values of $\xi_1$ and $\xi_2$.

Now $\eta_1$ and $\eta_2$ are the only irreducible characters of $G$ not fully known. Orthogonality relations in $G$ for $|C_G(x)|$ give the values $|\eta_1(x)|^2 + |\eta_2(x)|^2 = \eta_1(x)^2 + \eta_2(x)^2$. Since we also know $\eta_1(x) + \eta_2(x)$ by Step 7, and $\eta_1(zx) = -\eta_1(x)$, $\eta_2(zx) = -\eta_2(x)$, we easily complete the character table of $G$.

*Step 9*   The theorem holds if $\varepsilon = -1$.

*Proof*   $\varepsilon = -1$ means $q \equiv -1$ (mod 4). Hence $-1$, an element of $F^\#$ of order 2, is not a square in $F^\#$. Thus

$$c^{-1} = \begin{pmatrix} 1 & 0 \\ -1 & 1 \end{pmatrix} \quad \text{and} \quad c = \begin{pmatrix} 1 & 0 \\ 1 & 1 \end{pmatrix}$$

are not conjugate in $G$, forcing $c^{-1} \in (d)$, $\{zc\}^{-1} \in (zd)$, $d^{-1} \in (c)$, $\{zd\}^{-1} \in (zc)$. By Theorem 23.3, $G$ has four nonreal irreducible complex characters.

$1_G$, $\psi$ and the $\chi_i$ and $\theta_j$ are all real, so $\xi_1$, $\xi_2$, $\eta_1$, $\eta_2$ are not real and we must have $\bar{\xi}_1 = \xi_2, \bar{\eta}_1 = \eta_2$.

In this case, the only $x$'s in $G$ with $xz \in (x)$ are the $x$'s in $(b^{(q+1)/4})$. So again we know $|C_{G/\langle z\rangle}(x\langle z\rangle)|$. $\eta_1$ and $\eta_2$ are the only irreducible characters of $G/\langle z\rangle$ not fully known. The orthogonality relations 5.3(3) in the group $G/\langle z\rangle$ now give the values of

$$|\eta_1(x)|^2 + |\eta_2(x)|^2 = |\eta_1(x)|^2 + |\overline{\eta_1(x)}|^2 = 2|\eta_1(x)|^2.$$

Since we also know $\eta_1(x) + \eta_2(x) = \eta_1(x) + \overline{\eta_1(x)}$ by Step 7, and $(\eta_i, 1_G)_G = 0$, we complete the table of values of $\eta_1, \eta_2$.

Now $\xi_1$, $\xi_2 = \bar{\xi}_1$ are the only irreducible characters of $G$ not fully known. Orthogonality relations in $G$ for $|C_G(x)|$ give the values $|\xi_1(x)|^2 + |\xi_2(x)|^2 = 2|\xi_1(x)|^2$. Since we also know $\xi_1(x) + \xi_2(x) = \xi_1(x) + \overline{\xi_1(x)}$ by Step 5, and $\xi_i(zx) = -\xi_i(x)$, we easily complete the character table of $G$.

**Theorem 38.2**   *Let $F$ be the finite field of $q = 2^n$ elements, and let $v$ be a generator of the cyclic group $F^{\#} = F - \{0\}$. Denote*

$$1 = \begin{pmatrix} 1 & 0 \\ 0 & 1 \end{pmatrix}, \quad c = \begin{pmatrix} 1 & 0 \\ 1 & 1 \end{pmatrix}, \quad a = \begin{pmatrix} v & 0 \\ 0 & v^{-1} \end{pmatrix}$$

*in $G = SL(2, F)$. $G$ contains an element $b$ of order $q + 1$.*

*For any $x \in G$, let $(x)$ denote the conjugacy class of $G$ containing $x$. Then $G$ has exactly $q + 1$ conjugacy classes $(1), (c), (a), (a^2), \ldots, (a^{(q-2)/2})$, $(b), (b^2), \ldots, (b^{q/2})$, where*

| $x$ | $1$ | $c$ | $a^l$ | $b^m$ |
|---|---|---|---|---|
| $|(x)|$ | $1$ | $(q^2 - 1)$ | $q(q + 1)$ | $q(q - 1)$ |

*for $1 \leq l \leq (q - 2)/2, 1 \leq m \leq q/2$.*

*Let $\rho \in \mathbf{C}$ be a primitive $(q - 1)$th root of 1, $\sigma \in \mathbf{C}$ a primitive $(q + 1)$th root of 1. Then the character table of $G$ over $\mathbf{C}$ is*

| | $1$ | $c$ | $a^l$ | $b^m$ |
|---|---|---|---|---|
| $1_G$ | $1$ | $1$ | $1$ | $1$ |
| $\phi$ | $q$ | $0$ | $1$ | $-1$ |
| $\chi_i$ | $q+1$ | $1$ | $\rho^{il}+\rho^{-il}$ | $0$ |
| $\theta_j$ | $q-1$ | $-1$ | $0$ | $-(\sigma^{jm}+\sigma^{-jm})$ |

*for $1 \leq i \leq (q - 2)/2, \; 1 \leq j \leq q/2, \; 1 \leq l \leq (q - 2)/2, \; 1 \leq m \leq q/2$.*

*Proof, Step 1*   $G$ contains an element $b$ of order $q + 1$. If $x \in \langle b \rangle^{\#}$, then $C_G(x) = \langle b \rangle$; if $x$ is conjugate in $G$ to a power $x^r$, then $x^r \in \{x, x^{-1}\}$.

*Proof of Step 1*   Same as the proof of Step 1 of the previous theorem.

*Step 2*   $a = \begin{pmatrix} v & 0 \\ 0 & v^{-1} \end{pmatrix}$ has order $q - 1$. If $x \in \langle a \rangle^{\#}$, then $C_G(x) = \langle a \rangle$; if $x$ is conjugate in $G$ to $x^r$, then $x^r \in \{x, x^{-1}\}$.

*Proof*   Same as the proof of Step 2 of the previous theorem.

*Step 3*   The conjugacy classes of $G$ are as described.

*Proof*

$$C_G(c) = \left\{ \begin{pmatrix} \alpha & 0 \\ \beta & \alpha \end{pmatrix} \mid \alpha, \beta \in F, \alpha^2 = 1 \right\}$$

$$= \left\{ \begin{pmatrix} 1 & 0 \\ \beta & 1 \end{pmatrix} \mid \beta \in F \right\}$$

has order $q$, so $|(c)| = q^2 - 1$. Of course $|(1)| = 1$. In Steps 1–3 we have described conjugacy classes of $G$ containing

$$1 + q^2 - 1 + \frac{q-2}{2} q(q+1) + \frac{q}{2} q(q-1) = q(q^2 - 1) = |G|$$

elements, so we are done.

*Step 4*   The functions $\psi, \chi_1, \ldots, \chi_{(q-2)/2}$, as listed, are irreducible characters of $G$.

*Proof*   Let $S$ be the Sylow 2-subgroup

$$\left\{ \begin{pmatrix} 1 & 0 \\ \beta & 1 \end{pmatrix} \mid \beta \in F \right\} \text{ of } G, \qquad H = \left\{ \begin{pmatrix} \alpha & 0 \\ \beta & \alpha^{-1} \end{pmatrix} \mid \beta \in F, \alpha \in F^{\#} \right\};$$

then $S \lhd H$ and $H = S \langle a \rangle$, $S \cap \langle a \rangle = 1$, $|H| = |S||\langle a \rangle| = q(q-1)$. We see that $(c) \cap H = S^{\#}$ has order $q - 1$. If $1 \le l \le (q-2)/2$, then

$$(a^l) \cap H = \left\{ \begin{pmatrix} v^l & 0 \\ \alpha & v^{-l} \end{pmatrix}, \begin{pmatrix} v^{-l} & 0 \\ \alpha & v^l \end{pmatrix} \mid \alpha \in F \right\}$$

is a set of order $2q$. For any $1 \leq m \leq q/2$, $(b^m) \cap H = \emptyset$.

Using this information, we can compute $\lambda^G$, $\lambda$ any character of $H$. For $0 \leq i \leq q - 1$, let $\lambda_i$ be the linear character of $H$ satisfying

$$\lambda_i : \begin{pmatrix} v^t & 0 \\ \alpha & v^{-t} \end{pmatrix} \to \rho^{it}, \qquad \text{any } \alpha \in F;$$

thus $S \subseteq \ker \lambda_i$ and $\lambda_{q-1} = \lambda_0 = 1_H$. The values of the $\lambda_i^G$ are

| $x$ | $1$ | $c$ | $a^l$ | $b^m$ |
|---|---|---|---|---|
| $\lambda_i^G(x)$ | $q + 1$ | $1$ | $\rho^{il} + \rho^{-il}$ | $0$ |

If $1 \leq i \leq (q - 2)/2$, then $\chi_i = \lambda_i^G$ satisfies $(\chi_i, \chi_i)_G = 1$, so $\chi_i$ is an irreducible character of $G$. $(\lambda_0^G, \lambda_0^G)_G = 2$, and $(\lambda_0^G, 1_G)_G = (1_H^G, 1_G)_G = (1_H, 1_H)_H = 1$, so $\lambda_0^G = 1_G + \psi$, $\psi$ an irreducible character of $G$.

*Step 5*   The functions $\theta_1, \ldots, \theta_{q/2}$, as listed, are irreducible characters of $G$, so we are done.

*Proof*   Let $\varphi_j$ be the linear character of $\langle b \rangle$ satisfying

$$\varphi_j : b^t \to \sigma^{jt}.$$

Then $\varphi_j^G$ has the values

| $x$ | $1$ | $c$ | $a^l$ | $b^m$ |
|---|---|---|---|---|
| $\varphi_j^G(x)$ | $q(q - 1)$ | $0$ | $0$ | $\sigma^{jm} + \sigma^{-jm}$ |

For $1 \leq j \leq q/2$, let $\theta_j$ be the generalized character $\theta_j = \psi \lambda_j^G - \lambda_j^G - \varphi_j^G$. The values of $\theta_j$ are

| $x$ | $1$ | $c$ | $a^l$ | $b^m$ |
|---|---|---|---|---|
| $\theta_j(x)$ | $q - 1$ | $-1$ | $0$ | $-(\sigma^{jm} + \sigma^{-jm})$ |

$(\theta_j, \theta_j)_G = 1$ and $\theta_j(1) > 0$, so $\theta_j$ is an irreducible character of $G$.

The sum of the squares of the degrees of $1_G$, $\psi$, and the $\chi_i$ and $\theta_j$ is

$$1 \cdot 1 + 1 \cdot q^2 + \frac{q - 2}{2}(q + 1)^2 + \frac{q}{2}(q - 1)^2 = q(q^2 - 1) = |G|,$$

completing the proof.

*Remarks*  For the matrix representations of $SL(2, q)$, see Dietz [1] and Tanaka [1]. Major work on the representations of $GL(n, q)$ has been done in Green [2, 3], Morris [1], and Steinberg [1]; see also Boerner [1], Burrow [1], Farahat [1], Drobotenko [1], Morris [2], and Springer [2]. Steinberg [2, 4] has done important work on characters of other linear groups; see also Curtis [1], Dagger [1], and Ennola [1]. Srinivasan [1] gives a complete table of characters of $Sp(4, q)$.

EXERCISE

Give a new proof, independent of §35, that if $q \geq 4$ then $PSL(2, q)$ is simple.

# Bibliography

This bibliography emphasizes theory and applications of ordinary representations; Part B will contain a bibliography on modular representation theory. Reiner [2] contains an excellent bibliography for integral representation theory. Other good surveys of the literature appear in Curtis and Reiner [1], Berman [8], and Boerner [3].

Alperin, J. L.; Kuo, Tzee-Nan.
1. The exponent and the projective representations of a finite group. *Illinois J. Math.* **11** (1967), 410–413.

Asano, Keizo.
1. Einfacher Beweis eines Brauerschen Satzes über Gruppencharaktere. *Proc. Japan Acad.* **31** (1955), 501–503.

Basmaji, B. G.
1. Monomial representations and metabelian groups. *Nagoya Math. J.* **35** (1969), 99–107.

Bayar, Ergün.
1. Eine neue Einführung in die Darstellungstheorie symmetrischer Gruppen. *Mitt. Math. Sem. Giessen* **81** (1969), 45 pp.

Berger, T. R.
1. Class two *p*-groups as fixed point free automorphism groups. *Illinois J. Math.* **14** (1970), 121–149.

Berggren, J. L.
1. Finite groups in which every element is conjugate to its inverse. *Pacific J. Math.* **28** (1969), 289–293.

Berman, S. D.
1. On the theory of representations of finite groups. (Russian) *Dokl. Akad. Nauk SSSR* (*N. S.*) **86** (1952), 885–888.
2. p-adic ring of characters. (Russian) *Dokl. Akad. Nauk SSSR* (*N. S.*) **106** (1956), 583–586.
3. The number of irreducible representations of a finite group over an arbitrary field. (Russian) *Dokl. Akad. Nauk SSSR* (*N. S.*) **106** (1956), 767–769.
4. Groups of which all representations are monomial. (Ukrainian) *Dopovidi Akad. Nauk Ukrain. RSR* (1957), 539–542.
5. Representations of groups of order $2^m$ over an arbitrary field of zero characteristic. (Ukrainian) *Dopovidi Akad. Nauk Ukrain. RSR* (1958), 243–246.
6. On Schur's index. (Russian) *Uspehi Mat. Nauk* **16** (1961), no. 2 (98), 95–99.

7. The smallest field in which can be realized all complex representations of a *p*-group of odd order. (Russian) *Uspehi Mat. Nauk* **16** (1961), no. 3 (99), 151–153.

8. Representations of finite groups. (Russian) *Algebra 1964*, pp. 83–122. Akad. Nauk SSSR Inst. Naučn. Informacii, Moscow, 1966.

Bivins, Robert L.; Metropolis, N.; Stein, Paul R.; Wells, Mark B.

1. Characters of the symmetric groups of degree 15 and 16. *Math. Tables and Other Aids to Computation* **8** (1954), 212–216.

Blichfeldt, H. F.

1. *Finite Collineation Groups*. Univ. of Chicago Press, Chicago, Illinois, 1917.

Boerner, Hermann.

1. Über die rationalen Darstellungen der allgemeinen linearen Gruppe. *Arch. Math.* **1** (1948), 52–55.

2. *Representation of Groups with Special Consideration for the Needs of Modern Physics*. Wiley-Interscience, New York, 1963.

3. Darstellungstheorie der endlichen Gruppen. *Enzyklopädie der mathematischen Wissenschaften mit Einschluss ihrer Anwendungen*, Heft 6, Teil II, No. 15. B. G. Teubner, Stuttgart, 1967, 80 pp.

Brauer, Richard.

1. On the representation of a group of order *g* in the field of the *g*th roots of unity. *Amer. J. Math.* **67** (1945), 461–471.

2. Applications of induced characters. *Amer. J. Math.* **69** (1947), 709–716.

3. On the representations of groups of finite order. *Proceedings of the International Congress of Mathematicians, Cambridge, Mass., 1950*, vol. 2, pp. 33–36. Amer. Math. Soc., Providence, R. I., 1952.

4. A characterization of the characters of groups of finite order. *Ann. of Math.* (2) **57** (1953), 357–377.

5. On some conjectures concerning finite simple groups. *Studies in Mathematical Analysis and Related Topics*, pp. 56–61. Stanford Univ. Press, Stanford, Calif., 1962.

6. On finite groups and their characters. *Bull. Amer. Math. Soc.* **69** (1963), 125–130.

7. *Representations of Finite Groups*. Lectures on Modern Mathematics, Vol. I, pp. 133–175. Wiley, New York, 1963.

8. A note on theorems of Burnside and Blichfeldt. *Proc. Amer. Math. Soc.* **15** (1964), 31–34.

9. On quotient groups of finite groups. *Math. Zeitschr.* **83** (1964), 72–84.

10. Über endliche lineare Gruppen von Primzahlgrad. *Math. Ann.* **169** (1967), 73–96.

11. On pseudo groups. *J. Math. Soc. Japan* **20** (1968), 13–22.

12. On the order of finite primitive projective groups in a given dimension. Nachrichten der Akademie der Wissenschaften in Göttingen. Mathematisch-Physikalische Klasse (1969), No. 11, 1–4.

Brauer, Richard; Feit, Walter.

1. An analogue of Jordan's theorem in characteristic *p*. *Ann. of Math.* (2) **84** (1966), 119–131.

Brauer, Richard; Fowler, K. A.

1. On groups of even order. *Ann. of Math.* (2) **62** (1955), 565–583.

Brauer, Richard; Sah, Chih-han.

1. *Theory of Finite Groups: A Symposium*. Edited by Richard Brauer and Chih-Han Sah. Benjamin, New York, 1969.

Brauer, Richard; Suzuki, Michio.
1. On finite groups of even order whose 2-Sylow group is a quaternion group. *Proc. Nat. Acad. Sci. U.S.A.* **45** (1959), 1757–1759.

Brauer, Richard; Suzuki, Michio; Wall, G. E.
1. A characterization of the one-dimensional unimodular projective groups over finite fields. *Illinois J. Math.* **2** (1958), 718–745.

Brauer, Richard; Tate, John.
1. On the characters of finite groups. *Ann. of Math.* (2) **62** (1955), 1–7.

Burgoyne, P. Nicholas; Fong, Paul.
1. The Schur multipliers of the Mathieu groups. *Nagoya Math. J.* **27** (1966), 733–745.

Burnside, W.
1. On soluble irreducible groups of linear substitutions in a prime number of variables. *Acta Math.* **27** (1903), 217–224.
2. On groups of order $p^a q^b$. I, II. *Proc. London Math. Soc.* **2** (1904), 388–392, 432–437.

Burrow, Martin D.
1. A generalization of the Young diagram. *Canadian J. Math.* **6** (1954), 498–508.
2. *Representation Theory of Finite Groups.* Academic Press, New York-London, 1965.

Clifford, A. H.
1. Representations induced in an invariant subgroup. *Ann. of Math.* (2) **38** (1937), 533–550.

Cline, Edward T.
1. Stable Clifford theory. To appear.

Cooney, Sister Miriam P.
1. The non-solvable Hall subgroups of the general linear group. *Math. Zeitschr.* **114** (1970), 245–270.

Curtis, Charles W.
1. The Steinberg character of a finite group with a BN-pair. *J. Algebra* **4** (1966), 433–441.

Curtis, Charles W.; Benson, Clark.
1. On the degrees and rationality of certain characters of finite Chevalley groups. I, II. In Reiner [3], pp. 1–5.

Curtis, Charles W.; Fossum, Timothy V.
1. On centralizer rings and characters of representations of finite groups. *Math. Zeitschr.* **107** (1968), 402–406.

Curtis, Charles W.; Reiner, Irving.
1. *Representation Theory of Finite Groups and Associative Algebras.* Wiley-Interscience, New York, 1962.

Dade, Everett C.
1. Answer to a question of R. Brauer. *J. Algebra* **1** (1964), 1–4.
2. Lifting group characters. *Ann. of Math.* (2) **79** (1964), 590–596.
3. Characters and solvable groups. (Preprint) University of Illinois, Urbana, Ill., 1967.
4. Products of orders of centralizers. *Math. Zeitschr.* **96** (1967), 223–225.
5. Carter subgroups and Fitting heights of finite solvable groups. *Illinois J. Math.* **13** (1969), 449–514.

6.  Counterexamples to a conjecture of Tamaschke. *J. Algebra* **11** (1969), 353–358.
7.  Isomorphisms of Clifford extensions. (Preprint) Université de Strasbourg, Strasbourg, France, 1969.
8.  Compounding Clifford's theory. *Ann. of Math.* (2) **91** (1970), 236–290.

Dagger, S. W.
1.  A class of irreducible characters for certain classical groups. *J. London Math. Soc.* (2) **2** (1970), 513–520.

DeMeyer, Frank R.; Janusz, Gerald J.
1.  Finite groups with an irreducible representation of large degree. *Math. Zeitschr.* **108** (1969), 145–153.

Deskins, W. E.
1.  A note on the relationship between certain subgroups of a finite group. *Proc. Amer. Math. Soc.* **9** (1958), 655–660.

Dickson, Leonard E.
1.  *Linear Groups.* Dover, New York, 1958.

Dietz, Helmut.
1.  Zur Darstellungstheorie der binären projektiven Gruppe über einem Galoisfeld. *Math. Nachr.* **7** (1952), 219–256.

Dieudonné, Jean.
1.  *Sur les groupes classiques.* Actualités Sci. Ind., no. 1040. Hermann et Cie., Paris, 1948.
2.  *La géométrie des groupes classiques.* Seconde édition. Springer-Verlag, Berlin-Göttingen-Heidelberg, 1963.

Dixon, John D.
1.  High speed computation of group characters. *Numer. Math.* **10** (1967), 446–450.
2.  *Problems in Group Theory.* Blaisdell, Waltham, Mass., 1967.
3.  The Fitting subgroup of a linear solvable group. *J. Austral. Math. Soc.* **7** (1967), 417–424.
4.  Normal *p*-subgroups of solvable linear groups. *J. Austral. Math. Soc.* **7** (1967), 545–551.
5.  The solvable length of a solvable linear group. *Math. Zeitschr.* **107** (1968), 151–158.

Dornhoff, Larry.
1.  *M*-groups and 2-groups. *Math. Zeitschr.* **100** (1967), 226–256.
2.  Jordan's theorem for solvable groups. *Proc. Amer. Math. Soc.* **24** (1970), 533–537.
3.  Jordan's theorem for solvable groups. In Reiner [3], pp. 37–38.

Drobotenko, E. S.
1.  Irreducible complex representations of the group *GL*(3, *q*). (Russian) *Dopovidi Akad. Nauk Ukrain. RSR Ser. A* (1967), 104–109.

Ennola, Veikko.
1.  On the characters of the finite unitary groups. *Ann. Acad. Sci. Fenn. Ser. AI* No. 323 (1963), 35 pp.

Ernest, John A.
1.  Central intertwining numbers for representations of finite groups. *Trans. Amer. Math. Soc.* **99** (1961), 499–508.
2.  Embedding numbers for finite groups. *Proc. Amer. Math. Soc.* **13** (1962), 567–570.

Evens, Leonard.
1. On the Chern classes of representations of finite groups. *Trans. Amer. Math. Soc.* **115** (1965), 180–193.

Farahat, H. K.
1. On Schur functions. *Proc. London Math. Soc.* (3) **8** (1958), 621–630.

Fein, Burton.
1. The Schur index for projective representations of finite groups. *Pacific J. Math.* **28** (1969), 87–100.

Feit, Walter.
1. The degree formula for the skew-representations of the symmetric group. *Proc. Amer. Math. Soc.* **4** (1953), 740–744.
2. On a conjecture of Frobenius. *Proc. Amer. Math. Soc.* **7** (1956), 177–187.
3. On groups which contain Frobenius groups as subgroups. *Proc. Sympos. Pure Math.*, Vol. I, pp. 22–28. Amer. Math. Soc., Providence, R. I., 1959.
4. A characterization of the simple groups $SL(2, 2^a)$. *Amer. J. Math.* **82** (1960), 281–300; correction, ibid. **84** (1962), 201–204.
5. On a class of doubly transitive permutation groups. *Illinois J. Math.* **4** (1960), 170–186.
6. Group characters. Exceptional characters. *Proc. Sympos. Pure Math.*, Vol. VI, pp. 67–70. Amer. Math. Soc., Providence, R. I., 1962.
7. *Characters of Finite Groups.* Benjamin, New York-Amsterdam, 1967.

Feit, Walter; Hall, Marshall; Thompson, John G.
1. Finite groups in which the centralizer of any non-identity element is nilpotent. *Math. Zeitschr.* **74** (1960), 1–17.

Feit, Walter; Thompson, John G.
1. Groups which have a faithful representation of degree less than $(p - 1)/2$. *Pacific J. Math.* **11** (1961), 1257–1262.
2. A solvability criterion for finite groups and some consequences. *Proc. Nat. Acad. Sci. U.S.A.* **48** (1962), 968–970.
3. Finite groups which contain a self-centralizing subgroup of order 3. *Nagoya Math. J.* **21** (1962), 185–197.
4. Solvability of groups of odd order. *Pacific J. Math.* **13** (1963), 775–1029.

Fischer, Bernd.
1. Die Brauersche Charakterisierung der Charaktere endlicher Gruppen. *Math. Ann.* **149** (1962/3), 226–231.

Fong, Paul.
1. A note on splitting fields of representations of finite groups. *Illinois J. Math.* **7** (1963), 515–520.

Ford, Charles.
1. Some results on the Schur index of a representation of a finite group. *Canadian J. Math.* **22** (1970), 626–640.

Fossum, Timothy V.
1. Projective representations and induced linear characters. *Proc. Amer. Math. Soc.* **24** (1970), 106–111.

Foulser, David A.
1. Solvable primitive permutation groups of low rank. *Trans. Amer. Math. Soc.* **143** (1969), 1–54.

Frame, J. Sutherland.
  1.  Double coset matrices and group characters. *Bull. Amer. Math. Soc.* **49** (1943), 81–92.
  2.  Group decomposition by double coset matrices. *Bull. Amer. Math. Soc.* **54** (1948), 740–755.
  3.  An irreducible representation extracted from two permutation groups. *Ann. of Math.* (2) **55** (1952), 85–100.
  4.  The constructive reduction of finite group representations. *Proc. Sympos. Pure Math.*, Vol. VI, pp. 89–99. Amer. Math. Soc., Providence, R. I., 1962.
Frame, J. Sutherland; Tamaschke, Olaf.
  1.  Über die Ordnungen der Zentralisatoren der Elemente in endlichen Gruppen. *Math. Zeitschr.* **83** (1964), 41–45.
Frobenius, G.
  1.  Über Relationen zwischen den Charakteren einer Gruppe und denen ihrer Untergruppen. Sitzungsberichte Preuss. Akad. Wiss., Berlin (1898), 501–515.
Frobenius, G.; Schur, Issai.
  1.  Über die reellen Darstellungen der endlichen Gruppen. Sitzungsberichte Preuss. Akad. Wiss., Berlin (1906), 186–208.
Gabriel, J. R.
  1.  On the construction of irreducible representations of the symmetric group. *Proc. Cambridge Philos. Soc.* **57** (1961), 330–340.
Gallagher, Patrick X.
  1.  Group characters and commutators. *Math. Zeitschr.* **79** (1962), 122–126.
  2.  Group characters and normal Hall subgroups. *Nagoya Math. J.* **21** (1962), 223–230.
  3.  Group characters and Sylow subgroups. *J. London Math. Soc.* **39** (1964), 720–722.
  4.  Determinants of representations of finite groups. *Abh. Math. Sem. Univ. Hamburg* **28** (1965), 162–167.
  5.  The generation of the lower central series. *Canadian J. Math.* **17** (1965), 405–410.
  6.  Zeros of characters of finite groups. *J. Algebra* **4** (1966), 42–45.
  7.  Counting subgroups of linear groups. *Math. Zeitschr.* **115** (1970), 9–10.
  8.  The number of conjugacy classes in a finite group. In Reiner [3], pp. 51–52.
Gaschütz, Wolfgang.
  1.  Endliche Gruppen mit treuen absolut-irreduziblen Darstellungen. *Math. Nachr.* **12** (1954), 253–255.
Gel'fand, I. M.; Graev, M. I.
  1.  Construction of irreducible representations of simple algebraic groups over a finite field. (Russian) *Dokl. Akad. Nauk SSSR* **147** (1962), 529–532.
Glauberman, George.
  1.  Subgroups of finite groups. *Bull. Amer. Math. Soc.* **73** (1967), 1–12.
  2.  A characterization of the Suzuki groups. *Illinois J. Math.* **12** (1968), 76–98.
  3.  Correspondences of characters for relatively prime operator groups. *Canadian J. Math.* **20** (1968), 1465–1488.
  4.  Prime-power factor groups of finite groups. *Math. Zeitschr.* **107** (1968), 159–172.
  5.  On a class of doubly transitive permutation groups. *Illinois J. Math.* **13** (1969), 394–399.

Goldschmidt, David M.
1. A group-theoretic proof of the $p^a q^b$-theorem for odd primes. *Math. Zeitschr.* **113** (1970), 373–375.
Gorenstein, Daniel.
1. *Finite Groups.* Harper and Row, New York-London, 1968.
Gorenstein, Daniel; Walter, John H.
1. On finite groups with dihedral Sylow 2-subgroups. *Illinois J. Math.* **6** (1962), 553–593.
2. The characterization of finite groups with dihedral Sylow 2-subgroups. I, II, III. *J. Algebra* **2** (1965), 85–151, 218–270, 354–393.
Green, James A.
1. On the converse to a theorem of R. Brauer. *Proc. Cambridge Philos. Soc.* **51** (1955), 237–239.
2. The characters of the finite general linear groups. *Trans. Amer. Math. Soc.* **80** (1955), 402–447.
3. Les polynômes de Hall et les caractères des groupes $GL(n, q)$. *Colloque d'algèbre supérieure*, Centre Belge de Recherches Mathématiques, pp. 207–215. Gauthier-Villars, Paris, 1957.
Grove, Larry C.
1. Real representations of split metacyclic groups. In Reiner [3], pp. 65–66.
Gündüzalp, Yavuz.
1. Über die gewöhnlichen irreduziblen Charaktere der symmetrischen Gruppe. *Mitt. Math. Sem. Giessen* **81** (1969), 53 pp.
Hall, Marshall, Jr.
1. *The Theory of Groups.* Macmillan, New York, 1959.
Hamel, Ray O.
1. Idempotents in group algebras and exceptional characters. *J. Algebra* **15** (1970), 283–292.
Harada, Koichiro.
1. A characterization of the groups $LF(2, q)$. *Illinois J. Math.* **11** (1967), 647–659.
Herzog, Marcel.
1. On finite groups with a cyclic Sylow subgroup. *Illinois J. Math.* **14** (1970), 188–193.
Hestenes, Marshall D.
1. Singer groups. *Canadian J. Math.* **22** (1970), 492–513.
Higman, Donald G.
1. Focal series in finite groups. *Canadian J. Math.* **5** (1953), 477–497.
2. Finite permutation groups of rank 3. *Math. Zeitschr.* **86** (1964), 145–156.
3. Intersection matrices for finite permutation groups. *J. Algebra* **6** (1967), 22–42.
Higman, Graham.
1. Odd characterizations of finite simple groups. Lecture notes, University of Michigan, 1968.
Hoheisel, Guido.
1. Über Charaktere. *Monatsh. Math. Phys.* **48** (1939), 448–456.
Huppert, Bertram.
1. Monomiale Darstellungen endlicher Gruppen. *Nagoya Math. J.* **6** (1953), 93–94.
2. Normalteiler und maximale Untergruppen endlicher Gruppen. *Math. Zeitschr.* **60** (1954), 409–434.

3.  Lineare auflösbare Gruppen. *Math. Zeitschr.* **67** (1957), 479–518.
4.  Zweifach transitive auflösbare Gruppen. *Math. Zeitschr.* **68** (1957), 126–150.
5.  Gruppen mit modularer Sylow-Gruppe. *Math. Zeitschr.* **75** (1960/61), 140–153.
6.  *Endliche Gruppen. I.* Springer-Verlag, Berlin-New York, 1967.

Isaacs, I. Martin.
1.  Extensions of certain linear groups. *J. Algebra* **4** (1966), 3–12.
2.  Finite groups with small character degrees and large prime divisors. *Pacific J. Math.* **23** (1967), 273–280.
3.  Two solvability theorems. *Pacific J. Math.* **23** (1967), 281–290.
4.  Fixed points and characters in groups with non-coprime operator groups. *Canadian J. Math.* **20** (1968), 1315–1320.
5.  Extensions of group representations over nonalgebraically closed fields. *Trans. Amer. Math. Soc.* **141** (1969), 211–228.
6.  Groups having at most three irreducible character degrees. *Proc. Amer. Math. Soc.* **21** (1969), 185–188.
7.  Invariant and extendible group characters. *Illinois J. Math.* **14** (1970), 70–75.
8.  Systems of equations and generalized characters in groups. *Canadian J. Math.* **22** (1970), 1040–1046.
9.  Symplectic action and the Schur index. In Reiner [3], pp. 73–75.

Isaacs, I. Martin; Passman, Donald S.
1.  Groups whose irreducible representations have degrees dividing $p^e$. *Illinois J. Math.* **8** (1964), 446–457.
2.  Groups with representations of bounded degree. *Canadian J. Math.* **16** (1964), 299–309.
3.  A characterization of groups in terms of the degrees of their characters. I, II. *Pacific J. Math.* **15** (1965), 877–903; **24** (1968), 467–510.
4.  Groups with relatively few non-linear irreducible characters. *Canadian J. Math.* **20** (1968), 1451–1458.

Itô, Noboru.
1.  On the degrees of irreducible representations of a finite group. *Nagoya Math. J.* **3** (1951), 5–6.
2.  Note on $A$-groups. *Nagoya Math. J.* **4** (1952), 79–81.
3.  On a theorem of H. F. Blichfeldt. *Nagoya Math. J.* **5** (1953), 75–77.
4.  On monomial representations of finite groups. *Osaka Math. J.* **6** (1954), 119–127.
5.  On transitive simple permutation groups of degree $2p$. *Math. Zeitschr.* **78** (1962), 453–468.
6.  A note on transitive permutation groups of degree $p = 2q + 1$, $p$ and $q$ being prime numbers. *J. Math. Kyoto Univ.* **3** (1963), 111–113.
7.  On transitive permutation groups of prime degree. (Japanese) *Sûgaku* **15** (1963/4), 129–141.
8.  Über die Darstellungen der Permutationsgruppen von Primzahlgrad. *Math. Zeitschr.* **89** (1965), 196–198.
9.  On a conjecture of J. S. Frame. *Nagoya Math. J.* **30** (1967), 79–81.
10. On uniprimitive permutation groups of degree $2p$. *Math. Zeitschr.* **102** (1967), 238–244.

Iwahori, Nagayoshi.
1.  On a property of a finite group. *J. Fac. Sci. Univ. Tokyo Sect. I* **11** (1964), 47–64.

Iwahori, Nagayoshi; Matsumoto, Hideya.
  1. Several remarks on projective representations of finite groups. *J. Fac. Sci. Univ. Tokyo Sect. I* **10** (1964), 129–146.
Janko, Zvonimir.
  1. A new finite simple group with abelian 2-Sylow subgroups. *Proc. Nat. Acad. Sci. U.S.A.* **53** (1965), 657–658.
  2. A new finite simple group with abelian Sylow 2-subgroups and its characterization. *J. Algebra* **3** (1966), 147–186.
  3. A characterization of the Mathieu simple groups. I, II. *J. Algebra* **9** (1968), 1–19, 20–41.
  4. A characterization of the simple group $G_2(3)$. *J. Algebra* **12** (1969), 360–371.
  5. Some new simple groups of finite order. *Symposia Mathematica* (INDAM, Rome, 1967/68), Vol. 1, pp. 25–64. Academic Press, London, 1969.
Janusz, Gerald J.
  1. Some remarks on Clifford's theorem and the Schur index. *Pacific J. Math.* **32** (1970), 119–129.
Jordan, Camille.
  1. Mémoire sur les équations différentielles linéaires à intégrale algébrique. *J. Reine Angew. Math.* **84** (1878), 89–215.
Kanunov, N. F.
  1. On works of Theodor Molien in the theory of representations of finite groups. (Russian) *Istor.-Mat. Issled.* No. 17 (1966), 57–88.
Keller, Gordon.
  1. Concerning the degrees of irreducible characters. *Math. Zeitschr.* **107** (1968), 221–224.
Kerber, Adalbert.
  1. Zur Darstellungstheorie von Kranzprodukten. *Canadian J. Math.* **20** (1968), 665–672.
  2. Zur Darstellungstheorie von Symmetrien symmetrischer Gruppen. *Mitt. Math. Sem. Giessen* **80** (1969), 1–27.
  3. Zu einer Arbeit von J. L. Berggren über ambivalente Gruppen. *Pacific J. Math.* **33** (1970), 669–675.
  4. Zur Theorie der $M$-Gruppen. *Math. Zeitschr.* **115** (1970), 4–6.
Kilmoyer, Robert W., Jr.
  1. The reflection character of a finite group with a $BN$-pair. In Reiner [3], pp.91–94.
Kochendörffer, Rudolf.
  1. Über treue irreduzible Darstellungen endlicher Gruppen. *Math. Nachr.* **1** (1948), 25–39.
Kondo, Takeshi.
  1. The characters of the Weyl group of type $F_4$. *J. Fac. Sci. Univ. Tokyo Sect. I* **11** (1965), 145–153.
Kovács, L. G.; Neumann, B. H.
  1. *Proceedings of the International Conference on the Theory of Groups held at the Australian National University, Canberra, 10–20 August, 1965.* Edited by L. G. Kovács and B. H. Neumann. Gordon and Breach, New York, 1967.
Kronstein, Karl.
  1. Characters and systems of subgroups. *J. Reine Angew. Math.* **224** (1966), 147–163.

2. Character tables and the Schur index. In Reiner [3], pp. 97–98.

Lang, Serge.
   1. *Algebra*, Addison-Wesley, Reading, Mass., 1965.

Leonard, Henry S., Jr.
   1. On finite groups which contain a Frobenius factor group. *Illinois J. Math.* **9** (1965), 47–58.

Leonard, Henry S., Jr.; McKelvey, Katherine K.
   1. On lifting characters in finite groups. *J. Algebra* **7** (1967), 168–191.

Lindsey, John H., II.
   1. On the Suzuki and Conway groups. *Bull. Amer. Math. Soc.* **76** (1970), 1088–1090.
   2. On the Suzuki and Conway groups. In Reiner [3], pp. 107–109.

Littlewood, Dudley E.
   1. *The Theory of Group Characters and Matrix Representations of Groups.* 2nd edition, Oxford Univ. Press, Oxford, 1958.

Lorenz, Falko.
   1. *Bestimmung der Schurschen Indizes von Charakteren endlicher Gruppen.* Eberhard-Karls-Universität, Tübingen, 1966, 83 pp.

Mackey, George W.
   1. On induced representations of groups. *Amer. J. Math.* **73** (1951), 576–592.
   2. Symmetric and antisymmetric Kronecker squares and intertwining numbers of induced representations of finite groups. *Amer. J. Math.* **75** (1953), 387–405.

Mangold, Ruth.
   1. Beiträge zur Theorie der Darstellungen endlicher Gruppen durch Kollineationen. *Mitt. Math. Sem. Giessen* **69** (1966), 44 pp.

Mann, Avinoam.
   1. Simple groups having $p$-nilpotent 2nd-maximal subgroups. *Israel J. Math.* **6** (1968), 233–245.

Martineau, R. P.
   1. A characterization of Janko's simple group of order 175,560. *Proc. London Math. Soc.* (3) **19** (1969), 709–729.

McKay, J. K. S.
   1. A method for computing the character table of a finite group. *Computers in Mathematical Research*, pp. 140–148. North-Holland, Amsterdam, 1968.

Miller, G. A.; Blichfeldt, H. F.; Dickson, L. E.
   1. *Theory and Applications of Finite Groups.* Dover, New York, 1961.

Montague, Stephen.
   1. On rank 3 groups with a multiply transitive constituent. *J. Algebra* **14** (1970), 506–522.

Morris, Alun O.
   1. The characters of the group $GL(n, q)$. *Math. Zeitschr.* **81** (1963), 112–123.
   2. A note on lemmas of Green and Kondo. *Proc. Cambridge Philos. Soc.* **63** (1967), 83–85.

Murnaghan, Francis D.
   1. The dimensions of the irreducible representations of a finite group. *Proc. Nat. Acad. Sci. U.S.A.* **37** (1951), 441–442.
   2. On the characters of the symmetric group. *Proc. Nat. Acad. Sci. U.S.A.* **41** (1955), 396–398.

Nagao, Hirosi.
1.  On the groups with the same table of characters as symmetric groups. *J. Inst. Polytech. Osaka City Univ. Ser. A* **8** (1957), 1–8.
2.  A remark on the orthogonality relations in the representation theory of finite groups. *Canadian J. Math.* **11** (1959), 59–60.

Osima, Masaru.
1.  On the representations of groups of finite order. *Math. J. Okayama Univ.* **1** (1952), 33–61.
2.  On the induced characters of groups of finite order. *Math. J. Okayama Univ.* **3** (1953), 47–64.
3.  Some remarks on the characters of the symmetric group. *Canadian J. Math.* **5** (1953), 336–343.

Oyama, Tuyosi.
1.  On the groups with the same table of characters as alternating groups. *Osaka J. Math.* **1** (1964), 91–101.

Padzerski, Gerhard.
1.  *Über lineare auflösbare Gruppen.* Universität Halle, Halle, 1963, 85 pp.

Pahlings, Herbert.
1.  Beiträge zur Theorie der projektiven Darstellungen endlicher Gruppen. *Mitt. Math. Sem. Giessen* **77** (1968), 61 pp.
2.  Gruppen mit irreduziblen Darstellungen hohen Grades. *Mitt. Math. Sem. Giessen* **85** (1970), 27–44.

Passman, Donald S.
1.  Character kernels of discrete groups. *Proc. Amer. Math. Soc.* **17** (1966), 487–492.
2.  Groups whose irreducible representations have degrees dividing $p^2$. *Pacific J. Math.* **17** (1966), 475–496.
3.  Groups with normal solvable Hall $p'$-subgroups. *Trans. Amer. Math. Soc.* **123** (1966), 99–111.
4.  *Permutation Groups.* Benjamin, New York-Amsterdam, 1968.

Poljak, S. S.
1.  Irreducible complex representations of the group of triangular matrices over a finite prime field. (Ukrainian) *Dopovidi Akad. Nauk Ukrain. RSR* (1966), 13–16.
2.  On the characters of irreducible complex representations of the group of triangular matrices over a finite prime field. (Ukrainian) *Dopovidi Akad. Nauk Ukrain. RSR* (1966), 434–436.

Prokop, Wilfried.
1.  *Über eine Formel von Frobenius zur Berechnung der Charaktere endlicher Gruppen.* Eidgenössische Technische Hochschule, Zurich, 1948, 39 pp.

Puttaswamiah, B. M.
1.  On the reduction of permutation representations. *Canadian Math. Bull.* **6** (1963), 385–395.

Reiner, Irving.
1.  The Schur index in the theory of group representations. *Michigan Math. J.* **8** (1961), 39–47.
2.  A survey of integral representation theory. *Bull. Amer. Math. Soc.* **76** (1970), 159–227.
3.  *Proceedings of the Symposium on Representation Theory of Finite Groups and*

*Related Topics.* Irving Reiner, editor. Proceedings of Symposia in Pure Mathematics, Vol. XXI Amer. Math. Soc., Providence, R.I. 1971.

Reynolds, William F.
1. Projective representations of finite groups in cyclotomic fields. *Illinois J. Math.* **9** (1965), 191–198.
2. Sections, isometries, and generalized group characters. *J. Algebra* **7** (1967), 394–405.

Ribenboim, P.
1. Linear representation of finite groups. Queen's Papers in Pure and Applied Mathematics, No. 5. Queen's University, Kingston, Ont., 1966.

Rigby, J. F.
1. Primitive linear groups containing a normal nilpotent subgroup larger than the centre of the group. *J. London Math. Soc.* **35** (1960), 389–400.

Robinson, G. de B.
1. On the representations of the symmetric group. I, II, III. *Amer. J. Math.* **60** (1938), 745–760; **69** (1947), 286–298; **70** (1948), 277–294.
2. On the disjoint product of irreducible representations of the symmetric group. *Canadian J. Math.* **1** (1949), 166–175.
3. *Representation Theory of the Symmetric Group.* Univ. of Toronto Press, Toronto, 1961.

Roquette, Peter.
1. Arithmetische Untersuchung des Charakterringes einer endlichen Gruppe. *J. Reine Angew. Math.* **190** (1952), 148–168.
2. Realisierung von Darstellungen endlicher nilpotenter Gruppen. *Arch. Math.* **9** (1958), 241–250.

Sah, Chih-Han.
1. Existence of normal complements and extension of characters in finite groups. *Illinois J. Math.* **6** (1962), 282–291.
2. Automorphisms of finite groups. *J. Algebra* **10** (1968), 47–68.

Saksonov, A. I.
1. The integral ring of characters of a finite group. (Russian) *Vesci Akad. Navuk BSSR Ser. Fiz.-Mat. Navuk* (1966), no. 3, 69–76.
2. An answer to a question of R. Brauer. (Russian) *Vesci Akad. Navuk BSSR Ser. Fiz.-Mat. Navuk* (1967), no. 1, 129–130.
3. Group-albegras of finite groups over a number field. (Russian) *Dokl. Akad. Nauk BSSR* **11** (1967), 302–305.

Schur, Issai.
1. Über die Darstellungen der endlichen Gruppen durch gebrochene lineare Substitutionen. *J. Reine Angew. Math.* **127** (1904), 20–50.
2. Untersuchungen über die Darstellungen der endlichen Gruppen durch gebrochene lineare Substitutionen. *J. Reine Angew. Math.* **132** (1907), 85–137.

Scott, William R.
1. *Group Theory.* Prentice-Hall, Englewood Cliffs, N. J., 1964.

Seitz, Gary M.
1. Finite groups having only one irreducible representation of degree greater than one. *Proc. Amer. Math. Soc.* **19** (1968), 459–461.
2. $M$-groups and the supersolvable residual. *Math. Zeitschr.* **110** (1969), 101–122.

Seitz, Gary M.; Wright, C. R. B.
  1. On finite groups whose Sylow subgroups are modular or quaternion-free. *J. Algebra* **13** (1969), 374–381.
Sekino, Kaoru.
  1. Über die Zerlegung der Gruppencharaktere. *J. Fac. Sci. Univ. Tokyo. Sect. I* **7** (1954), 255–263; **8** (1960), 195–228, 333–362; **9** (1962), 249–260.
Serre, Jean-Pierre.
  1. *Représentations linéaires des groupes finis.* Hermann, Paris, 1967.
Shamash, Jack; Shult, Ernest.
  1. On groups with cyclic Carter subgroups. *J. Algebra* **11** (1969), 564–597.
Solomon, Louis.
  1. On the sum of the elements in the character table of a finite group. *Proc. Amer. Math. Soc.* **12** (1961), 962–963.
  2. The representation of finite groups in algebraic number fields. *J. Math. Soc. Japan* **13** (1961), 144–164.
  3. On Schur's index and the solutions of $G^n = 1$ in a finite group. *Math. Zeitschr.* **78** (1962), 122–125.
  4. The orders of the finite Chevalley groups. *J. Algebra* **3** (1966), 376–393.
  5. A decomposition of the group algebra of a finite Coxeter group. *J. Algebra* **9** (1968), 220–239.
  6. The Steinberg character of a finite group with BN-pair. In Brauer and Sah [1], pp. 213–221.
  7. On the affine group over a finite field. In Reiner [3], pp. 145–147.
Spitznagel, Edward L., Jr.
  1. Hall subgroups of certain families of finite groups. *Math. Zeitschr.* **97** (1967), 259–290.
Springer, Tonny A.
  1. On induced group characters. *Indagationes Math.* **10** (1948), 250–258.
  2. Generalization of Green's polynomials. In Reiner [3], pp. 149–153,
Srinivasan, Bhama.
  1. The characters of the finite symplectic group Sp(4, $q$). *Trans. Amer. Math. Soc.* **131** (1968), 488–525.
Steinberg, Robert.
  1. A geometric approach to the representations of the full linear group over a Galois field. *Trans. Amer. Math. Soc.* **71** (1951), 274–282.
  2. Prime power representations of finite linear groups. I, II. *Canadian J. Math.* **8** (1956), 580–591; **9** (1957), 347–351.
  3. Complete sets of representations of algebras. *Proc. Amer. Math. Soc.* **13** (1962), 746–747.
  4. Lectures on Chevalley groups. Yale University, New Haven, Conn., 1967.
Suprunenko, D. A.
  1. *Soluble and Nilpotent Linear Groups.* Amer. Math. Soc., Providence, R. I., 1963.
  2. On the theory of solvable linear groups. (Russian) *Dokl. Akad. Nauk SSSR* **184** (1969), 47–50.
  3. On the theory of solvable linear groups. (Russian) *Sibirsk. Mat. Ž.* **10** (1969), 1161–1172.
Suzuki, Michio.
  1. On finite groups with cyclic Sylow subgroups for all odd primes. *Amer. J. Math.*

77 (1955), 657–691.

2.   The nonexistence of a certain type of simple groups of odd order. *Proc. Amer. Math. Soc.* **8** (1957), 686–695.

3.   *Applications of Group Characters.* Proc. Sympos. Pure Math., Vol. I, pp. 88–99. Amer. Math. Soc., Providence, R. I., 1959.

4.   On characterizations of linear groups. I-II, III, IV. *Trans. Amer. Math. Soc.* **92** (1959), 191–219; *Nagoya Math. J.* **21** (1962), 159–183; *J. Algebra* **8** (1968), 223–247.

5.   A new type of simple groups of finite order. *Proc. Nat. Acad. Sci. U.S.A.* **46** (1960), 868–870.

6.   Finite groups with nilpotent centralizers. *Trans. Amer. Math. Soc.* **99** (1961), 425–470.

7.   On a finite group with a partition. *Arch. Math.* **12** (1961), 241–254.

8.   On a class of doubly transitive groups. I, II. *Ann. of Math.* (2) **75** (1962), 105–145; **79** (1964), 514–589.

9.   On the existence of a Hall normal subgroup. *J. Math. Soc. Japan* **15** (1963), 387–391.

10.  Characterizations of linear groups. *Bull. Amer. Math. Soc.* **75** (1969), 1043–1091.

Tachikawa, Hiroyuki.

1.   A remark on generalized characters of groups. *Sci. Rep. Tokyo Kyoiku Daigaku. Sect. A* **4** (1954), 332–334.

Taketa, K.

1.   Über die Gruppen, deren Darstellungen sich sämtlich auf monomiale Gestalt transformieren lassen. *Proc. Jap. Imp. Acad.* **6** (1930), 31–33.

Tamaschke, Olaf.

1.   Ringtheoretische Behandlung einfach transitiver Permutationsgruppen. *Math. Zeitschr.* **73** (1960), 393–408.

2.   S-Ringe und verallgemeinerte Charaktere auf endlichen Gruppen. *Math. Zeitschr.* **84** (1964), 101–119.

3.   S-rings and the irreducible representations of finite groups. *J. Algebra* **1** (1964), 215–232.

4.   A generalized character theory on finite groups. In Kovács and Neumann [1], pp. 347–355.

5.   A generalization of normal subgroups. *J. Algebra* **11** (1969), 338–352.

Tanaka, Shun'ichi.

1.   Construction and classification of irreducible representations of special linear group of the second order over a finite field. *Osaka J. Math.* **4** (1967), 65–84.

Tate, John.

1.   Nilpotent quotient groups. *Topology* **3** (1964), 109–111.

Tazawa, Masatada.

1.   Über die isomorphe Darstellung der endlichen Gruppe. *Tôhoku Math. J.* **47** (1940), 87–93.

Thompson, John G.

1.   Normal *p*-complements for finite groups. *Math. Zeitschr.* **72** (1960), 332–354.

2.   Normal *p*-complements for finite groups. *J. Algebra* **1** (1964), 43–46.

3.   Characterizations of finite simple groups. *Proc. Internat. Congr. Math.* (*Moscow*, 1966), pp. 158–162. Izdat. "Mir," Moscow, 1968.

4. A non-duality theorem for finite groups. *J. Algebra* **14** (1970), 1–4.
5. Normal *p*-complements and irreducible characters. *J. Algebra* **14** (1970), 129–134.

Tsuzuku, Tosiro.
1. On multiple transitivity of permutation groups. *Nagoya Math. J.* **18** (1961), 93–109.

Wales, David B.
1. Finite linear groups in seven variables. *Bull. Amer. Math. Soc.* **74** (1968), 197–198.
2. Finite linear groups of prime degree. *Canadian J. Math.* **21** (1969), 1025–1041.
3. Finite linear groups of degree seven. I, II. *Canadian J. Math.* **21** (1969), 1042–1056; *Pacific J. Math.* **34** (1970), 207–235.

Wall, C. T. C.
1. On groups consisting mostly of involutions. *Proc. Cambridge Philos. Soc.* **67** (1970), 251–262.

Walter, John H.
1. On the characterization of linear and projective linear groups. I, II. *Trans. Amer. Math. Soc.* **100** (1961), 481–529; **101** (1961), 107–123.

Ward, Harold N.
1. An extension of a theorem on monomial groups. *Amer. Math. Monthly* **76** (1969), 534–535.

Weidman, Donald R.
1. The character ring of a finite group. *Illinois J. Math.* **9** (1965), 462–467.

Weisner, Louis.
1. Condition that a finite group be multiply isomorphic with each of its irreducible representations. *Amer. J. Math.* **61** (1939), 709–712.

Weyl, Hermann.
1. *The Classical Groups. Their Invariants and Representations.* Princeton Univ. Press, Princeton, N. J., 1939.

Wielandt, Helmut.
1. Primitive Permutationsgruppen vom Grad 2*p*. *Math. Zeitschr.* **63** (1956), 478–485.
2. *Finite Permutation Groups.* Academic Press, New York-London, 1964.

Wigner, Eugene P.
1. On representations of certain finite groups. *Amer. J. Math.* **63** (1941), 57–63.

Winter, David L.
1. Finite groups having a faithful representation of degree less than $(2p + 1)/3$. *Amer. J. Math.* **86** (1964), 608–618.
2. Finite *p*-solvable linear groups with a cyclic Sylow *p*-subgroup. *Proc. Amer. Math. Soc.* **18** (1967), 341–343.
3. On the restrict-induce map of group characters. *Proc. Amer. Math. Soc.* **19** (1968), 246–248.
4. Solvability of certain *p*-solvable linear groups of finite order. *Pacific J. Math.* **34** (1970), 827–835.

Witt, Ernst.
1. Die algebraische Struktur des Gruppenringes einer endlichen Gruppe über einem Zahlkörper. *J. Reine Angew. Math.* **190** (1952), 231–245.

Wong, Warren J.
1. A characterization of the alternating group of degree 8. *Proc. London Math. Soc.* (3) **13** (1963), 359–383.
2. Linear groups analogous to permutation groups. *J. Austral. Math. Soc.* **3** (1963), 180–184.
3. A characterization of the Mathieu group $M_{12}$. *Math. Zeitschr.* **84** (1964), 378–388.
4. On linear *p*-groups. *J. Austral. Math. Soc.* **4** (1964), 174–178.

Yamazaki, Keijiro.
1. A note on projective representations of finite groups. *Sci. Papers College Gen. Ed. Univ. Tokyo* **14** (1964), 27–36.
2. On projective representations and ring extensions of finite groups. *J. Fac. Sci. Univ. Tokyo Sect. I* **10** (1964), 147–195.

Zassenhaus, Hans J.
1. An equation for the degrees of the absolutely irreducible representations of a group of finite order. *Canadian J. Math.* **2** (1950), 166–167.

Žmud', È. M.
1. On isomorphic linear representations of finite groups. (Russian) *Mat. Sb.* (*N. S.*) **38** (80) (1956), 417–430.
2. A group-theoretic Möbius-Delsarte function and the theory of linear representations of finite groups. (Russian) *Izv. Vysš. Učebn. Zaved. Matematika* (1957), no. 1, 133–141.
3. On kernels of homomorphisms of linear representations of finite groups. (Russian) *Mat. Sb.* (*N. S.*) **44** (86) (1958), 353–408.
4. Isomorphisms of irreducible projective representations of finite groups. (Russian) *Zap. Meh.-Mat. Fak. i Har'kov Mat. Obšc.* (4) **26** (1960), 333–372.
5. Structure-theoretic properties of the kernels of homomorphisms of irreducible representations of finite groups. (Russian) *Algebra and Math. Logic: Studies in Algebra* (Russian), pp. 98–110. Izdat. Kiev. Univ., Kiev, 1966.